OXFORD REVIEWS OF REPRODUCTIVE BIOLOGY

Volume 17
1995

OXFORD REVIEWS OF REPRODUCTIVE BIOLOGY EDITORIAL BOARD

Professor R B Heap (chairman), Institute of Animal Physiology and Genetics Research, Cambridge Research Station, Babraham Hall, Cambridge CB2 4AT, UK.

Professor D T Baird, Department of Obstetrics and Gynaecology, University of Edinburgh, 37 Chalmers Street, Edinburgh EH3 9EW, UK.

Professor F Bazer, 442D Kleberg Center, Department of Animal Science, Texas A & M University, College Station, TX 77843–2471, USA.

Professor H Beier, Institut für Anatomie und Reproduktionsbiologie Medizinische Fakultät der RWTW Aachen, Wendlingweg 2, D–W–5100 Aachen, Germany.

Professor A R Bellvé, Department of Anatomy and Cell Biology and Center of Cellular and Molecular Urology, Columbia Presbyterian Medical Center, New York NY 10032, USA.

Dr H M Charlton (editor), Department of Human Anatomy, University of Oxford, South Parks Road, Oxford OX1 3QX, UK.

Professor Hilary Dobson, University of Liverpool, Department of Veterinary Preclinical Sciences, Leahurst, Neston, Wirral L64 7TE, UK.

Professor A P F Flint, Animal Physiology Section, Dept of Physiology and Environmental Science, University of Nottingham, Sutton Bonington, Loughborough, Leics, LE12 5RD, UK.

Professor M Johnson, University of Cambridge, Department of Anatomy, Downing Street, Cambridge, CB2 3DY, UK.

Professor D de Kretser (director), Institute of Reproduction and Development, Monash University Level 3, Block E, Monash Medical Centre, 246, Clayton Road, Clayton, Victoria 3168, Australia.

Professor A McNeilly, MRC Reproductive Biology Unit, Centre for Reproductive Biology, 37 Chalmers Street, Edinburgh, EH3 9EH, UK.

Dr S R Milligan, Physiology Section, Division of Biomedical Sciences, King's College, Strand, London WC2R 2LS, UK.

Professor R Ozon, Laboratoire de Physiologie de la Reproduction, Université Pierre et Marie Curie, Bâtiment B, 7ème étage–Case 13, 9 quai St Bernard, 75005 Paris, France.

Professor R V Short, Department of Physiology, Monash University, Clayton, Victoria 3168, Australia.

Professor S K Smith, University of Cambridge Clinical School, Department of Obstetrics and Gynaecology, The Rosie Maternity Hospital, Robinson Way, Cambridge, CB2 2SW, UK.

OXFORD REVIEWS OF REPRODUCTIVE BIOLOGY

EDITED BY
H.M. CHARLTON

Volume 17
1995

Oxford   New York   Tokyo
OXFORD UNIVERSITY PRESS
1995

Oxford University Press, Walton Street, Oxford OX2 6DP
Oxford   New York
Athens Auckland Bangkok Bombay
Calcutta Cape Town Dar es Salaam Delhi
Florence Hong Kong Istanbul Karachi
Kuala Lumpur Madras Madrid Melbourne
Mexico City Nairobi Paris Singapore
Taipei Tokyo Toronto
and associated companies in
Berlin Ibadan

Oxford is a trade mark of Oxford University Press

Published in the United States
by Oxford University Press Inc., New York

© Oxford University Press, 1995

All rights reserved. No part of this publication may be
reproduced, stored in a retrieval system, or transmitted, in any
form or by any means, without the prior permission in writing of Oxford
University Press. Within the UK, exceptions are allowed in respect of any
fair dealing for the purpose of research or private study, or criticism or
review, as permitted under the Copyright, Designs and Patents Act, 1988, or
in the case of reprographic reproduction in accordance with the terms of
licences issued by the Copyright Licensing Agency. Enquiries concerning
reproduction outside those terms and in other countries should be sent to
the Rights Department, Oxford University Press, at the address above.

This book is sold subject to the condition that it shall not,
by way of trade or otherwise, be lent, re-sold, hired out, or otherwise
circulated without the publisher's prior consent in any form of binding
or cover other than that in which it is published and without a similar
condition including this condition being imposed
on the subsequent purchaser.

A catalogue record for this book is available from the British Library

Library of Congress Cataloging in Publication Data
Library of Congress Card No 80–648347

ISBN 0 19 262629 9

Typeset by The Electronic Book Factory Ltd, Fife, Scotland
Printed bound in Great Britain by
Biddles Ltd, Guildford and King's Lynn

Contents

	List of contributors	vii
	List of abbreviations	ix
1	Mechanisms of autoimmune disease in the testis and ovary	1
	Kenneth S. K. Tung and Cory Teuscher	
2	Adaptive strategies regulating energy balance in human pregnancy	33
	Andrew M. Prentice, Sally D. Poppitt, Gail R. Goldberg, and Ann Prentice	
3	Angiogenic growth factor expression in the uterus	61
	Steven K. Smith	
4	Ovarian endocrine control of sperm progression in the Fallopian tubes	85
	Ronald H. F. Hunter	
5	The origin of genetic defects in the human and their detection in the pre-implantation embryo	125
	Joy D. A. Delhanty and Alan H. Handyside	

6 Regulation of pituitary
   gonadotrophin gene expression   159
   Julie E. Mercer and William W. Chin

7 Neural and endocrine mechanisms
   underlying the synchrony of sexual
   behaviour and ovulation in the
   sheep   205
   Dominique Blache and Graeme B. Martin

8 The gonadotrophin-releasing
   hormone receptor: structural
   determinants and regulatory
   control   255
   Stuart C. Sealfon and Robert P. Millar

   Index   285

Contributors

Dominique Blache: School of Agriculture (Animal Science), The University of Western Australia, Nedlands, WA 6009, Australia.

William W. Chin: Division of Genetics, Department of Medicine, Brigham and Women's Hospital; Howard Hughes Medical Institute, Harvard Medical School, Boston, MA, 02115, USA.

Joy D. A. Delhanty: Department of Genetics and Biometry, Galton Laboratory, University College London, Wolfson House, 4 Stephenson Way, London NW1 2HE, UK.

Gail R. Goldberg: MRC Dunn Clinical Nutrition Centre, Hills Road, Cambridge CB2 2DH, UK.

Alan H. Handyside: Institute of Obstetrics and Gynaecology, Royal Postgraduate Medical School, Hammersmith Hospital, Du Cane Road, London W12 0NN, UK.

Ronald H. F. Hunter: University of Edinburgh, 32 Gilmour Road, Edinburgh EH16 5NT, UK.

Graeme B. Martin: CSIRO, Division Animal Production, Private Bag, P. O. Wembley, WA 6014, Australia.

Julie E. Mercer: Prince Henry's Institute of Medical Research, P. O. Box 152, Clayton 3168, Australia.

Robert P. Millar: MRC Regulatory Peptides Research Unit, Department of Chemical Pathology, University of Cape Town Medical School, South Africa.

Sally D. Poppitt: MRC Dunn Clinical Nutrition Centre, Hills Road, Cambridge CB2 2DH, UK.

Ann Prentice: MRC Dunn Nutrition Centre, Milton Road, Cambridge CB4 1XJ, UK.

Andrew M. Prentice: MRC Dunn Clinical Nutrition Centre, Hills Road, Cambridge CB2 2DH, UK.

Stuart C. Sealfon: Fishberg Research Center for Neurobiology and Department of Neurology, Mount Sinai School of Medicine, New York, NY 10029, USA.

Steven K. Smith: Department of Obstetrics and Gynaecology, University of Cambridge, Rosie Maternity Hospital, Cambridge CB2 2SW, UK.

Cory Teuscher: Department of Microbiology, Brigham Young University, Provo, UT 84602, USA.

Kenneth S. K. Tung: Department of Pathology, University of Virginia, Charlottesville, VA 22908, USA.

Abbreviations

AChR	acetylcholine receptor
bFGF	basic fibroblast growth factor
BME	bovine microvascular endothelium
BMR	basal metabolic rate
CAT	chloramphenicol acetyltransferase
CF	cystic fibrosis
CNS	central nervous system
CSF	colony-stimulating factor
D3TX	day 3 thymectomized
DIT	diet-induced thermogenesis
EAO	experimental allergic orchitis
FGF	fibroblast growth factor
FISH	fluorescent *in situ* hybridization
FSH	follicle-stimulating hormone
GnRH	gonadotrophin-releasing hormone
HIP-70	hormone-induced protein (70 KDa)
HuVE	human umbilical vein endothelium
IFNγ	interferon-γ
IL	interleukin
IVF	*in vitro* fertilization
LH	luteinizing hormone
OVLT	organum vasculosum of the lamina terminalis
OVX/HPD	ovariectomized, hypothalamus–pituitary disconnected
PCR	polymerase chain reaction
PDGF	platelet-derived growth factor
PGE_1	prostaglandin E_1
PGE_2	prostaglandin E_2
RNase	ribonuclease
SCF	stem cell factor
TEE	total energy expenditure
TGF-β	transforming growth factor-β
TNF	tumour necrosis factor
TRH	thyrotrophin-releasing hormone
TSH	thyroid-stimulating hormone
VEGF	vascular endothelial growth factor

1 Mechanisms of autoimmune disease in the testis and ovary

KENNETH S.K. TUNG and CORY TEUSCHER

I **Introduction**

II **Local and systemic regulatory mechanisms that prevent autoimmune orchitis and oophoritis**
   1 Tissue barriers and antigen sequestration are important but not sufficient to protect male germ cell antigens and prevent experimental autoimmune orchitis
   2 Local immunoregulatory environment in the testis
   3 Suppressor T cells for testis antigen are demonstrated by induction of tolerance transferable by T cells and establishment of suppressor T cell line
   4 Suppressor T cells are demonstrable by the analysis of T cells from normal mice with capacity to induce or prevent ovarian, testicular, and gastric autoimmune diseases

III **Events that can bring about autoimmune disease of ovary and testis**
   1 Autoimmune oophoritis and orchitis occur when regulatory T cells are depleted
   2 Molecular mimicry can cause autoimmune oophoritis through pathogenic T cell activation
      i Murine autoimmune oophoritis can be induced by a peptide from the zona pellucida ZP3 protein
      ii T cell epitope mimicry as a mechanism that triggers autoimmune oophoritis

IV **Pathogenesis of experimental autoimmune orchitis and oophoritis**
   1 Pathogenetic role of inflammatory CD4+ T cells
   2 Pathogenic role of antibodies
   3 A novel mechanism of ovarian autoantibody induction by the T cell peptide of the ZP3 mini-autoantigen

V **Immunogenetic approach to autoimmune oophoritis and orchitis**

VI **Summary and conclusions**

I INTRODUCTION

Autoimmune diseases of the testis and the ovary are responsible for some spontaneous infertilities seen in animals, and are likely causes of some human infertilities as well (Table 1.1) (Tung and Lu 1991). Testicular biopsies from infertile men have been found to contain immunopathology of granulomatous orchitis, testicular immune complexes, and epididymal granulomas of noninfectious origin, changes which are also found in experimental autoimmune disease of the testis. On the other hand,

Table 1.1 *Experimental autoimmune diseases of testis and ovary*

Experimental autoimmune (allergic) orchitis and oophoritis that result from immunization with tissue antigen
- Classical experimental autoimmune orchitis induced by immunization with testis antigen with adjuvant[a]
- Experimental autoimmune orchitis induced by immunization with testis antigen without adjuvant[b]
- Autoimmune oophoritis induced by immunization with a mini-autoantigenic peptide from murine ZP3[c]

Autoimmune diseases of ovary, testis, and/or other organs that result from manipulations of the normal immune system
- Mice thymectomized on days 1–4 (D3TX) after birth[d]
- Adult murine CD5[low] spleen T cells transferred to athymic mice, to mice without T cells, or to SCID mice[e]
- Neonatal mice treated with cyclosporin A[f]
- Fetal rat thymus grafted in athymic mice[g]
- Neonatal mouse thymus grafted in athymic mice[h]
- Normal murine T cells from adult or neonatal thymus, or neonatal spleen injected into athymic mice[i]
- Mice with a transgenic Vα protein of the T cell receptor[j]
- RT6-depleted rat spleen T cells injected into athymic rats[k]
- OX22[high] (or CD45RC[high]) rat spleen T cells injected into athymic rats[l]

Other models of autoimmune orchitis
- Spontaneous autoimmune orchitis in dog[m], mink[n], rat[o], and human[p]
- Post-vasectomy autoimmune orchitis[q]
- Orchitis in rats with the transgenic HLA/B27 molecule[r]

[a] Voisin et al. 1951; Freund et al. 1953; Waksman 1959; Sato et al. 1981; Kohno et al. 1983.
[b] Sakamoto et al. 1985; Itoh et al. 1991a. [c] Rhim et al. 1992. [d] Nishizuka and Sakakura 1969; Kojima and Prehn 1981; Taguchi and Nishizuka 1981; Tung et al. 1987b, 1987c. [e] Sakaguchi et al. 1985; Sugihara et al. 1988; Smith et al. 1992. [f] Sakaguchi and Sakaguchi 1989; [g] Taguchi et al. 1986. [h] Sakaguchi and Sakaguchi 1990. [i] Smith et al. 1992. [j] Sakaguchi et al. 1994. [k] McKeever et al. 1990. [l] Fowell et al. 1991. [m] Fritz et al. 1976. [n] Tung et al. 1981. [o] Furbeth et al. 1989. [p] Morgan 1976. [q] Bigazzi et al. 1976; Tung 1978. [r] Hammer et al. 1990.

both ovarian autoantibodies and ovarian inflammation (oophoritis) have been documented in women with premature ovarian failure. By far the strongest evidence for an autoimmune basis of these diseases is the concurrence of autoimmune orchitis or autoimmune oophoritis in patients with autoimmunity syndromes involving multiple endocrine organs (LaBerbera et al. 1988).

There are two major experimental approaches for induction of autoimmune diseases of the gonads (Table 1.1). The first is by immunization with testis or ovarian antigen, usually with adjuvant, giving rise to the classical experimental autoimmune diseases of the testis (autoimmune orchitis or allergic orchitis and the ovary (autoimmune oophoritis). The second is by the deliberate alteration of the immune system in normal animals, without injecting antigen or adjuvants.

This chapter will review recent research on autoimmune diseases of the testis and ovary based on the two experimental approaches. The purpose is to review and attempt to understand the control mechanisms that normally prevent these gonadal autoimmune diseases from occurring, the events that overcome these control mechanisms, and the mechanisms that amplify the disease process once it is initiated.

II   LOCAL AND SYSTEMIC REGULATORY MECHANISMS THAT PREVENT AUTOIMMUNE ORCHITIS AND OOPHORITIS

1 Tissue barriers and antigen sequestration are important but not sufficient to protect male germ cell antigens and prevent experimental autoimmune orchitis

The unique haploid testis autoantigens have a late ontogeny and are not available to interact with lymphocytes early in life. Immunological tolerance might require the interaction between self antigens and developing lymphocytes early in development. Consequently, it has been assumed that systemic tolerance of the germ cell antigens does not exist, and the responsibility of testis antigen protection is assigned to the blood–testis barrier as a complete immunological barrier.

However, male germ cell autoantigens are not completely sequestered. Autoimmunogenic antigens exist on preleptotene spermatocytes, located outside the blood–testis barrier (Yule et al. 1988). Additionally, testis-unique peptides are presumably recognized by orchitogenic CD4+ T cells outside the straight tubules that link the seminiferous tubules to the rete testis (Tung et al. 1987a; Yule and Tung 1993). Lastly, the seasonal establishment of the mink blood–testis barrier has been found to coincide with tubular lumen formation, preceded by the appearance of antigenic germ cells (Pelletier 1986).

There is now strong evidence for active regional and systemic regulation against autoimmune response to testis antigens. The testicular interstitial space may be an immunologically privileged site, providing an immunochemical barrier. Suppressor T cells relevant to experimental autoimmune orchitis (EAO) have been identified. Moreover, thymectomy of mice within three days after birth (D3TX) leads to ablation of the systemic tolerance mechanism, and, importantly, autoimmune diseases in the D3TX animals are prevented by the infusion of normal T cells (Taguchi and Nishizuka 1987; Smith et al. 1991).

Thus, the protection of testicular autoantigens is probably dependent on two mechanisms: the confinement of most of the germ cell antigens by the strong but regionally incomplete tissue barrier; and the dynamic systemic (and the less well defined, regional) tolerance mechanisms that regulate the autoreactive lymphocytes. This dual mechanism concept is supported by the experimental finding of D3TX-induced autoimmune diseases (Taguchi and Nishizuka 1981). While over 95 per cent of female D3TX mice develop autoimmune oophoritis, less than 30 per cent of male D3TX mice develop autoimmune orchitis. However, the prevalence of orchitis increased to over 90 per cent when the D3TX mice were vasectomized. Inasmuch as vasectomy causes leakage of sperm antigens, the difference in the incidence of the two diseases might reflect the relative sequestration of testis, but not of ovarian antigens in the normal animals.

2 Local immunoregulatory environment in the testis

Organ and tissue allografts, including parathyroid and pancreatic islets, which would be rapidly rejected under the renal capsule, survived for prolonged periods when engrafted inside the testis (Head et al. 1983; Selawry et al. 1987).

T cell response elicited by mitogen or antibody to the T cell receptor was shown to be suppressed by proteins in fluids drained from the testicular interstitial space (Pollanen et al. 1988) or in the supernatant of cultured Sertoli cells (Wyatt et al. 1988; De Cesaris et al. 1992). Also, Leydig cells form rosettes with thymocytes, and lymphocytes, and, in their presence, the T cell proliferative responses to alloantigens or mitogen were reduced (Born and Wekerle 1981, 1982).

Another approach has been to identify cytokines within the testis known to regulate lymphocyte responses to mitogen. These have included: interleukin-1-like factor, transforming growth factor-β (TGF-β), fibroblast growth factor, transglutaminase, and prostaglandin E_2 (Hedger et al. 1990). Inhibin and activin produced in the testes also respectively accentuate and reduce thymocyte proliferation in response to mitogen (Hedger et al. 1989; Lee et al. 1989). Proteins which are known to inhibit complement activation exist in the testis, where they may reduce complement-mediated inflammation and complement-dependent cytolysis (Tarter and Alexander 1984). Finally, a major synthetic product of the Sertoli cell is sulphated glycoprotein 2, which has been shown to inhibit cell lysis by the complement attack complex, C56789 (Griswold et al. 1986; Jenne and Tschopp 1989).

In summary, some testicular factors with the capacity to affect lymphocyte functions *in vitro* have been identified. However, their physiological role has not yet been fully evaluated.

3 Suppressor T cells for testis antigen are demonstrated by induction of tolerance transferable by T cells and establishment of suppressor T cell line

Classical murine EAO is elicited by immunization with testis antigen in complete Freund's adjuvant and *Bordetella pertussis* toxin (Sato et al. 1983; Kohno et al. 1983). More Recently, EAO has also been induced by two subcutaneous injection, two weeks apart, of viable, dissociated testicular cells, without adjuvants (Sakamoto et al. 1985; Itoh et al. 1991a). This second disease model, which is characterized by severe orchitis and aspermatogenesis in the absence of vasitis or epididymitis, is transferrable to normal mice by CD4+ T cells (Itoh et al. 1991b) or by a CD4+ T cells line (Itoh et al. 1992a).

Antigen-specific tolerance has been demonstrated in this novel model following repeated intravenous injection of deaggregated, soluble testis antigens (Mukasa et al. 1992). C3H/He mice treated in this manner developed long-lasting resistance to the induction of EAO from subsequent challenge with testicular cells. Moreover, splenic T cells from mice given intravenous soluble testis antigen endowed recipients with resistance to EAO in cell-transfer experiments. In this model of tolerance, the suppressor T cells were defined as CD4− CD8+.

Also based on this model, a T cell line with the capacity to suppress experimental autoimmune orchitis has been derived and characterized. The T cells were obtained from the spleens of C3H/He mice which had received 10 subcutaneous injections of testis cells over a five-month period, followed by repeated stimulations *in vitro* with solubilized testis antigens (Itoh et al. 1992b). The CD4+ T cell line that emerged did not transfer EAO to normal recipients. Instead, disease suppression was apparent when the cell line was transferred to recipients during autoimmune orchitis induction. When cells were injected seven days after the second subcutaneous testicular cell injection, orchitis did not occur. Delayed hypersensitivity and antibody responses to testis antigens were also abrogated by the T cell line. This data suggested that suppression is manifest in the efferent phase of disease induction, presumably before the onset of testicular pathology.

Suppressor T cells have also been demonstrated in the classic murine

EAO model induced by testis homogenate in adjuvants. Sub-lines of the BALB/c mice differ in EAO susceptibility (Teuscher et al. 1987a), BALB/c By are susceptible whereas BALB/c Jare highly resistant. When spleen cells from the resistant BALB/cJ line, immunized with testis antigen in adjuvant, were transferred to BALB/c By mice, the recipients became resistant to EAO induction (Mahi-Brown and Tung 1990; Teuscher et al. 1990). T cells responsible for EAO suppression were isolated and typed as CD4+ CD8− T cells (Teuscher et al. 1990). Studies on suppressor T cells in EAO illustrate the complex nature of cellular immunoregulation in autoimmune disease.

4 Suppressor T cells are demonstrable by the analysis of T cells from normal mice with capacity to induce or prevent ovarian, testicular, and gastric autoimmune disease

The existence of pathogenic T cells in normal individuals, and their regulation by suppressor T cells, has also been demonstrated in a series of experiments based on a remarkably simple design (Table 1.2) (Sakaguchi and Sakaguchi 1990; Smith et al. 1992).

When CD4+ CD8− thymocytes from normal female adult or neonatal BALB/C mice were transferred to athymic BALB/c mice, approximately 75 per cent of the recipients developed significant oophoritis, autoimmune gastritis, or both two months later. The tissue pathology was associated with detectable serum autoantibody to the oocytes and the gastric autoantigen H^+K^+-ATPase. Since the CD4+ CD8− subset represents mature thymocytes that have passed beyond the stage of T cell development in which the deletion of self-reactive T cells should occur, mature pathogenic self-reactive T cells are not deleted in normal thymuses. This conclusion is supported by a study on mice with the transgenic expression of H^+K^+-ATPase β chain in the thymus. Thymocytes from these mice transferred oophoritis but not gastritis to athymic recipients (Alderuccio et al. 1993).

The next experiment uncovered the dichotomy that oophoritis would develop in athymic syngeneic recipients of neonatal, but not adult splenic CD4+ T cells. This finding is consistent with the results of an earlier study (Smith et al. 1989) that tracked Vβ11+ T cells that recognize the endogenous superantigen of MHC class II, I-E, and endogenous retroviral peptides (Dyson et al. 1991; Woodland et al. 1991). Endogenous superantigens are recognized by T cell receptors in a Vβ-specific manner. The study showed that Vβ11+ T cells, undeleted in neonatal thymus, were enriched in the neonatal spleen (Smith et al. 1989).

Therefore, tolerance for self antigens, including those relevant to autoimmune oophoritis, is ontogenetically regulated. Specifically, the neonatal repertoire is enriched in self-reactive T cells, which, for reasons that remain unclear, do not require regulation. Since thymectomy soon after

Table 1.2 *Induction of autoimmune oophoritis or autoimmune gastritis by transfer of normal BALB/c T cells to syngeneic athymic nu/nu mice or scid mice*

Cell source	Cell number (×10^{-6})	Autoimmune oophoritis or autoimmune gastritis (no. positive/total (%))	Ovarian or gastric autoantibodies (no. positive/total (%))
None	–	0/6 (0)	0/6 (0)
Day 3-old thymocytes	20	7/9 (78)	3/5 (60)
Day 3-old CD8-depleted spleen cells	20	4/5 (80)	4/5 (80)
Adult CD4+CD8− thymocytes	20	8/8 (100)	8/8 (100)
Adult spleen cells	20	0/17 (0)	0/17 (0)
CD5-depleted adult spleen cells, and D3 or D7 spleen cells	10+10	1/22 (5)	0/22 (0)

Lymphocytes from female BALB/c Nu/+ mice were untreated, or selected by appropriate monoclonal antibody and complement. Cell purity was verified by fluorescence-activated cell sorter analysis, then cells were counted and injected intravenously into female BALB/c nu/nu or BALB/c SCID recipients. Pathology and serum antibodies in recipients were analysed two to three months later. (Data from Smith *et al.* 1992.)

birth should limit the T cell repertoire to that of the neonate, and skew it to one enriched in self-reactive T cells, this could explain why autoimmune diseases occur spontaneously in D3TX mice. Indeed, Vβ11+ self-reactive T cells have been found to be enriched in adult I-E+ D3TX mice (Smith *et al.* 1989; Jones *et al.* 1990). This is one of the two possible mechanisms for autoimmune disease occurrence in the D3TX mice. However, it should be emphasized that the fate in the thymus of T cells capable of transferring oophoritis and gastritis to athymic nude (nu/nu) mice differs in one major respect from the earlier finding on Vβ11 T cells that recognize the endogenous superantigen (Smith *et al.* 1989); whereas endogenous superantigen-specific T cells are deleted in the adult but not the neonatal thymus, the deletion of oophoritiogenic T cells is not apparent in the adult or neonatal thymus.

Although adult spleen T cells did not transfer oophoritis and gastritis to athymic recipients, the fraction of adult splenic CD4+ T cells that expresses a low level of cell-surface CD5 molecules were able to do so. Splenic T cells were treated with CD5 antibody and complement, and the residual T cells were then injected into recipients (Sakaguchi *et al.* 1985). Although all T cells are known to be CD5+, this treatment eliminated 95 per cent of the CD4+ T cells, as demonstrable by flow cytometry. The residual 5 per cent, which expressed a low level of CD5 (CD4+ CD5^{low} T cells), were responsible for disease transfer (Smith *et al.* 1992). These studies clearly establish the existence of oophoritogenic, mature T cells within the peripheral T cell pool of normal adult mice. These T cells appear to have a unique phenotype that includes low cell-surface expression of the CD5 molecule, a known T cell activation marker (Spertini *et al.* 1991). In a similar study of normal rats, self-reactive, pathogenic CD4+ T cells were reported to produce interleukin-2 (Il-2) but not Il-4 upon activation (Fowell *et al.* 1991), thus they are CD4+ T cells of the T_H1 phenotype (Mossman *et al.* 1986). Collectively, the data indicate that the potentially pathogenic T cells present in adult spleens are present but normally non-functional. However, they can cause autoimmune oophoritis when the regulators are removed which indicates that the non-functional status of the autoreactive T cells is reversible.

The existence of regulatory T cells with the capacity to suppress oophoritogenic T cells has also been demonstrated. When adult splenic T cells and neonatal T cells were co-injected into athymic recipients, oophoritis and gastritis that would be induced by the neonatal T cells was abrogated by the adult T cells (Table 1.2). At present, the nature of the regulatory T cells in the oophoritis model is not well defined. However, in analogous autoimmune models of athymic rats, the regulatory T cells were reported to produce IL-4 but not IL-2 (Fowell *et al.* 1991); that is, they are CD4+ T cells of the T_H2 functional phenotype (Mossman *et al.* 1986). A second study, on autoimmune thyroiditis and diabetes, has indicated that the regulatory T cells might bear the RT6 allotypic marker (McKeever *et al.* 1990).

This series of experiments has emphasized the crucial role of regulatory or suppressor T cells with the capacity to render oophoritogenic T cells non-functional in the normal individual. While our discussion has focused so far on ovarian and gastric autoimmunity, similar data have also been obtained for orchitogenic T cells (Smith and Tung, unpublished).

III EVENTS THAT CAN BRING ABOUT AUTOIMMUNE DISEASE OF OVARY AND TESTIS

The analysis of normal murine T cells described above indicates that pathogenic self-reactive T cells exist in normal mice, but are controlled by regulatory T cells. This is consistent with the concept of T cell balance based on results in autoimmune thyroiditis research (Rose *et al.* 1981). It follows that if the balance is tipped in favour of effector T cell activity, autoimmune diseases could occur. This scenario has been substantiated in two models of autoimmune oophoritis: the depletion of regulatory cells by D3TX or other manipulations of the normal immune system; and the activation of oophoritogenic T cells by non-ovarian peptides that mimic an oophoritogenic T cell peptide.

1 Autoimmune oophoritis and orchitis occur when regulatory T cells are depleted

Since 1967, several important models of autoimmune endocrinopathy that include orchitis and oophoritis have been investigated systematically by Nishizuka and his colleagues (Nishizuka and Sakakura 1969; Kojima and Prehn 1981; Taguchi *et al.* 1986; Taguchi and Nishizuka 1987; Sakaguchi and Sakaguchi 1989; Sakaguchi *et al.* 1994; reviewed in Tung *et al.*, 1994), (Table 1.1).

D3TX (C57BL/6 × A/J)F1 mice developed autoimmune oophoritis spontaneously, and the disease was readily transferrable to young recipients by CD4+ (but not CD8+) T cells from diseased individuals (Taguchi and Nishizuka, 1980; Smith *et al.* 1991). Of particular significance was the observation that autoimmune oophoritis that followed D3TX was preventable by the infusion of T cells from normal adult mice, provided the D3TX recipients had not reached 10–12 days of age (Sakaguchi *et al.* 1991; Smith *et al.* 1991). Both adult thymocyte and adult spleen cells suppressed oophoritis in the D3TX mice, and among adult spleen T cells, the regulatory function was assigned to CD4+ T cells that expressed high levels of CD5 (Smith *et al.* 1991). The finding of disease suppression by normal T cells indicates that autoimmune diseases occur in D3TX mice because of a deprivation of regulatory T cells. This conclusion is corroborated by similar findings in all the autoimmune models based on the perturbation of the normal immune system (Table 1.1), in which the common feature was the prevention of diseases by normal T cells.

Therefore, two mechanisms may be involved in the pathogenesis of D3TX autoimmunity: the enrichment and expansion of the self-reactive, neonatal T cell repertoire as discussed above; and the deprivation of regulatory T cells that may have a relatively late ontogeny.

2 Molecular mimicry can cause autoimmune oophoritis through pathogenic T cell activation

Autoimmune oophoritis also develops when the balance is tipped in favor of the expansion of pathogenic T cells. Although peptides from endogenous self antigens in the normal host do not spontaneously activate T cells, this evidently can occur through molecular mimicry when the host is immunized by peptides that mimic the self peptide with adjuvant. This possibility has been documented in a new murine autoimmune oophoritis model.

i Murine autoimmune oophoritis can be induced by a peptide from the Zona pellucida ZP3 protein

Autoimmune oophoritis was first induced in rats immunized with heterologous or homologous ovarian antigens in complete Freund's adjuvant (Jankovic et al. 1973; Vajnstangl et al. 1979). Research in the 1980s described ovarian failure in several animal species immunized with heterologous zona pellucida or ZP3 (Wood et al. 1981; Mahi-Brown et al. 1982; Gulyas et al. 1983; Sacco et al. 1987), and an autoimmune basis for the ovarian failure was subsequently established by Rhim et al. (1992).

The zona pellucida is the acellular matrix surrounding developing and ovulated oocytes, and it exists as degraded proteins within atretic follicles. ZP3 is a major zona pellucida glycoprotein composed of 424 amino acids (Dean 1992). A 15-mer peptide of murine ZP3 (328–342), which possesses a known B cell epitope of ZP3 (335–342), was originally selected as an experimental contraceptive vaccine antigen (Millar et al. 1989). This peptide was subsequently found to elicit a high incidence of severe autoimmune oophoritis in mice (Fig. 1.1). The immunized mice developed ZP3-specific T cell responses and antibodies to ZP3. Antibody was detected in treated mice both in the serum and bound to the ovarian zona pellucida. By studying a series of truncated ZP3 328–342 peptides, the shortest oophoritogenic epitope was mapped to the octamer sequence, ZP3 330–337. This peptide overlaps by two residues with the known heptamer B cell epitope (335–342) (Table 1.3). Because the well characterized 13-mer ZP3 peptide possesses all the functional attributes of a regular self protein

Table 1.3 *Structural and functional domains of the murine ZP3 mini-autoantigen*

ZP3 peptide	Amino acid sequence	Biological functions
330–342	NSSSSQFQIHGPR	Mini-autoantigen
330–338	NSSSSQFQ	Minimum T cell epitope
		Pathogenic epitope
		Induction of amplified ZP3 antibody
335–342	QFQIHGPR	Native B cell epitope

antigen, it is a mini-autoantigen eminently suited for studies of complex autoimmune responses.

ii T cell epitope mimicry as a mechanism that triggers autoimmune oophoritis

We have investigated molecular mimicry based on the model of autoimmune oophoritis induced by the ZP3 mini-autoantigen. We asked whether non-ovarian peptides could be recognized by ZP3-specific T cells and induce autoimmune oophoritis. And if this occurs, what mechanism may be responsible for the molecular mimic? We searched the protein sequence library for peptides with amino acid sequences that simulate the nanomer T cell epitope 330–339 in the ZP3 mini-autoantigen and uncovered one from the δ chain and one from the γ chain of the murine acetylcholine receptor (AChR) (Luo *et al*. 1993). Despite the limited homology between these peptides and ZP3 330–339, we investigated the AChR peptides because of reported clinical association between premature ovarian failure and myasthenia gravis (LaBerbera *et al*. 1988).

Autoimmune oophoritis developed in mice immunized with the AChR δ peptide that shares four amino acid residues with the ZP3 mini-autoantigenic T cell epitope. The cross-reaction between these T cell peptides was established by the finding that both the ZP3 peptide and the AChR δ peptide were recognized by a ZP3-specific T cell clone, both restricted to the class II MHC molecule, IA-$\alpha^k\beta^b$ (Table 1.4, study 1). In contrast, mice immunized with the AChR γ peptide that shares two common residues with the ZP3 peptide did not develop ovarian pathology.

To investigate the mechanism of antigen mimicry, we mapped the residues in the ZP3 mini-autoantigen and the AChR δ peptides which were critical for disease induction and for recognition by pathogenic T cells. Of the nine residues in the ZP3 peptide, four were critical and three of these four were shared between the ZP3 peptide and the AChR δ peptide

(Table 1.4, study 2). To provide direct evidence for molecular mimicry at these critical residues, we studied nanomer polyalanine peptides into which selected residues of the ZP3 or the AChR δ peptides were inserted (Table 1.4, study 3). The peptide with the four critical residues of the ZP3 peptide, and the peptide with the residues common to ZP3 and AChR δ peptides both elicited severe oophoritis and stimulated the ZP3-specific T cell clone to proliferate (Luo *et al.* 1993). This study provides conclusive evidence that non-ovarian peptides sharing sufficient critical residues with ZP3, a self peptide, can potentially stimulate ZP3-specific T cell clones. Through the mechanism of T cell peptide mimicry, a non-ovarian peptide has elicited autoimmune oophoritis by clonal expansion and activation of ZP3-specific, pathogenic T cells.

IV PATHOGENESIS OF EXPERIMENTAL AUTOIMMUNE ORCHITIS AND OOPHORITIS

1 Pathogenetic role of inflammatory CD4+ T cells

The importance of the T cell-mediated mechanism has been established by experiments involving the adoptive transfer of disease by lymphocytes of known function and antigen specificity. T cells that had been activated *in vitro* transferred severe orchitis and vasitis to normal euthymic mice (Mahi-Brown *et al.* 1987; Mahi-Brown and Tung 1989; Feng *et al.* 1990; Itoh *et al.* 1991*b* Itoh *et al.* 1992*a*). Similarly, T cells from mice immunized with oophoritogenic ZP3 peptides transfered oophoritis to syngeneic recipients within two days (Rhim *et al.* 1992). In both cases, CD4+ (not CD8+) T cells were responsible for disease (Mahi-Brown *et al.* 1987; Itoh *et al.* 1991*b*; Rhim *et al.* 1992). The importance of CD4+ T cells in disease transfer was subsequently confirmed by studies based on CD4+ T cell lines (Itoh *et al.* 1992*a* Rhim *et al.* 1992; Yule and Tung 1993).

A recent study based on T cell clones has further defined the pathogenetic mechanism of EAO (Yule and Tung 1993). Despite the use of

Fig. 1.1 Histopathology of murine autoimmune oophoritis. (A) Normal murine Graafian antral follicle. (B) Severe oophoritis with inflammatory cells invading Graafian follicles in mouse 10 days after immunization with 50 nmol ZP3$^{328-342}$. Note lymphocyte infiltration in the granulosa cell layer and accumulation of neutrophils and macrophages in the region of the oocyte (*arrows*). (C) Granuloma in interfollicular area of B6AF$_1$ mouse ovary after immunization with ZP3$^{328-342}$. (D) 21 days after immunization with ZP3$^{328-342}$. There is loss of large follicles and ovarian atrophy. *Arrows* point to small ovarian follicles without oocytes. Sections stained with hematoxylin and eosin. Panels A–C are at eight times higher magnification than panel D. (Reproduced from Rhim *et al.* 1992, with permission of the publisher.)

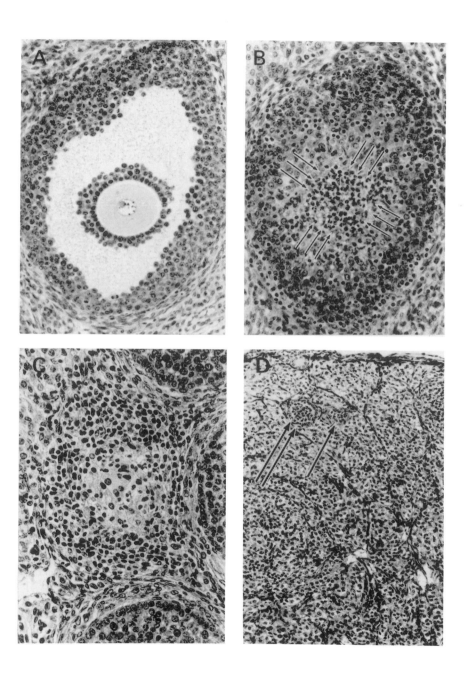

Table 1.4 Mechanism of autoimmune oophoritis induced by murine acetylcholine receptor γ chain peptide 96–104

Study	Peptide sequence	Oophoritis incidence (%)	severity 1 2 3 4	Zona-bound IgG	Proliferation of J3 Clone (δCPM) 0.1	1 (Peptide, μM)	10	100
1	CFA ONLY	0/15	— — — —	0/15	NS	NS	250	NS
	NSSSSQFQI	13/15 (87)	2 4 7 0	14/14	NS	NS	140 633	NS
	NNNDGSFQI	8/14 (57)	3 4 1 0	9/14	NS	NS	54 348	NS
	NNVDGVFEV	0/5	— — — —	0/5	NS	NS	−2 097	NS
2	NSSSSQFQI	9/10 (90)	1 3 5 0	10/10	169	12 216	32 874	249 315
	ASSSSQFQI	0/9 (0)	0 0 0 0	0/9	−185	518	600	46 739
	NASSSQFQI	8/9 (90)	2 3 2 1	6/9	−56	25 924	51 738	236 450
	NSASSQFQI	10/10 (100)	2 2 6 0	4/5	1 756	34 433	85 815	219 133
	NSSASQFQI	5/5 (100)	0 1 3 1	5/5	835	44 120	79 919	167 197
	NSSSAQFQI	6/10 (60)	2 2 2 0	3/5	−167	1 046	34 234	141 256
	NSSSSAFQI	0/9 (0)	0 0 0 0	0/5	−194	−206	637	15 810
3	NAAAAAQFQA	8/10 (80)	3 3 2 0	8/10	27 284	109 783	158 031	89 808
	NAAAAAFQI	13/15 (87)	3 7 3 0	15/15	30 769	64 960	124 713	121 605
	NAAAAAFAA	0/10 (0)	— — — —	0/7	−2 615	−2 039	1 295	11 456

Adult female B6AFI mice immunized with 50 μg of the peptide in adjuvant, or adjuvant alone, were studied 14 days later. Antibody to zona pellucida is detected by direct immunoflourescence as intense IgG bound to ovarian zona pellucida. The proliferative responses of the ZP3 330–342-specific, oophoritogenic T cell clone J3 to the peptides (30 μM) in the presence of mitomycin-treated L cells transfected with IA($\alpha^k\beta^b$), are expressed as δcpm. The amino acid sequence of ZP3 (330–338) is NSSSSQFQI, and that of ACRd96–104 is NNNDGSFQI. NS, not studied. (Data from Luo et al. (1993), with permission of the publisher.)

crude tissue antigens, both T cell lines and all of 16 independent T cell clones transferred EAO to normal syngeneic mice with pathology that affected one or more of the testis, epididymis, or the vas deferens. Thus, orchitogenic peptides of restricted number, dominant immunogenicity or both are likely to occur. Although testis antigen-derived T cell clones responded preferentially to testis antigen, and sperm antigen-derived clones responded more to sperm antigens, each of the 16 clones responded to both antigens. Thus common orchitogenic antigens exist in these germ cell populations, though the quantity of target antigens may differ in distribution. The fact that sperm-specific T cell clones elicited disease clearly indicates that germ cells alone can elicit EAO. All orchitogenic T cell clones expressed CD4 and the $\alpha\beta$ T cell receptor. When activated, they produced IL-2, interferon γ (IFNγ), and tumor necrosis factor (TNF), but not IL-4. This cytokine profile characterizes the T_H1 CD4+ T cell subset, known to be responsible for the delayed-type immunological reaction including granulomatous inflammation. Importantly, disease transfer was significantly and reproducibly attenuated when recipients were given neutralizing antibody to TNF, but not when they were given neutralizing antibody to IFNγ. Hence TNF has been defined as a cytokine important in the pathogenesis of this autoimmune disease.

The main target antigen in EAO recognized by CD4+ T cells has been clearly demonstrated by the unique distribution of testicular pathology in recipients of activated orchitogenic T cell lines or T cell clones (Tung *et al.* 1987*a*; Yule and Tung 1993). The predominant testicular pathology invariably begins and focuses around the straight tubules and, to a lesser extent, the rete testis and the ductus efferentes (Fig. 1.2). It is known that CD4+ T cells recognize peptides in association with class II MHC molecules (Unanue 1989), and in normal mouse testis, class II MHC-positive macrophages are sparse, but form a dense cuff around the straight tubules, where orchitogenic T cells preferentially encounter germ cell peptides to initiate EAO (Tung *et al.* 1987*a*). The inflammatory infiltrate then blocks the passage of tubular spermatozoa and fluids to cause dilatation of the seminiferous tubules. Severe orchitis spreads centripetally to involve peripheral seminiferous tubules, and ultimately testicular atrophy and necrosis. The distribution of orchitis differs between passive and active EAO (Tung *et al.* 1987*a*). In active EAO, the inflammatory infiltrates are random, with initial accumulations of lymphoid cells around blood vessels in the interstitium, outside the seminiferous tubules.

Studies based on T cell clones against the ZP3 mini-autoantigen have obtained results for autoimmune oophoritis identical to those described for the orchitogenic T cell clones. They were also CD4+, and they produced IL-2, TNF and IFNγ, but not IL-4, when activated *in vitro*. Moreover, oophoritis transfer was also blocked by antibody to TNF but not to IFNγ (Luo *et al.* 1993; Tung *et al.*, unpublished). A main target

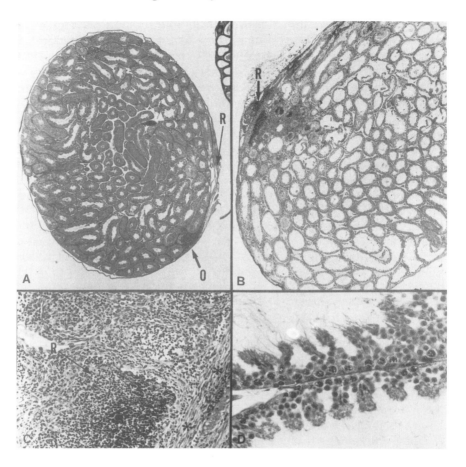

Fig. 1.2 Predominant histopathological lesions in active EAO (panel A) and passive EAO (panels B, C, and D). (A) A common histopathological finding in active EAO, illustrated in two testes, is subcapsular orchitis involving peripheral seminiferous tubules. (B) In passive EAO, inflammation in the regions of straight tubules, rete testis (R) is common, and this leads to severe dilation of seminiferous tubules. Compare this testis with a testis with active EAO (Panel A) taken at the same magnification. (C) At higher magnification, inflammation in the rete regions is severe and obstructs the lumen. (D) Dilated seminiferous tubules have an attenuated but intact germinal epithelium, without evidence of orchitis (=O). (Reproduced from Tung et al. 1987a, with permission of the publisher.)

antigenic structure in the ovaries is the atretic follicles which contain ZP3 antigens from degenerating oocytes, often surrounded by class II MHC-positive macrophages. When oophoritogenic T cell lines or T cell clones were transferred to normal recipients, they induced granulomatous inflammation, which began in the hilar region and affected predominantly the atretic follicles in the ovarian medulla (Tung *et al.*, unpublished).

2 Pathogenic role of antibodies

As discussed at the beginning of this chapter, there is conclusive evidence for the presence of autoimmunogenic germ cells outside the blood–testis barrier that react with circulating autoantibodies *in vivo* (Yule *et al.* 1988). When frozen testis sections at different times after immunization with testis antigens in adjuvant were studied, intense deposits of IgG were detectable on cells of the seminiferous tubules outside the blood-testis barrier (*Fig. 1.3*). The IgG deposits were characterized as antibodies bound to the preleptene spermatocytes, and were detected as early as seven days after immunization, five to six days before the onset of orchitis.

Testis IgG deposits were elicited by immunization with testis homogenate in incomplete Freund's adjuvant, which does not elicit EAO. Thus the immune deposits are not sufficient to cause EAO. The lack of pathogenicity is probably related to the unique immunoglobulin subclass of the IgG deposits. Only IgG1 and IgG3 were detected, and complement components were absent. However, the antigenic molecules in these complexes may provide the orchitogenic peptides, presented by macrophages or dendritic cells to stimulate the effector T cells to induce EAO. If this was proven correct, it could explain how T cells recognize germ cell antigens sequestered behind the blood–testis barrier at the level of the seminiferous tubules.

In EAO, antibodies were also detected as immune complexes around aspermatogenic tubules free of orchitis. Although focal deposits of immune complexes were detectable in EAO of rabbits (Tung and Woodroffe 1978) and mice (Kohno *et al.* 1983; Itoh *et al.* 1991*a*), abundant immune complexes were found surrounding the aspermatogenic tubules in spontaneous EAO of the dark mink (Tung *et al.* 1981), aging Brown Norway rats (Furbeth *et al.* 1989), vasectomized rabbits (Bigazzi *et al.* 1976), and possibly testes from infertile men (Salomon *et al.* 1982; Lehmann *et al.* 1987). However, the pathogenic role of testicular immune complexes in EAO has not yet been explored.

3 A novel mechanism of ovarian autoantibody induction by the T cell peptide of the ZP3 mini-autoantigen

The characteristic features of organ-specific autoimmune disease include the detection of: T cell response to the self peptide; pathology in the

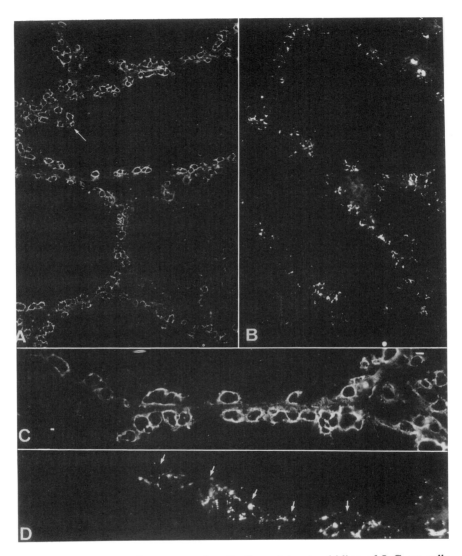

Fig. 1.3 (A) Immunofluorescence localization of *in vivo* biding of IgG on cells located at the periphery of seminiferous tubules. (B) The deposits appear granular in some testes. (C) Similar deposits are transferred passively to normal mice by sera of orchidectomized donors immunized with testis in adjuvants. (D) Co-localization of IgG and tritiated thymidine, detected by autoradiography in mice immunized with testis in adjuvant 10 days earlier, and injected with tritiated thymidine two hours before immunization. The latter treatment labels exclusively germ cells located outside the Sertoli cell barrier. (Reproduced from Yule *et al.* 1988, with permission of the publisher.)

target organ; serum autoantibodies to the target antigen; and binding of autoantibodies to the target antigen that may affect the physiological function of the antigen. In the study on autoimmune oophoritis induced by the ZP3 mini-autoantigen, we have so far demonstrated the importance of a pathogenic T cell response to the peptide. An unexpected finding was that the activation of effector T cells *per se* can lead to the spontaneous occurence of autoantibodies against the native ZP3 protein (Lou and Tung 1993).

The production of autoantibody in response to a pure T cell epitope was first suggested by the detection of antibody to the zone pellucida in mice immunized with the truncated ZP3 mini-autoantigen (330–338) that lacked the intact B epitope 335–342 (Table 1.3) (Lou and Tung 1993). Later, similar zona pellucida antibodies were also detected in mice immunized with the AChR δ peptide that mimicked the ZP3 peptide, and even the polyalanine peptide that contained the critical residues of either the ZP3 peptide or the AChR δ chain peptide (Table 1.4) (Luo *et al.* 1993). That an exclusively T cell epitope of murine ZP3 elicited autoantibodies against ZP3 outside the T cell peptide was confirmed by the following observations.

1. The truncated ZP3 peptides that induced antibody to the zona pellucida did not contain any additional B cell epitopes that would cross-react with native ZP3.
2. The zona pellucida-bound IgG was 400-fold enriched in antibody activity against the zona pellucida, so the zona-bound IgG represented tissue-bound antibodies.
3. The antibodies produced by mice immunized with a T cell epitope reacted with ZP3 by immunoblot and with native zona pellucida *in vivo*.
4. Perhaps most importantly, endogenous ovarian antigens were required for the induction of zona pellucida antibodies. Thus, autoantibodies against the zona pellucida were not detected in ovariectomized mice immunized two weeks later with the ZP3 peptides that lacked the B epitope 336–342.

More recently, we have documented that ovarian pathology is not a prerequisite for the induction of autoantibody to the zona pellucida (Lou and Tung, unpublished). Thus, the autoantibody response is not merely a response to antigens released from diseased ovaries. This phenomenon has been called autoantibody amplification.

Autoantibody amplification can be explained by the following sequence of events. In normal mice, ovarian antigenic macromolecules including ZP3 normally reach the regional lymph nodes where they encounter antigen-specific B cells, bind to the antigen receptors, become internalized, and are processed within the endosomes (Unanue 1989). Processed antigenic peptides of ZP3 associated with class II MHC molecules are recycled to the B cell surface. In normal mice, the series of events ends here. However, in mice immunized with T cell peptide, the activated ZP3-specific T cells may encounter the peptide–MHC complexes on the ZP3-specific B cells. Thereupon, they can activate the B cells to produce antibodies. The antibody specificity should coincide with the specificity of the antigen receptor of the B cells that capture the ovarian antigen, different from that of the ZP3 mini-autoantigen.

An implication of the finding is that self-reactive B cells in normal female mice can respond to self ovarian antigen and, therefore, they are probably not tolerized. In addition, the occurrence of amplified autoantibodies indicates that serum autoantibodies need not mirror the immunogens that initiate an autoimmune disease. As such, investigation of molecular mimicry in autoimmunity based on autoantibodies could be misleading. However, we have found that autoantibodies elicited by the endogenous antigens preferentially recognize conformational determinants, can bind to target antigens at high affinity, and are potentially important in autoimmune disease pathogenesis (Lou and Tung 1993).

The phenomenon of autoantibody amplification also supports an important concept in autoimmunity that is emerging. Autoimmune T cell response and antibody response with specificity other than the immunogen or the immunogenic epitope can occur frequently. The phenomenon has also been documented in immune response to murine myelin basic peptide in experimental autoimmune encephalomyelitis (Lehmann *et al.* 1992), murine gastric parietal cell K^+H^+-dependent ATPase in autoimmune gastritis of the D3TX mice (Alderuccio *et al.* 1993), and, most recently, for the pancreatic islet β-cell autoantigen, glutamic acid decarboxylase, in the diabetic NOD mice (Kaufman *et al.* 1993; Tisch *et al.* 1993). This phenomenon of antigen spreading is expected to influence the design of the antigen-based immunotheraphy of autoimmune diseases.

V IMMUNOGENETIC APPROACH TO AUTOIMMUNE OOPHORITIS AND ORCHITIS

A molecular immunogenetic approach offers an alternative strategy to explore the pathogenesis of autoimmune diseases of the gonads. Identification of susceptible and resistant strains of inbred mice and the phenotyping of segregating populations derived therefrom allows an estimation of the number of genes involved. With the recent development of simple sequence polymorphisms, or microsatellites (Love *et al.* 1990; Deitrich *et al.* 1992), and random amplified polymorphic markers (Woodward *et al.* 1992; Wardell *et al.* 1993), both of which show DNA variations between inbred strains of mice, it is now feasible to localize the genes controlling even complex traits to specific chromosomal regions (Cornall *et al.* 1991;

Todd *et al.* 1991; Ghosh *et al.* 1993). This approach may reveal linkage to candidate genes that can be sequenced, or it may provide the first step in the positional cloning of the genes of interest (Prins *et al.* 1993). It is particularly useful in studying known polygenic diseases such as EAO and autoimmune oophoritis involving both MHC-linked and non-MHC-linked genes (Kojima and Prehn 1981; Teuscher 1985; Teuscher *et al.* 1985; Teuscher *et al.* 1987*a*; Tokunaga *et al.* 1993*a,b*).

Our first study addressed the genetic control of differential susceptibility to EAO of the testis, the epididymis, and the vas deferens. Mice with the H-2^s haplotype developed mainly orchitis, whereas mice of the H-2^k haplotype developed pathology mainly in the epididymis and the vas deferens (epididymovasitis) (Teuscher *et al.* 1985; Tung *et al.* 1985). This difference in response was also demonstrated using the BXH series of recombinant inbred lines derived from the EAO-susceptible C57BL/6J and the EAO-resistant C3H/HeJ strains (Person *et al.* 1992). A comparison of the strain distribution pattern of autoimmune orchitis with that of the typed alleles segregating in the BXH RI lines has not demonstrated definitive linkage. However, the data suggest that the locus controlling orchitis resistance tends to segregate with the *D17Tu20* or *D17Tu52* loci on chromosome 17. A similar comparison of autoimmune epididymovasitis indicated maximal concordance with *MARK-1* (Watson and Paigen 1990), and the androgen-regulated locus on chromosome 7.

Further genetic studies on EAO have uncovered four additional orchitis susceptibility genes: *Orch-1*, which maps within *H-2*, *and the Orch-2*, *Bphs*, *and Orch-3* genes, which reside outside *H-2*. Studies based on intra-MHC recombinant inbred mice have indicated that *Orch-1* is an orchitis-susceptibility gene mapping within the *H-2S/H-2D* interval (Teuscher *et al.* 1985; Teuscher *et al.* 1990), between the *IR* gene controlling antibody responsiveness to trinitrophenol-Ficoll (*TNP-Ficol*) and the locus encoding TNF-α (Teuscher *et al.* 1990). This has established the gene order in the region as: *H-2S—TNP-Ficoll—Orch-1—Tnfα—H-2D*. Recent molecular analysis has further mapped *Orch-1* within a 50–60 kb segment from *Hsp 70.1* to a region proximal to *G7* (Person *et al.* 1992; Snoek *et al.*1993), that encompasses the *Hsp 70.1*, *Hsp70.3*, *Hsc70t*, *G7b*, and *G7a/Bat6* genes.

Genes outside the MHC strongly influence EAO susceptibility. DBA/2J (H-2^D) mice are resistant to EAO (Teuscher 1985; Teuscher *et al.* 1985), and BALB/cByJ(H-2^D) mice are susceptible to EAO, whereas BALB/cByJ X DBA/2J F1 (CD2F1) mice are resistant. Among a (BALB/cBY X DBA/2J)F1 XBALB/cByJ (BC1) population ($n = 172$), 54 per cent of the animals were classified as resistant, and 46 per cent as susceptible. Thus dominant resistance appears to be controlled by a single gene, *Orch-2* (Teuscher *et al.* 1985). Using DNAs isolated from the phenotyped BCI population and previously mapped microsatellite markers distinguishing DBA/2J and BALB/cByJ, and we have linked the *Orch-2* locus to

$D11Mit8$ (χ^2 = 17.2, df = 1, p < 0.001; t-test = 4.12, df = 100, p < 0.001). The maximal support interval on chromosome 11 places $Orch$-2 within a 10 cM region which encodes a number of candidate loci of immunological relevance. These include: nu, Idd-4, H-3, Tca-3, $Mip1a$, and $Mip1b$ (Buchberg and Camper 1993).

Induction of classical murine EAO requires pertussis toxin as an adjuvant. The $Bphs$ locus controls the phenotypic expression of susceptibility to $Bordetella\ pertussis$-induced histamine sensitization (Wardlaw 1970). Mice which possess a susceptible allete at this locus die from hypovolemic shock following histamine challenge, and those with a resistant allele do not. Susceptibility to experimental autoimmune encephalomyelitis (Linthicum and Frelinger 1982) and EAO (Teuscher 1985) has been shown to be associated with, and in the latter case linked to, a susceptible allele at this locus. Using microsatellites and RAPD markers, we have mapped the $Bphs$ locus to mouse chromosome 6, telomeric of $Tcrb$ and centromeric of Prp (Sudweeks $et\ al.$ 1993). Our mapping studies placed the $Bphs$ locus within a region encoding a number of immunologically important candidate genes, including Igk, ly-2, IL-$5r$, Ly-4, and $Tnfr$-2 (Moore and Elliot 1993). Additionally, the locus encoding the guanine nucleotide-binding protein, β-polypeptide-3 (Gnb-3), which may be relevant to pertussis toxin action, also resides within the same support interval.

The role of disease-resistance genes has also become evident from studies on substrains of BALB/c mice. Disease resistance, inherited as a recessive trait in the BALB/cJ mice, has been shown by segregation analysis to be associated with a single genotypic difference, $Orch$-3 (Person $et\ al.$ 1992). $Orch$-3 is not related to elevated levels of serum α-fetoprotein in adult BALB/cJ animals (Olsson $et\ al.$ 1977; Teuscher $et\ al.$ 1987b),or to the $Bphs$ locus (Teuscher $et\ al.$ 1987b).

Finally, we have studied the genes which control susceptibility to autoimmune oophoritis in D3TX mice. A previous study by Kojima and Prehn (1981) designed to identify such genes suggested that H-2-linked genes play little, if any, role in determining disease outcome. In addition, in a recent study on autoimmune oophoritis and other polyendocrinopathy that developed in mice with the transgenic T cell receptor Vα protein, it was found that the organ distribution of pathology was strictly controlled by the genetic background of the mice that carried the transgene (Sakaguchi $et\ al.$ 1994). In a multi-center collaborative study, 144 D3TX (C57BL/6J X A/J) F1 X C57BL/6J backcross (BCI) mice were crossbred with the disease susceptible A/J strain (90 per cent disease incidence), (C57BL/6J X A/J)F1 hybrid (100 per cent disease incidence), and the disease-resistant C57BL/6J (8 per cent disease incidence) (Wardell $et\ al.$, in press). Histological analysis of the ovaries at 60 days of age revealed that 77 exhibited oophoritis, whereas 67 were resistant. The ratio of 1.15 between susceptible and resistant animals is consistent with that expected for a single locus.

Microsatellites have been employed to generate a genomic exclusion map that localized the autoimmune oophoritis susceptibility gene (*AOD-1*) to chromosome 16, linked to *D16Mit4* ($\chi^2 = 12.5$, df = 1, $p < 0.001$). Several candidate genes of immunological relevance reside on chromosome 16, including *Ly-7, lfgt, lfrc, lgl-1, Mls-3*, and *VpreB* (Reeves *et al.* 1993). Fine mapping of the region encoding *AOD-1* will be necessary for positional cloning and characterization of the gene function.

VI SUMMARY AND CONCLUSIONS

The local testicular immunoregulatory environment, provided by the strong blood–testis barrier and perhaps intratesticular cellullar and humoral factors, partially impedes autoimmune responses to testis antigens of late ontogeny. In addition, dynamic systemic tolerance mechanisms acting on the autoreactive T cells play a critical protective role. Recent studies indicate that pathogenic T cells capable of eliciting autoimmune diseases in these organs develop in the normal thymus and persist in the normal peripheral immune system. However, the function of the pathogenic T cells is normally controlled by regulatory T cells. Some important phenotypic differences between these two T cell subsets are being defined. When the clonal balance of these CD4+ T cell subsets is tipped in favour of pathogenic T cells, autoimmune diseases of the gonads could ensue.

Loss of regulatory T cells may occur through aberrant T cell development. This has been amply documented in several exciting models of autoimmune oophoritis and orchitis following the deliberate perturbation of the normal immune system, including neonatal thymectomy. On the other hand, oophoritogenic T cells can be activated by stimulation with non-ovarian peptides that cross-react with self peptides at the level of the T cell receptor. This novel form of antigen mimicry depends in part on the sharing between unrelated peptides of the few critical amino acids that are required for activation of pathogenic T cells.

The inflammatory CD4+(T_H1) T cell mechanism has been established to be a critical pathogenetic pathway for autoimmune orchitis and autoimmune oophoritis. In both cases, the pathogenic T cell clones secrete Il-2, IFNγ, and TNF; and TNF has been shown to be required for amplification of the pathogenic T cell response. The tissue locations wherein pathogenic T cells encounter testicular and ovarian target antigens have been suggested by the consistent localization of histopathology elicited by the pathogenic T cell clones.

Antibodies bind to both testicular and ovarian target antigens during the development of autoimmune orchitis and autoimmune oophoritis. However, the precise role of antibody in these autoimmune diseases has not been critically explored. This is an important research problem. Its resolution will influence the future of contraceptive vaccine development based on ovarian antigens, as well as the clarification of mechanism whereby autoantibody may access ejaculated human spermatozoa to cause infertility.

In the case of ovarian disease, a novel mechanism of autoantibody induction has been uncovered. Immunization of female mice with a pure T cell peptide from ZP3 can lead to the production of antibodies against ZP3 domains outside the immunogenic peptide. Evidently, endogenous antigens from normal and pathological ovaries may reach peripheral immune tissues, and provide the antigenic stimulus to trigger an autoantibody response. This occurs at the same time as activation of ZP3-specific T cells is detected, and is not a consequence of tissue injury. Importantly, these autoantibodies react with native antigenic determinants, and are potentially important in autoimmune disease pathogenesis.

Exciting research based on molecular linkage analysis of inbred mice has been embarked on for mapping genes that influence the susceptibility and resistance to autoimmune diseases of the testis and ovary. This work, based on contemporary technology which can be extended to human diseases in the future, is expected to elucidate new and unanticipated mechanisms that underlie the complex gonadal autoimmune diseases.

In conclusion, contemporary research on experimental testicular and ovarian autoimmune diseases have emphasized similarities between them as well as their similarities with autoimmune diseases affecting other organs. Because of this trend, it is expected that research based on the unique models of gonadal autoimmune diseases and the well-defined self antigens will clarify not only the autoimmune diseases of the gonads, but also the autoimmune process in general.

REFERENCES

Alderuccio, F., Toh, B. H., Tan, S. S., Gleeson, P., and Driel, I. (1993). An autoimmune disease with multiple molecular targets abrogated by the transgenic expression of a single autoantigen in the thymus. *Journal of Experimental Medicine*, **178**, 419–26.

Bigazzi, P. E., Kosuda, L. L., Hsu, K. C., and Andres, G. A. (1976). Immune complex orchitis in vasectomized rabbits. *Journal of Experimental Medicine*, **143**, 382–404.

Born, W. and Wekerle, H. (1981). Selective, immunologically nonspecific adherence of lymphoid and myeloid cells to Leydig cells. *European Journal of Cell Biology*, **25**, 76–81.

Born, W. and Wekerle, H. (1982). Leydig cells nonspecifically suppress lymphoproliferation *in vitro*: implications for the testis as an immunologically privileged site. *American Journal of Reproductive Immunology*, **2**, 291–95.

Buchberg, A. M. and Camper, S. A. (1993). Mouse chromosome 11. *Mammalian Genome*, **4**, S164–75.

Cornall, R. J., Prins, J. B., Todd, J. A., Pressey, A., DeLarato, N. H., Wicker, L. S., *et al.* (1991). Type 1 diabetes in mice is linked to the interleukin-1 receptor and *Lsh/Ity/Bcg* genes on chromosome 1. *Nature*, **353**, 262–5.

De Cesaris, P., Filippini, A., Cervelli, C., Riccioli, A., Muci, S., Starace, G., *et al.* (1992). Immunosuppressive molecules produced by Sertoli cells cultured in vitro: biological effects on lymphocytes. *Biochemical and Biophysical Research Communications*, **186**, 1639–46.

Dean, J. (1992). Biology of mammalian fertilization: Role of the zone pellucida. *Journal of Clinical Investigation*, **89**, 1055–9.

Deitrich, W., Katz, H., Lincoln, S. E., Shin, H. S., Friedman, J., Dracopoli, N. C., *et al.* (1992). A genetic map of the mouse suitable for typing intraspecific crosses. *Genetics*, **131**, 423–47.

Dyson, P. J., Knight, A. M., Fairchild, S., Simpson, E., and Tomonair, K. (1991). Genes encoding ligands for deletion of Vβ11 T cells cosegregate with mammary tumor virus genomes. *Nature*, **349**, 531–2.

Feng, Z. Y., Ming, L. D., Louis, L. J., and Fei, W. Y. (1990). Adoptive transfer of murine autoimmune orchitis with spermspecific cases of male infertility. *Archives of Andrology*, **24**, 51–9.

Fowell, D., McKnight, A. J., Powrie, F., Dyke, R., and Mason, D. (1991). Subsets of CD4 T cells and their roles in the induction and prevention of autoimmunity. *Immunological Review*, **123**, 37–64.

Freund, J., Lipton, M. M., and Thompson, G. E. (1953). Aspermatogenesis in the guinea pig induced by testicular tissue and adjuvant. *Journal of Experimental Medicine*, **97**, 711–25.

Fritz, T. E., Lombard, L. A., Tyler, S. A., and Norris, W. P., (1976). Pathology and familial incidence of orchitis and its relation to thyroiditis in a closed beagle colony. *Experimental Molecular Pathology*, **24**, 142–58.

Furbeth, C., Hubner, G., and Thoenes, G. H. (1989). Spontaneous immune complex orchitis in Brown Norway rats. *Virchows Archives (Cell Pathology)*, **57**, 37–45.

Ghosh, S., Palmer, S. M., Rodrigues, N. R., Cordell, H. J., Hearne, C. M., Cornall, R.J., *et al.* (1993). Polygenic control of autoimmune diabetes in nonobese diabetic mice. *Nature Genetics*, **4**, 404–9.

Griswold, M. D., Robert, K., and Bishop, P. (1986). Purification and characterization of a sulfated glycoprotein secreted by Sertoli cells. *Biochemistry*, **25**, 7265–70.

Gulyas, B. J., Gwatkin, R. B. L., and Yuan, L. C. (1983). Active immunization of cynomolgus monkeys (*Macaca facicularis*) with porcine zonae pellucidae. *Gamete Research*, **4**, 299–307.

Hammer, R. E., Maika, S. D., Richardson, J. A., Tang, J. Y. P., and Taurog, J. D. (1990). Transgenic rats expressing HLA-B27 and human β2 microglobulin with spontaneous inflammatory disease in multiple organ systems: an animal model of HLA-B27-associated disease. *Cell*, **63**, 1099–112.

Head, J. R., Neaves, W. B., and Billingham, R. E. (1983). Immune privilege in the testis. I. Basic parameters of allograft survival. *Transplantation*, **36**, 423–31.

Hedger, M. P., Drummond, A. E., Robertson, D. M., Risbridger, G. P., and de

Krester, D. M. (1989). Inhibin and activin regulate [³H]thymidine uptake by rate thymocytes and 3T3 cells *in vitro*. *Molecular and Celluar Endocrinology*, **61**, 133–8.

Hedger, M. P., Qin, J. X., Robertson, D. M., and de Kretser, D. M. (1990). Intragonadal regulation of immune system functions. *Reproduction and Fertility Development*, **2**, 263–80.

Itoh, M., Hiramine, C., and Hojo, K. (1991a). A new murine model of autoimmune orchitis induced by immunization with viable syngeneic germ cells alone. 1. Immunological and histological studies. *Clinical Experimental Immunology*, **83**, 137–42.

Itoh, M., Mukasa, A., Tokunaga, Y., Hiramine, C., and Hojo, K. (1991b). New experimental model for adoptive transfer of murine autoimmune orchitis. *Andrologia*, **23**, 415–20.

Itoh, M., Hiramine, C., Mukasa, A., Tokunaga, Y., Fukui, Y., Takeuchi, Y., *et al.* (1992a). Establishment of an experimental model of autoimmune epididymoorchitis induced by the transfer of a T-cell line in mice. *International Journal of Andrology*, **15**, 170–81.

Itoh, M., Mukasa, A., Tokunaga, Y., Hiramine, C., and Hojo, K. (1992b). Suppression of efferent limb of testicular autoimmune response by a regulatory CD4+ T cell line in mice. *Clinical Experimental Immunology*, **87**, 455–60.

Jankovic, B. D., Markovic, B. M., Petrovic, S., and Isakovic, K. (1973). Experimental autoimmune oophoritis in the rat. *European Journal of Immunology*, **3**, 375–7.

Jenne, D. E. and Tschopp, J. (1989). Molecular structure and functional characterization of a human complement cytolysis inhibitor found in blood and seminal plasma: identity to sulfated glycoprotein 2, a constituent of rat testis fluid. *Proceedings of the National Academy of Sciences of the USA*, **86**, 7123–7.

Jones, L. A., Chin, T., Merriam, G. R., Nelson, L. M., and Kruisbeck, A. M. (1990). Failure of clonal deletion in neonatally thymectomized mice: tolerance is preserved through clonal anergy. *Journal of Experimental Medicine*, **139**, 1277–85.

Kaufman, D., Clare-Salzler, M., Tian, J., Forsthuber, T., Ting, G., Robinson, P., *et al.* (1993). Spontaneous loss of T-cell tolerance to glutamic acid decarboxylase in murine insulin-dependent diabetes. *Nature*, **366**, 69–72.

Kohno, S., Munoz, J. A., Williams, T. M., Teuscher, C., Bernard, C. C. A., and Tung, K. S. K. (1983). Immunopathology of murine experimental allergic orchitis. *Journal of Immunology*, **130**, 2675–82.

Kojima, A. and Prehn, R. T. (1981). Genetic susceptibility to post-thymectomy autoimmune disease in mice. *Immunogenetics*, **14**, 15–27.

LaBerbera, A. R., Miller, M. M., Ober, C., and Rebar, R. W. (1988). Autoimmune etiology in premature ovarian failure. *American Journal of Reproductive Immunology and Microbiology*, **16**, 115–22.

Lee, W., Mason, A. J., Schwall, R., Szonyi, E., and Mather, J. P. (1989). Secretion of activin by interstitial cells in the testis. *Science*, **243**, 396–8.

Lehmann, D., Temminch, D., Da Rugna, D., Leibundgut, B., Sulmoni, A., and Muller, H. L. (1987). Role of immunological factors in male infertility: immunohistochemical and serological evidence. *Laboratory Investigations*, **57**, 21–8.

Lehmann, P. V., Forsthuber, T., Miller, A., and Sercarz, E. (1992). Spreading of T-cell autoimmunity to cryptic determinants of an autoantigen. *Nature*, **358**, 155–7.

Linthicum, D. S. and Frelinger, J. A. (1982). Acute autoimmune encephalomyelitis is mice II: Susceptibility is controlled by the combination of *H-2* and histamine sensitization genes. *Journal of Experimental Medicine*, **156**, 31–40.

Lou, Y. H. and Tung, K. S. K. (1993). T cell peptide of a self protein elicits autoantibody to the protein antigen: implications for specificity and pathogenetic role of antibody in autoimmunity. *Journal of Immunology*, **151**, 5790–99.

Love, J. M., Knight, A. M., McAleer, M. A., and Todd, J. A. (1990). Towards construction of a high resolution map of the mouse genome using PCR-analysed microsatellites. *Nucleic Acids Research*, **18**, 4123–30.

Luo, A. M., Garza, K. M., Hunt, D., and Tung, K. S. K. (1993). Antigen mimicry in autoimmune disease: sharing of amino acid residues critical for pathogenic T cell activation. *Journal of Clinical Investigation*, **92**, 2117–23.

Mahi-Brown, C. A. and Tung, K. S. K. (1989). Activation requirements of donor T cells and host T cell recruitment in adoptive transfer of murine experimental autoimmune orchitis (EAO). *Cellular Immunology*, **124**, 368–79.

Mahi-Brown, C. A. and Tung, K. S. K. (1990). Transfer of susceptibility to experimental autoimmune orchitis from responder to non-responder substrains of BALB/c mice. *Journal of Reproductive Immunology*, **18**, 247–57.

Mahi-Brown, C. A., Huang, T. T. F., and Yanagimachi, R. (1982). Infertility in bitches induced by active immunization with porcine zonae pellucidae. *Journal of Experimental Zoology*, **222**, 89–95.

Mahi-Brown, C. A., Yule, T. D., and Tung, K. S. K. (1987). Adoptive transfer of murine autoimmune orchitis to naive recipients with immune lymphocytes. *Cellular Immunology*, **106**, 408–19.

McKeever, U., Mordes, J. P., Greiner, D. L., Apple, M. C., Rozing, J. R., Handler, E. S., et al. (1990). Adoptive transfer of autoimmune diabetes and thyroiditis to athymic rats. *Proceedings of the National Academy of Sciences of The USA*, **87**, 7618–22.

Millar, S. E., Chamow, S. M., Baur, A. W., Oliver, C., Robey, F., and Dean, J. (1989). Vaccination with a synthetic zona pellucida peptide produces long-term contraception in female mice. *Science*, **246**, 935–8.

Moore, K. J. and Elliot, R. W. (1993). Mouse chromosome 6. *Mammalian Genome*, **4**, S88–109.

Morgan, A. D. (1976). Inflammation and infestation of the testis and paratesticular structures. In *Pathology of the testis* (ed. R. C. B. Pugh), pp. 79–138. Blackwell, Oxford.

Mossman, T. R., Cherwinski, H., Bond, M. W., Giedlin, M. A., and Coffman, R. L. (1986). Two types of murine helper T cell clones. I. Definition according to profiles of lymphokine activities and secreted proteins. *Journal of Immunology*, **136**, 2348–57.

Mukasa, A., Itoh, M., Tokunaga, Y., Hiramine, C., and Hojo, K. (1992). Inhibition of a novel model of murine experimental autoimmune orchitis by intravenous administration with a soluble testicular antigen: participation of CD8+ regulatory T cells. *Clinical Immunology and Immunopathology*, **62**, 210–9.

Nishizuka, Y. and Sakakura, T. (1969). Thymus and reproduction: sex-linked dysgenesis of the gonad after neonatal thymectomy. *Science*, **166**, 753–5.

Olsson, M., Lindahl, G., and Ruoslahti, E. (1977). Genetic control of alpha-fetoprotein synthesis in the mouse. *Journal of Experimental Medicine*, **145**, 819–27.

Pelletier, R. M. (1986). Cyclic formation and decay of the blood-testis barrier in the mink (*Mustela vison*), a seasonal breeder. *American Journal of Anatomy*, **175**, 91–117.

Person, P. L., Snoek, M., Demant, P., Woodward, S. R., and Teuscher, C. (1992). The immunogenetics of susceptibility and resistance to murine experimental allergic orchitis. *Regional Immunology*, **4**, 284–97.

Pollanen, P., Soder, O., and Uksila, J. (1988). Testicular immunosuppressive protein. *Journal of Reproductive Immunology*, **14**, 125–28.

Prins, J. B., Todd, J. A., Rodrigues, N. R., Ghosh, S., Hogarth, P. M., Wicker, L. S., *et al*, (1993). Linkage on chromosome 3 to autoimmune diabetes and defective Fc receptor IgG in NOD mice. *Science*, **260**, 695.

Reeves, R. H., Irving, N. G., and Miller, R. D. (1993). Mouse chromosome 16. *Mammalian Genome* **4**, S223–9.

Rhim, S. H., Millar, S. E., Robey, F., Luo, A. M., Lou, Y. H., Yule, T., *et al.* (1992). Autoimmune diseases of the ovary induced by a ZP3 peptide from the mouse zona pellucida. *Journal of Clinical Investigation*, **89**, 28–35.

Rose, N. R., Kong, Y-C. M., Okayasu, I., Giraldo, A. A., Beisel, K., and Sundick, R. S. (1981). T cell regulation in autoimmune thyroiditis. *Immunological Review*, **55**, 299–314.

Sacco, A. G., Pierce, D. L., Subramanian, M. G., Yurewicz, E. C., and Dukelow, W. R. (1987). Ovaries remain functional in squirrel monkeys (*Samiri sciureus*) immunized with porcine zona pellucida 55,000 macromolecule. *Biology of Reproduction*, **36**, 484–90.

Sakaguchi, S. and Sakaguchi, N. (1989). Organ-specific autoimmune disease induced in mice by elimination of T cell subsets. V. National administration of cyclosporin A causes autoimmune disease. *Journal of Immunology*, **142**, 471–80.

Sakaguchi, S. and Sakaguchi, N. (1990). Thymus and autoimmunity: capacity of the normal thymus to produce self-reactive T cells and conditions required for their induction of autoimmune disease. *Journal of Experimental Medicine*, **172**, 537–45.

Sakaguchi, S., Takahashi, and T., Nishizuka, Y. (1982). Study on cellular events in post-thymectomy autoimmune oophoritis in mice. II. Requirement of Lyt-1 cells in normal female mice for the prevention of oophoritis. *Journal of Experimental Medicine*, **156**, 1577–86.

Sakaguchi, S. K., Fukuma, K., Kuribayashi, and Masuda, T. (1985). Organ-specific autoimmune diseases induced in mice by elimination of T cell subsets. I. Evidence for the active participation of T cells in natural self-tolerance: deficit of a T cell subset as a possible cause of autoimmune disease. *Journal of Experimental Medicine*, **161**, 72–87.

Sakaguchi, S., Ermak, T. H., Toda, M., Berg, L. J., Ho, W., de St. Groth, B. F., *et al.*, (1994). Induction of autoimmune disease in mice by germline alteration of the T cell receptor gene expression. *Journal of Immunology*, **152**, 1471–84.

Sakamoto, Y., Himeno, K., Sanui, H., Yoshida, S., and Nomoto, K. (1985). Experimental allergic orchitis in mice. I. A new model induced by immunization without adjuvants. *Clinical Immunology and Immunopathology*, **37**, 360–8.

Salomon, F., Saremaslani, P., Jakob, M., and Hedinger, C. F. (1982). Immune complex orchitis in infertile men. *Laboratory Investigations*, **47**, 555–67.

Sato, K., Hirokawa, K., and Hatakeyama, S. (1981). Experimental allergic orchitis in mice. Histopathological and immunological studies. *Virchows Archives (Pathology Anatomy)*, **392**, 147–58.

Selawry, H., Fojaco, R., and Whittington, K. (1987). Extended survival of MHC-compatible islet grafts from diabetes-resistant donors in spontaneously diabetic BB/W rat. *Diabetes*, **36**, 1061–7.

Smith, H., Chen, I. M., Kubo, R., and Tung, K. S. K. (1989). Neonatal thmectomy results in a repertoire enriched in T cells deleted in adult thymus. *Science*, **245**, 749–52.

Smith, H., Sakamoto, Y., Kasai, K., and Tung, K. S. K. (1991). Effector and regulatory cells in autoimmune oophoritis elicited by neonatal thymectomy. *Journal of Immunology*, **147**, 2928–33.

Smith, H., Lou, Y. H., Lacy, P., and Tung, K. S. K. (1992). Tolearance mechanism in ovarian and gastric autoimmune disease. *Journal of Immunology*, **149**, 2212–18.

Snoek, M., Jansen, M., Olavessen, M. G., Campbell, D. R., Teuscher, C., and van Vugt, H. (1993). The *Hsp7–* genes are located in the *Cr-H-2D* region: possible candidates for the *Orch-1* locus. *Genomics*, **15**, 350–56.

Spertini, F., Stohl, W., Ramesh, N., Moody, C., and Geha, R. S. (1991). Induction of human T-cell proliferation by a monoclonal antibody to CD5. *Journal of Immunology*, **146**, 47–52.

Sudweeks, J. D., Todd, J. A., Blankenhorn, E. P., Wardell, B. B., Woodward, S. R., Meeker, N. D., et al. (1993). Locus controlling *Bordetella pertussis*-induced histamine sensitization *9Bphs*, an autoimmune disease-susceptibility gene, maps distal to T-cell receptor B-chain gene on mouse chromosome 6. *Proceedings of the National Academy of Sciences of the USA*, **90**, 3700–4.

Sugihara, S. Y., Izumi., T., Yoshioka, H., Yagi, T., Tsujimura, Y., Kohno, Y., et al. (1988). Autoimmune thyroiditis induced in mice depleted of particular T cell subsets. I. Requirement of Lyt-1^{dull} L3T4^{bright} normal T cells for the induction of thyroiditis. *Journal of Immunology*, **141**, 105–13.

Taguchi, O. and Nishizuka, Y. (1980). Autoimmune oophoditis in the thymectomized mice: T-cell requirement in the adoptive cell transfer. *Clinical Experimental Immunology*, **42**, 324–31.

Tagchi, O. and Nishizuka, Y. (1981). Experimental autoimmune orchitis after neonatal thymectomy in the mouse. *Clinical Experimental Immunology*, **46**, 425–34.

Taguchi, O. and Nishizuka, Y. (1987). Self-tolerance and localized autoimmunity: mouse models of autoimmune disease that suggest that tissue-specific suppressor T-cells are involved in tolerance. *Journal of Experimental Medicine*, **165**, 146–56.

Taguchi, O., Takahashi, T., Masao, S., Namikawa, R., Matsuyama, M., and Nishizuka, Y. (1986). Development of multiple organ localized autoimmune disease in nude mice after reconstitution of T cell function by rat fetal thymus graft. *Journal of Experimental Medicine*, **164**, 60–71.

Tarter, T. H. and Alexander, N. J. (1984). Complement-inhibiting activity of seminal plasma. *American Journal of Reproductive Immunology*, **6**, 28–32.

Teuscher, C. (1985). Experimental allergic orchitis in mice II. Association of disease susceptibility with the locus controlling *Bordetella pertussis*-induced sensitivity to histamine. *Immunogenetics*, **22**, 417–25.

Teuscher, C., Smith, S. M., Goldberg, E. H., Shearer, G. M., and Tung, K. S. K. (1985). Experimental allergic orchitis in mice i. Genetic control of susceptibility and resistance to induction of autoimmune orchitis. *Immunogenetics*, **22**, 323–33.

Teuscher, C., Smith, S. M., and Tung, K. S. K. (1987*a*). Experimental allergic orchitis in mice. III. Differential susceptibility in BALB/c sublines. *Journal of Reproductive Immunology*, **10**, 219–30.

Teuscher, C., Blankenhorn, E. P., and Hickey, W. F. (1987*b*). Differential susceptibility to actively induced experimental allergic encephalomyelitis (EAE) and experimental allergic orchitis (EAO) among BALB/c substrains. *Cellular Immunology*, **110**, 294–304.

Teuscher, C., Gasser, D. L., Woodward, S. R., and Hickey, W. F. (1990). Experimental allergic orchitis in mice. VI. Recombinations within the H-2S/H-2D interval define the map position of the *H-2*-associated locus controlling disease susceptibility. *Immunogenetics*, **32**, 337–44.

Tisch, R., Yang, X-D, Singer, S., Liblau, R., Fugger, L., and McDevitt, H. (1993). Immune response to glutamic acid decarboxylase correlates with insulitis in non-obese diabetic mice. *Nature*, **366**, 72–5.

Todd, J. A., Aitman, T. J., Cornall, R. J., Ghosh, S., Hall, J. R. S., Hearne, C. M., *et al.* (1991). Genetic analysis of autoimmune type 1 diabetes mellitus in mice. *Nature*, **351**, 542–7.

Tokunaga, Y., Hiramine, C., Itoh, M., Mukasa, A., and Hojo, K. (1993*a*). Genetic susceptibility to the induction of murine experimental autoimmune orchitis (EAO) without adjuvant. I. Comparison of pathology, delayed-type hypersensitivity, and antibody. *Clinical Immunology and Immunopathology*, **66**, 239–47.

Tokunaga, Y., Hiramine, C., and Hojo, K. (1993*b*). Genetic susceptibility to the induction of murine experimental autoimmune orchitis (EAO) without adjuvant. II. Analysis of susceptibility to EAO induction using F1 hybrid mice and adoptive transfer system. *Clinical Immunology and Immunopathology*, **66**, 248–53.

Tung, K. S. K. (1978). Allergic orchitis lesions are adoptively transferred from vasoligated guinea pigs to syngeneic recipients. *Science*, **201**, 833–5.

Tung, K. S. K. and Woodroffe, A. (1978). Immunopathology of experimental allergic orchitis in the rabbit. *Journal of Immunology*, **120**, 320–8.

Tung, K. S. K., Ellis, L., Teuscher, C., Meng, A., Blaustein, J. C., Kohno, S., *et al.* (1981). The back mink *(Mustl a vision)*: a natural model of immunologic male infertility. *Journal of Experimental Medicine*, **154**, 1016–32.

Tung, K. S. K., Teuscher, C., Smith, S., Ellis, L., and Dufau, M. L. (1985). Factors that regulate the development of testicular autoimmune diseases. In *Hormone action and testicular function* (ed. K. J. Catt and M. L. Dufau). *Annals of the New York Academy of Science*, **438**, 171–88.

Tung, K. S. K., Yule, T. D., Mahi-Brown, C. A., and Listrom, M. B. (1987a). Distribution of histopathology and Ia positive cells in actively-induced and adoptively-transferred experimental autoimmune orchitis. *Journal of Immunology*, **138**, 752–9.

Tung, K. S. K., Smith, S., Teuscher, C., Cook, C., and Anderson, R. E. (1987b). Murine autoimmune oophoritis, epididymo-orchitis and gastritis induced by day-3 thymectomy: immunopathology. *American Journal of Pathology*, **126**, 293–302.

Tung, K. S. K., Smith, S., Matzner, P., Kasai, K., Oliver, J., Feuchter, F., and Anderson, R. E. (1987c). Murine autoimmune oophoritis, epididymo-orchitis and gastritis induced by day-3 thymectomy: autoantibodies. *American Journal of Pathology*, **126**, 303–14.

Tung, K. S. K. and Lu, C. Y. (1991). Immunologic basis of reproductive failure. In *Pathology of reproductive failure*, (ed. F. T. Kraus, I. Damjanov, and N. Kaufman), pp. 308–33. Williams and Wilkins, New York.

Tung, K. S. K., Taguchi, O., and Teuscher, C. (1994). Testicular and ovarian autoimmune diseases. In *Autoimmune disease models, a guidebook* (ed. I. R. Cohen and A. Millers), pp. 267–90. Academic Press, New York.

Unanue, E. R. (1989). Macrophages, antigen-presenting cells, and the phenomenon of antigen handling and presentation. In *Fundamental immunology*, (2nd edn), (ed. W. E. Paul), pp. 95–115. Raven Press, New York.

Vajnstangl, M., Petrovic, S., and Jankovic, B. D. (1979). Autoimmune oophoritis in the rat induced with isologous ovarian tissue. *Periodicum Biologorum*, **81**, 249–51.

Voisin, G. A., Delaunay, A., and Barber, M. (1951). Lesions testiculaires provoqúees chez le cobay par injection d'extrait de testicule homologue. CR Hedb, Seances (III). *Academy of Science*, **232**, 48–63.

Waksman, B. H. (1959). A histologic study of auto-allergic testis lesion in the guinea pig. *Journal of Experimental Medicine*, **109**, 311–23.

Wardell, B. B., Sudweeks, J. D., Meeker, N. D., Estes, S. S., Woodward, S. R., and Teuscher, C. (1993). The identification of Y chromosome-linked markers with random sequence oligonucleotide primers. *Mammalian Genome*, **4**, 109–12.

Wardell, B. B., Michael, S. D., Tung, K. S. K., Todd, J. A., Blankenhorn, E. P., Entee, K., *et al.* (1995) Aod1, the immunoregulatory locus controlling abrogation of tolerance in neonatal thymectomy-induced autoimmune ovarian dysgenesis, maps to mouse chromosome 16. *Proceedings of National Academy of Sciences of the USA*. (In press.)

Wardlaw, A. C. (1970). Inheritance of responsiveness to pertussis HSF in mice. *International Archives of Allergy*, **38**, 573–89.

Watson, G. and Paigen, K. (1990). Progressive induction of mRNA synthesis for androgen-responsive gene in mouse kidney. *Molecular Cellular Endocrinology*, **68**, 67–74.

Wood, D. M., Liu, C., and Dunbar, B. S. (1981). Effect of alloimmunization and heteroimmunization with zonae pellucidae on fertility in rabbits. *European Journal of Immunology*, **25**, 439–50.

Woodland, D. L., Happ, M. P., Gollob, K. J., and Palmer, E. (1991). An endogenous retrovirus mediating deletion of $\alpha\beta$ T cells? *Nature*, **349**, 529–30.

Woodward, S. R., Sudweeks, J., and Teuscher, C. (1992). Random sequence oligonucleotide primers detect polymorphic DNA products which segregate in inbred strains of mice. *Mammalian Genome*, **3**, 73–8.

Wyatt, C. R., Law, L., Magnuson, J. A., Griswold, M. D., and Magnuson, N.S. (1988). Suppression of lymphocyte proliferation by proteins secreted by cultured Sertoli cells. *Journal of Reproductive Immunology*, **14**, 27–40.

Yule, T. D., and Tung, K. S. K. (1993). Experimental autoimmune orchitis induced by testis and sperm antigen-specific T cell clones: an important pathogenic cytokine is tumor necrosis factor. *Endocrinology*, **133**, 1098–107.

Yule, T. D., Montoya, G. D., Russell, L. D., Williams, T. M., and Tung, K. S. K. (1988). Autoantigenic germ cells exist outside the blood-testis barrier. *Journal of Immunology*, **141**, 1161–7.

2 Adaptive strategies regulating energy balance in human pregnancy

ANDREW M. PRENTICE, SALLY D. POPPITT,
GAIL R. GOLDBERG, and ANN PRENTICE

I **Introduction**
  1 Nutritional stresses on women: evolutionary background
  2 A zoological perspective on human pregnancy
  3 Natural selection of energy-sparing traits

II **Energy needs for pregnancy: classical assumptions**
  1 Accretion in the products of conception
  2 Maintenance of the products of conception
  3 Maternal fat storage
  4 Physical activity
  5 Energy intake

III **Energy needs for pregnancy: new prospective studies**
  1 Description of datasets
  2 Adaptive adjustments in maintenance requirements
  3 Alterations in diet-induced thermogenesis
  4 Differences in fat storage
  5 Total metabolic costs of pregnancy
  6 Alterations in physical activity and total energy expenditure

IV **Implications of adaptive strategies for maternal and child health**
  1 Likely beneficial effects
  2 Possible detrimental effects

V **Conclusions**

I INTRODUCTION

1 Nutritional stresses on women: evolutionary background

The act of conception commits a woman to a substantial future investment of energy in her offspring. This obligation must be met by a combination of increased food intake and use of existing body stores. To guard against imprudent commitments which would end in non-viable pregnancies, fertility is suppressed in both under-nourished women (Frisch 1978) and, to a lesser extent, men (Keys *et al.* 1950). It is usually assumed that the body uses its level of adiposity as a measure of environmental conditions and of its own fitness for reproduction (Frisch 1978). However, examination of data from the Dutch famine of 1944–5 shows that this safeguard (a reduction to one-third of the expected number of births and 50 per cent amenorrhoea) only switches in when conditions have become quite severe (Stein *et al.* 1975).

The protective strategy of suppressed fertility has no way of anticipating the mother's future nutritional circumstances; in The Netherlands women who conceived at the beginning of the famine, when daily rations were estimated to be about 77 per cent of a pregnant woman's needs, had to carry their pregnancies through the depths of the winter blockade when rations fell to only 29 per cent of requirements (Smith 1947). Such variability in food supply is likely to have been widespread in our evolutionary past owing to winters in areas of high latitude, and to dry seasons in tropical zones. Droughts, severe winter freezes, harvest blights and wars create other desperate nutritional conditions in which women reproduce. There are over 450 recorded instances of major famine in written history alone (Keys *et al.* 1950), and many others will have passed unrecorded or have occurred in our pre-history.

Many communities throughout the world still have to contend with variable food supplies. Figure 2.1 shows the annual cycle of weight gain and loss in women of reproductive age in three rural Gambian villages—a consequence of seasonal changes in food availability and work load (Cole 1993). At the peak of the Sahelian drought years in the late 1970s and early 1980s an average woman was losing and then regaining about half of her body fat stores annually; in many women the fluctuation was even more

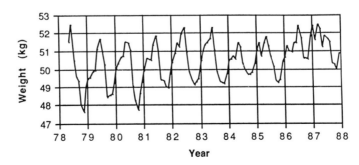

Fig. 2.1 Seasonal weight changes in rural Gambian women. (Reproduced with permission from Cole 1993.)

pronounced (Prentice *et al.* 1992). The time-course of these seasonal swings is such that a conception in the period of greatest food availability and highest fecundity (harvest season) places the third trimester of pregnancy in the worst part of the hungry season when birthweights are lowest (Prentice *et al.* 1992). Under these circumstances, suppressed fertility offers little protection.

In spite of these marginal, and sometimes severely inadequate, nutritional conditions human reproduction is remarkably successful if judged by the simple biological criteria of propagation of the species. Indeed it is in the poorest countries of Africa that population growth rates are highest with doubling times of close to 20 years despite the best efforts of family planning initiatives.

This review offers some clues as to how human reproduction is so successful by examining how women in diverse nutritional settings manage to balance their energy budget during gestation. We present recent data which demonstrate a very high level of variability in the metabolic responses to pregnancy. This plasticity appears to be governed by the mother's energy status and acts in a protective manner with under-nourished women showing significant energy-sparing adaptive strategies.

2 A zoological perspective on human pregnancy

Allometric scaling studies indicate complex size and species effects governing the energetics of mammalian reproduction (Payne and Wheeler 1967; Payne and Wheeler 1968; Robbins 1983; Blaxter 1989; Reiss 1989). We will focus here on the costs of human gestation. The size of the average human baby at birth fits the general cross-species relationship for mammals fairly well (Martin 1984). At around 6 per cent of maternal body weight, it is in line with generalized mammalian predictions based on the mother's metabolic body size (maternal weight$^{\sim 0.75}$ (Leutenegger 1976; Millar 1980). However, because the scaling factor is less than 1.0, large animals have proportionately smaller fetuses than small animals (Payne and Wheeler 1967; Payne and Wheeler 1968; Blaxter 1989). On the scale of all mammals, humans are on the large side and therefore have proportionately rather small babies compared with species such as mice and rats. For instance, a rat litter averages around 25 per cent of maternal weight at birth, representing a much greater stress on the dam (Spray, 1950; Naismith 1969).

Furthermore, human gestation is unusually slow. At 280 days it is almost twice as long as that for a ewe (150 days) in spite of a similar maternal body size. Human gestation is slow even compared with other primates for whom the general relationship is:

$$\text{gestation period (d)} = 31.3 \times \text{adult weight (g)}^{0.16}$$

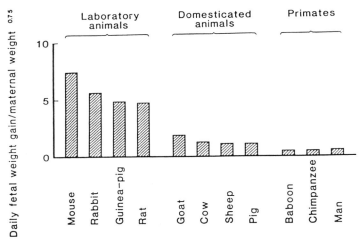

Fig. 2.2 Fetal growth rate as an indicator of the stress of gestation per unit time. (Adapted from Payne and Wheeler 1967, and reproduced with permission from Prentice and Whitehead 1987.)

(Kihlstrom 1972; Leutenegger 1973; Robbins 1983). For a 55 kg woman the predicted gestation of 179 days is 100 days shorter than the true value. It is usually assumed that our slow gestation and altricial state at birth are adaptations designed to maximize the time available for the growth and organization of a large brain (Martin 1980).

This combination of a small, usually singleton, neonate and a slow gestation places primates, particularly humans, in a special position with regard to the energetic stress of pregnancy. Figure 2.2 compares women with a number of other species in terms of the stress of pregnancy per unit time. The use of litter weight/maternal weight$^{0.75}$ as the unit of comparison was adopted from Payne and Wheeler (1967) and gives a crude measure of the relative energy costs of depositing new fetal tissue. By this measure the daily energy needs of a human pregnancy are very small because the cost is spread over a longer period. Figure 2.3 puts this into perspective by comparing the current internationally recommended increments in energy intake for pregnant (and lactating) women with those of a sheep with two lambs and a rat with eight pups (Prentice and Whitehead 1987).

Although the costs of gestation are small when measured per unit time the perspective is different when the total costs are analysed. The slow pace of human gestation means that the costs of maintaining the products of conception have to be borne for longer. If we exclude maternal fat deposition from the calculation, human maintenance costs average 400 per cent of the production costs for the products of conception

Adaptive strategies regulating energy balance in human pregnancy

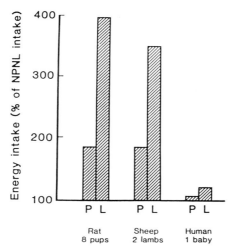

Fig. 2.3 Examples of the theoretical increases in food intake necessary to sustain pregnancy (P) and lactation (L). NPNL, non-pregnant, non-lactating. (Compiled from Cowan *et al.* 1980; Steingrimsdottir *et al.* 1980; Black *et al.* 1986. Reproduced with permission from Prentice and Whitehead 1987.)

whereas in the rat the figure is only 50 per cent (Kleiber *et al.* 1943; Spray 1950).

Human gestation also differs from most other species because of the greater fat stores available to buffer any short- or medium-term changes in energy supply from the diet. Figure 2.4 shows that even undernourished women have a considerably greater amount of energy reserved as fat than most other (non-Arctic and non-aquatic) mammals. The reasons for this are two-fold: humans as a species tend to be fat; and there is a marked sexual dimorphism giving women twice as much body fat as men in proportion to body weight. The reasons for this dimorphism are controversial (Pond 1991). One possible explanation is that larger fat stores confer a greater reproductive fitness by allowing women to withstand periods of food shortage.

As will be discussed below, these particular traits of human reproductive energetics have important implications for the metabolic and behavioural strategies adopted during pregnancy.

3 Natural selection of energy-sparing traits

In small species such as the rat, the energy required for accretion of fetal tissues is so much greater than the dam's usual maintenance costs that any energy-sparing adjustments in metabolic rate or physical activity, or

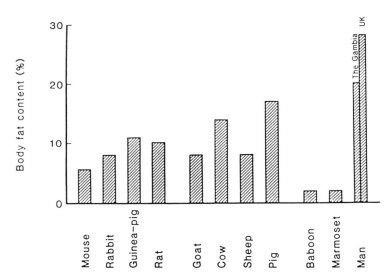

Fig. 2.4 Examples of maternal body fat content in the pre-gravid state. (Compiled from Spray 1950; Searle *et al.* 1972; Cowan *et al.* 1980; Steingrimsdottir *et al.* 1980; Pond 1984; Prentice 1984. Reproduced with permission from Prentice and Whitehead 1987.)

any attempts to subsidize fetal growth from existing fat stores, would be futile. Acquisition of a large amount of extra energy, with the consequent hypertrophy of the digestive system, is the only viable strategy. Failure to find extra food results in extreme responses such as abortion, resorption of fetuses, or infanticide after birth. This hastens the opportunity for a subsequent pregnancy when food supplies may be favourable and therefore has only limited effects on lifetime reproductive efficiency in a species with a rapid reproductive cycle. Under such conditions there would be minimal selective drive in favour of parsimonious metabolic adjustments.

The situation for humans is quite different. First, we have a slow reproductive cycle and tend to protect each gestational investment very carefully. Second, the marginal daily costs of gestation are so low that metabolic or behavioural adjustments, or both, in energy usage could realistically spare all of the extra energy needed. Third, the costs of maintenance are great compared with the costs of accretion so any modifications in maintenance could have a very significant effect on reproductive fitness and are likely to have been the subject of natural selection. The selective pressure in favour of any energy-sparing traits would probably have been very great since women who can continue to procreate successfully under conditions of marginal energy supply would soon dominate the gene pool.

A logical premise arising from the above considerations is that we

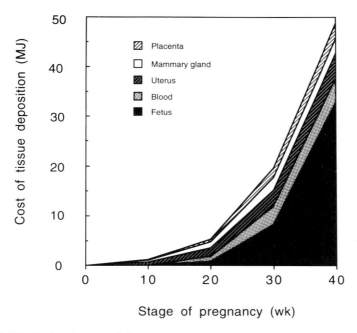

Fig. 2.5 Classical estimates of the energy deposited in the products of conception. (Adapted from Hytten and Chamberlain 1980.)

might expect to see some auto-regulatory adjustments in women's energy metabolism of sufficient magnitude to make a significant impact on fetal outcome. The remainder of this review summarizes recent data supporting this prediction.

II ENERGY NEEDS FOR PREGNANCY: CLASSICAL ASSUMPTIONS

1 Accretion in the products of conception

Figure 2.5 illustrates the classical assumptions for energy deposited as new tissue in the products of conception (except for maternal fat gain—see below) (Hytten and Leitch 1971). The needs of the fetus dominate because human neonates have an atypically high proportion of fat (Widdowson 1950; Robbins 1983). The total costs average about 49 MJ (11 500 kcal) of which 27 MJ (6500 kcal) is for protein and 22 MJ (5000 kcal) for fat. This is rather a small amount and represents only four or five days' energy intake for the mother.

These estimates have been obtained from tissue analysis performed many years ago including that of full-term stillborn babies (Widdowson 1950; Zeigler *et al.* 1976). More recent data are not available but there is no reason to question the validity of these former values as estimates of average deposition. There is little information available on the range of accretion costs under different nutritional conditions, and we therefore have to assume that they vary in proportion to the size of the baby and the amount of lean tissue (uterus, placenta, mammary secretory cells, and erythrocytes) gained by the mother. Fortunately, any errors in these assumptions have little bearing on the overall energetic calculations for pregnancy since the accretion costs of the conceptus are only 15 per cent of the total costs in a normal pregnancy.

2 Maintenance of the products of conception

The increase in metabolically active lean tissue in the mother and fetus requires an increase in the energy supply to support it. The energy costs of this process, and the question of whether the new tissue was more or less

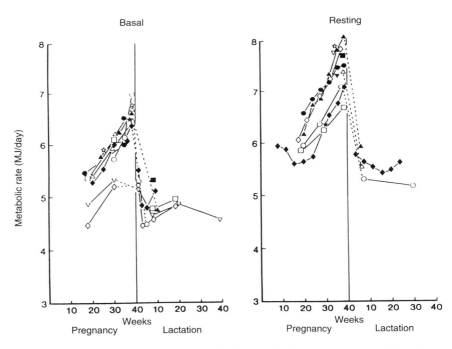

Fig. 2.6 Early published data on metabolic rate during pregnancy and lactation. (Publications span from 1910 to 1976. Reproduced with permission from Prentice and Whitehead (1987) where original citations can be found.)

active than the existing maternal cell mass, fascinated early physiologists and resulted in numerous investigations into the rise in basal metabolic rate in pregnancy. These studies yielded a consistent finding that metabolism was considerably elevated towards term (Fig. 2.6). Hytten and Leitch (1971) used these observations, together with other estimates of the likely costs of supporting the conceptus, to derive an estimated value of 152 MJ (36 000 kcal) as the maintenance costs for an average human pregnancy. As we shall show later, this value is an excellent estimate of the average cost of maintaining pregnancy in a well-nourished woman, but is associated with a very wide variance. Likewise the average curves in Fig. 2.6 conceal biologically important variability, and have missed aspects of the adaptive responses by only commencing the measurements in mid-pregnancy.

3 Maternal fat storage

In the past it has been assumed that the average amount of fat gained during a well-nourished pregnancy was about 3.5 kg with an energy content of about 135 MJ (FAO/WHO/UNU 1985). Since it is rather difficult to make direct measurements of fat deposition in human pregnancy, this estimate was derived as the residual after subtracting from the total pregnancy weight gain the weight of the gravid uterus, increases in breast and blood volume, and increased hydration (Hytten and Chamberlain 1980). It has always been recognized that there is very great variability in the amount of fat gained by different women.

4 Physical activity

Classical estimates of the energy costs of physical activity in pregnancy have been based largely on guesswork (FAO/WHO/UNU 1985). It has long been recognized that the gross costs of both weight-dependent and weight-independent activity increase in pregnancy particularly towards term, but it has frequently been assumed that a behavioural reduction in physical activity counter-balances this increase and may actually lead to net savings of energy during pregnancy, at least in women who have no external constraints on activity patterns such as the need to do agricultural work (FAO/WHO/UNU 1985). In practice it has proved quite difficult to identify any substantial decreases in activity using techniques such as activity diaries (van Raaij *et al.* 1990; Durnin 1992; Spaaij 1993). A possible explanation for this might be that many affluent women are so inactive in the non-pregnant state (Prentice *et al.* 1985) that there is little scope to become even less active during pregnancy. Another explanation might be that the intensity of self-paced work decreases and that this is not detected by activity diaries.

5 Energy intake

Estimates of energy intake in pregnancy frequently reveal a serious gap between the generally accepted calculations of energy requirements (FAO/WHO/UNU 1985) and the observed increase in food intake. The latter is frequently much lower even in affluent women who have no external constraints on food availability. Figure 2.7 summarises data from nine longitudinal studies covering over 850 pregnancies. The average increment in energy intake for all studies (shown by the bold line) is less than 0.5 MJ/day (120 kcal/day) even in late pregnancy, and represents an increase of only about 0.3 MJ/day (70 kcal/day) aggregated over the whole of pregnancy. There are certain factors which might bias this result, such as women becoming progressively less careful in recording all that they eat as they tire of the measurement process, but analysis of other longitudinal datasets in non-pregnant subjects suggests that declining compliance is unlikely to account for more than about 0.2 MJ/day (50 kcal/day). This paradox between calculated requirements and observed intakes remains a challenge to human nutritionists, and is only partly resolved by the adaptive changes in energy expenditure described below. Studies which have attempted to make careful simultaneous inventories of both the supply and

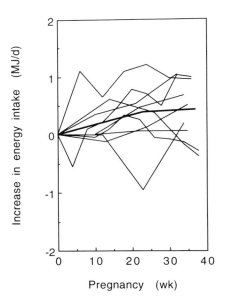

Fig. 2.7 Longitudinal estimates of changes in energy intake during pregnancy. Bold line shows average of all studies. (Compiled from Beal 1971; Doyle *et al.* 1982; Papoz *et al.* 1982; van Raaij *et al.* 1987; Truswell *et al.* 1988; Allen *et al.* 1992; Durnin 1992; Goldberg *et al.* 1993; Spaaij 1993.)

the demand of energy in human pregnancy generally still find a substantial discrepancy between the two, even using sophisticated techniques such as whole-body calorimetry and doubly labelled water ($^2H_2^{18}O$) to assess energy expenditure (Goldberg et al. 1993; Spaaij 1993).

III ENERGY NEEDS FOR PREGNANCY: NEW PROSPECTIVE STUDIES

1 Description of datasets

The past decade has seen the publication of 10 new longitudinal datasets covering 360 pregnancies from seven countries representing a broad range of maternal nutritional status: Sweden (Forsum et al. 1988), England (Goldberg et al. 1993), the Netherlands (van Raaij et al. 1987), Scotland (Durnin et al. 1987), the Philippines (Tuazon et al. 1987), Thailand (Thongprasert et al. 1987), and the Gambia (Lawrence et al. 1987; Poppitt et al. 1993). The importance of the new data arises from the fact that the women were first studied in the pre-pregnant state and were then followed longitudinally to the 36th week of gestation (except in the Philippines and Thailand where measurements were started at 10–16 weeks' gestation). The studies also used well standardized methods which were relatively similar between countries. Some studies measured basal metabolic rate (BMR) in whole-body calorimeters and others used ventilated hoods—a difference which should not introduce any interpretational difficulties particularly as each study used only one method for the serial measurements. Further methodological details are summarized elsewhere (Poppitt et al. 1994). These serial measurements allow the total integrated costs of gestation to be carefully quantified.

Table 2.1 lists maternal weight and body composition, weight gain, fat gain, and birth weight for the different populations. The range in average maternal pre-gravid weights (44.5–62.6 kg) provides a measure of the large differences in long-term nutritional status. The wide variability in acute nutritional status during pregnancy itself is indicated by the two-fold range in average gestational weight gains (6.4–13.6 kg).

2 Adaptive adjustments in maintenance requirements

Each of the studies measured BMR at six-week intervals in order to assess changes in maintenance energy requirements. Figure 2.8 illustrates the average changes in BMR relative to the pre-pregnant baseline in the 10 groups of women. The countries have been divided into developed (left hand side) and developing nations (right hand side). This rather subjective division accords with the basic measures of nutritional status

Table 2.1 Basic indicators of nutritional status in the 10 new longitudinal datasets

Country	Citation	Height (m)	Weight (kg)	Fat mass (kg)	Fat-free mass (kg)	Weight gain (kg)	Fat gain (kg)	Birth weight (kg)
Sweden	Forsum et al. (1988)	1.65	61.0	17.2	43.8	13.6	5.8	3.56
England	Goldberg et al. (1993)	1.64	61.7	18.6	43.1	13.2	3.4	3.77
Netherlands II	Spaaij (1993)	1.69	62.6	17.3	45.3	11.7	2.1	3.52
Netherlands I	van Raaij et al. (1987)	1.69	62.5	17.7	44.7	11.6	2.0	3.46
Scotland	Durnin et al. (1987)	1.62	57.3	14.9	42.4	11.7	2.3	3.37
Thailand	Thongprasert et al. (1987)	1.52	47.6	11.4	36.2	8.9	1.3	2.98
Philippines	Tuazon et al. (1987)	1.51	44.5	11.3	33.2	8.4	1.3	2.89
Gambia II	Poppitt et al. (1993)	1.57	52.0	12.2	39.8	8.7	2.3	3.02
Gambia I (S)	Lawrence et al. (1987)	1.57	51.2	9.9	41.3	7.9	2.0	2.99
Gambia I (U)	Lawrence et al. (1987)	1.58	51.6	11.0	40.6	6.4	−0.5	2.96

Height, weight, fat mass, and fat-free mass values refer to the pre-gravid state for all studies except Thailand and the Philippines, where first measurements were at 10 weeks' gestation. S, supplemented; U, unsupplemented.

Adaptive strategies regulating energy balance in human pregnancy 45

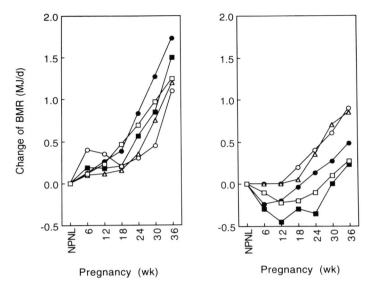

Fig. 2.8 Average changes in BMR relative to the pre-pregnant baseline in women studied longitudinally. Left-hand panel contains developed countries: Sweden (*solid circles*); England (*solid squares*); Netherlands I (*open circles*); Netherlands II (*open squares*); and Scotland (*open triangles*). Right-hand panel contains less developed countries: Thailand (*open triangles*); Philippines (*open circles*); Gambia I, supplemented (*open squares*); Gambia I, unsupplemented (*solid squares*); and Gambia II (*solid circles*). (Adapted from Poppitt *et al.* 1994.)

which are significantly different between the two groups. For example, maternal weight was 61.0 ± 0.9 kg (SE) vs 49.4 ± 1.3 kg, $p < 0.001$; and pregnancy weight gain was 12.4 ± 0.4 vs 8.0 ± 0.4 kg, $p < 0.001$. When divided in this manner there is a clear distinction between the metabolic responses to pregnancy: well-nourished women tend to show an immediate and progressive 'energy profligate' increase in BMR whereas under-nourished women show an 'energy sparing' suppression of metabolic rate in early gestation which offsets the later increase (Poppitt *et al.* 1994). The total maintenance cost of pregnancy in each of these studies is obtained by integrating the area under the curves with extrapolation to 40 weeks. This yields figures ranging from 210 MJ (50 000 kcal) in Sweden to −45 MJ (−11 000 kcal) in unsupplemented Gambian women (see Fig. 2.9).

In addition to the wide variation in metabolic responses to pregnancy between different populations there are also wide variations within populations. Figure 2.10 illustrates individual changes in BMR in women from England (Prentice *et al.* 1989; Goldberg *et al.* 1993) and the Gambia (Poppitt *et al.* 1993), and shows that energy-sparing and energy-profligate

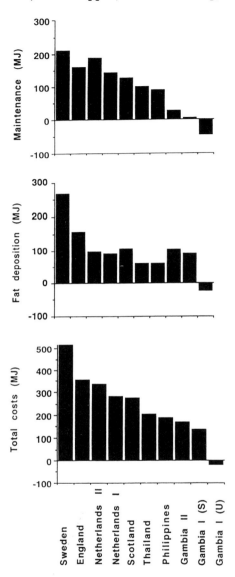

Fig. 2.9 Total metabolic cost of pregnancy and its major components in 10 recent longitudinal datasets. The energy deposited in the conceptus (not shown) was computed by scaling the value 49 MJ for a 3.4 kg baby. (Reproduced with permission from Poppitt *et al.* 1994.)

Adaptive strategies regulating energy balance in human pregnancy 47

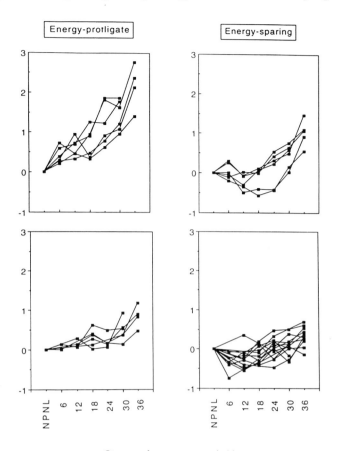

Fig. 2.10 Changes in BMR in individual English and Gambian subjects arbitrarily divided into energy-profligate and energy-sparing responders. NPNL, non-pregnant, non-lactating. (English data (upper-half) from Goldberg et al. 1993; Gambian data (lower half) from Poppitt et al. 1993.)

responses can be found in both communities, but with a preponderance of energy-sparing women in the Gambia and of energy-proligate women in England. The only other study which has looked closely at differences between individuals is the Netherlands II dataset from Spaaij (1993). This confirms our own observations of widely divergent responses. The integrated maintenance costs of pregnancy for the Netherlands data have been combined with the English and Gambian data in Fig. 2.11 which provides a graphic illustration of the plasticity in maintenance energy needs between individual women.

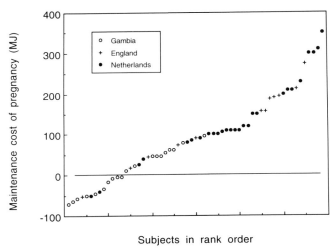

Fig. 2.11 Integrated maintenance costs of pregnancy in individual women from three different studies. (Data from Spaaij 1993; Goldberg et al. 1993; Poppitt et al. 1993.)

In both the English (Prentice et al. 1989) and the Netherlands II data sets, (Spaaij 1993), the best predictor of a woman's maintenance energy needs in pregnancy was found to be her pre-pregnant body fatness. In the larger of the two datasets (Netherlands II) the correlation between cumulative maintenance costs and pre-pregnant body fat percentage was $+0.50$ ($p < 0.01$) (Spaaij 1993).

3 Alterations in diet-induced thermogenesis

Possible changes in diet-induced thermogenesis (DIT) during pregnancy have been investigated in two longitudinal studies which have measured the thermic response to a standardized liquid meal (Illingworth et al. 1987; Spaaij 1993) and two whole-body calorimetry studies using normal meals (Schutz et al. 1988; Prentice et al. 1989; Schutz 1989). Illingworth et al. (1987) observed a significantly lower DIT in the second trimester (−28 per cent) but not in the first or third trimesters (−1 per cent and −15 per cent) compared with measurements made post-partum. Spaaij (1993) observed no significant differences in the three trimesters (−1 per cent, −5 per cent, and −5 per cent). Prentice et al. (1989) derived estimates of the sum of minor physical movement and DIT from 24 hour whole-body calorimeter measurements, and concluded that any changes were extremely small and non-significant. Finally, Schutz et al. (1988) and Schutz (1989) recorded DIT during pregnancy to be 9 ± 1 per cent of the energy consumed compared with 13.5 ± 1 per cent in the same women after lactation ($p < 0.05$). Elsewhere we have stressed that any changes in DIT during pregnancy are most unlikely to represent biologically significant effects (Prentice *et al.* 1987).

4 Differences in fat storage

Calculations of maternal fat gain must be interpreted with some caution because the normal assumptions underlying most methods for assessing body fat become progressively less valid during pregnancy (van Raaij *et al.* 1988). However the estimates in most of the new studies have been derived by combining several techniques and by making appropriate adjustments to the methodological assumptions. They are therefore likely to be fairly reliable.

The maternal fat gains listed in Table 2.1 reveal a very wide range from a loss of 0.5 kg in unsupplemented Gambian women (equivalent to a saving of about 20 MJ (4500 kcal)) to a gain of 5.8 kg in the Swedish women (at a cost of about 226 MJ (54 000 kcal)). Once again the wide variation between populations is mirrored by a wide variation between different women within each population. For instance, the study from England recorded a standard deviation of 3.2 kg with a range of −2 kg to +8 kg (Goldberg *et al.* 1993), and in the Netherlands II dataset the standard deviation was around 2.5 kg depending on the method used (Spaaij 1993). A study from Sweden which used magnetic resonance imaging to assess fat gains recorded a standard deviation of 3.8 kg (Sohlstrom 1993).

5 Total metabolic costs of pregnancy

The metabolic costs of pregnancy (as distinct from any costs associated with physical activity) can be computed by adding the costs of new tissue accretion, fat deposition, and maintenance. For the purposes of these calculations we have estimated accretion costs by making a pro rata adjustment (according to maternal body weight) of the accepted average value of 49 MJ (11 500 kcal) (see page 39). The total metabolic costs calculated in this manner for the 10 new longitudinal datasets are shown in Fig. 2.9. The pronounced covariance between the costs of maintenance and fat deposition is clear, and creates an exceptionally large range in the total metabolic costs, from 523 MJ (125 000 kcal) in the Swedish women to −30 MJ (−7 000 kcal) in the unsupplemented Gambian women. The average costs in the well nourished groups was close to the current international assumption of 336 MJ (80 000 kcal) (FAO/WHO/UNU 1985).

In an attempt to understand the mechanisms responsible for these wide differences we have examined possible cross-study relationships between pre-pregnant nutritional status and the metabolic costs of pregnancy (Poppitt et al. 1994). Maternal height, which may reflect early and long-term nutritional influences, was not correlated with any of the costs of pregnancy. Pre-pregnant weight was not related to the amount of fat deposited, but there were marginally significant correlations with both the maintenance ($r = +0.67$, $p < 0.05$) and the total costs of pregnancy ($r = +0.67$, $p < 0.05$). In contrast the women's pre-gravid fat mass was strongly correlated with both the maintenance costs ($r = +0.84$, $p < 0.01$) and the total metabolic costs ($r = +0.80$, $p < 0.01$) of pregnancy (Fig. 2.12). Perhaps surprisingly, there was no relationship between pre-gravid fat-free mass and any of the costs of pregnancy.

The correlation between the metabolic costs of pregnancy and women's initial fatness suggests the existence of a mechanism which can monitor the mother's pre-gravid nutritional status and adjust the homeorrhetic changes in maternal metabolism accordingly. Such a protective mechanism seems biologically plausible and would confer considerable survival benefits to the mother and her genes, but must be interpreted in the light of a number of possible confounding factors. First, as can be seen from Fig. 2.12, the costs of pregnancy were also highly correlated with the amount of weight gained in pregnancy (maintenance ($r = +0.92$, $p < 0.001$); fat deposition ($r = +0.81$, $p < 0.01$); and total costs ($r = +0.94$, $p < 0.001$)), and pre-pregnant fatness and weight gain were themselves highly correlated ($r = +0.88$, $p < 0.001$), making it difficult to establish which factor is dominant. However, within-study analyses comparing different individual women, performed by ourselves with English data (unpublished) and by Spaaij with the Netherlands II data (personal communication), found no association between pregnancy weight gain and maintenance costs. This contrasts with the positive association between pre-gravid fatness and maintenance costs discussed above. The inter-individual analysis therefore tends to support the cross-study analysis in suggesting the existence of a feedback signal generated from maternal fat reserves. Second, pre-pregnant fatness may simply be acting as a proxy measure for overall nutritional status and may be representative of energy balance during pregnancy itself. The difference in metabolic costs between the supplemented and unsupplemented Gambian women (Lawrence et al. 1987) certainly suggests that current intake in pregnancy plays a role in determining the response.

Irrespective of whether the signals modulating energy metabolism in pregnancy originate from maternal fat stores, the mother's current state of energy balance or a combination of the two, the resultant plasticity does appear to be energy sensitive and operates in a direction which tends to normalize energy balance.

Adaptive strategies regulating energy balance in human pregnancy 51

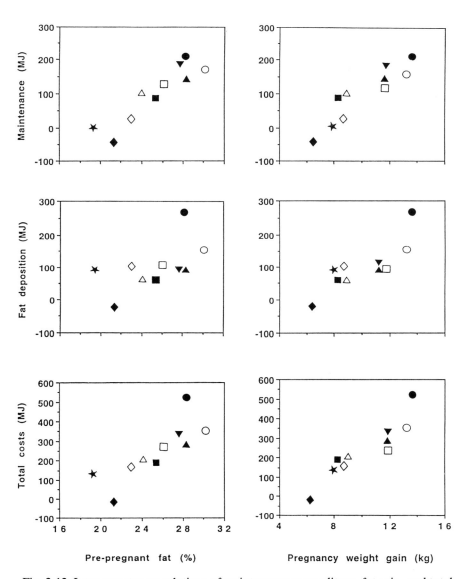

Fig. 2.12 Inter-country correlations of maintenance expenditure, fat gain, and total costs against pre-pregnant percentage fatness and pregnancy weight gain. Sweden (*solid circles*); England (*open circles*); Netherlands I (solid triangles); Netherlands II (*solid inverted triangles*); Scotland (*open squares*); Thailand (*open triangles*); Philippines (*solid squares*); Gambia I supplemented (*solid stars*); Gambia I, unsupplemented (*solid diamonds*); and Gambia II (*open diamonds*). (Reproduced with permission from Poppitt *et al.* 1994.)

6 Alterations in physical activity and total energy expenditure

In view of the low marginal costs of human pregnancy relative to the mother's normal maintenance needs, any energy saved through behavioural reductions in physical activity could potentially make very significant contributions to balancing the overall energy budget. If activity patterns remain constant total energy expenditure (TEE) will increase because the net energy cost of weight-independent exercise (for example cycling) and weight-dependent exercise (for example walking, climbing stairs) both increase towards term, by about 10 per cent and 15 per cent, respectively. If the amount of activity performed is small the effect of these changes is minimal and total energy expenditure remains rather constant when expressed as a multiple of BMR (that is as physical activity level equal to TEE/BMR). This is well illustrated in studies using whole-body calorimeters in which activity levels tend to be much lower than in real life. For instance, Prentice *et al.* (1989) demonstrated that 24 hour energy expenditure in a calorimeter was highly correlated with BMR throughout pregnancy ($r = +0.98$, $p < 0.001$). Another way of expressing this is that, when subjects are confined, changes in the energy cost of physical activity (and thermogenesis) contribute little to the variance already accounted for by differences in BMR. Figure 2.13 confirms this by showing that the

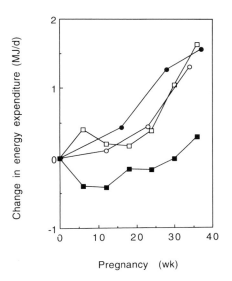

Fig. 2.13 Changes during pregnancy in 24 hour energy expenditure within a whole-body calorimeter. (Data from Schutz *et al.* 1988 and Schutz 1989; (*solid circles*); Prentice *et al.* 1989 (*open squares*); Poppitt *et al.* 1993 (*solid squares*) and de Groot *et al.*, in press (*open circles*)).

Adaptive strategies regulating energy balance in human pregnancy

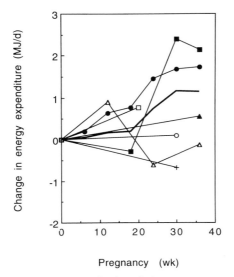

Fig. 2.14 Changes during pregnancy in free-living energy expenditure measured by the doubly labelled water method. (Data from Singh *et al.* 1989 (*open squares*); Goldberg *et al.* 1991 (*solid triangles*); Forsum *et al.* 1992 (*solid squares*); Heini *et al.* 1991 (*open triangles*); Goldberg *et al.* 1993 (*solid circles*), Bronstein *et al.* in press, normal weight (*open circles*) and obese (plus)).

differences observed between Gambian women and Europeans in terms of BMR are also present over the full day.

In the free-living state there is much more scope for behavioural modification. This has been investigated using the doubly labelled water technique which measures total free-living energy expenditure (Prentice 1990). Figure 2.14 summarizes all the available data from both longitudinal and cross-sectional studies. The average increase seems to be somewhat lower than that observed in calorimetry studies, indicating that there has been a slight suppression of physical activity. Expressing the results as TEE − BMR (in order to factor out the effect of changes in BMR) confirms this impression. However, any savings are small and the two studies which made true longitudinal measurements in the same women showed a slight increase in the energy cost of activity (Forsum *et al.* 1992; Goldberg *et al.* 1993).

We conclude that, although there is scope for modulating energy balance by altering levels of physical activity, there is little evidence of this being a major factor in practice. One explanation for this, as previously discussed, might be that in affluent women physical activity is already low, leaving little room for further reduction, and that in under-nourished women (in whom there is a greater scope for reduction in activity) the obligatory nature of many tasks, such as farming, childcare, drawing water, collecting firewood, pounding grain, or working in a factory, may override any energetically desirable decrease in activity.

IV IMPLICATIONS OF ADAPTIVE STRATEGIES FOR MATERNAL AND CHILD HEALTH

1 Likely beneficial effects

We have already alluded to the extraordinary efficiency of human reproduction when measured in terms of population growth rates. The fetus also appears well protected in other respects. For instance, in the absence of pathological defects in the fetus or placenta and considering only full-term births, human birthweight varies over a much smaller range than in many other mammalian species. This is well illustrated using the birthweight results from the 10 new longitudinal studies discussed above. The top portion of Fig. 2.15 shows the average birthweights for each population, with a span of about 20 per cent. However, much of this can be accounted for by the fact that the mothers are of very different body size. The middle portion of Fig. 2.15 shows birthweight expressed relative to maternal metabolic body size (weight$^{0.75}$) and reveals a remarkable consistency, with a between-study coefficient of variation of only 3.3 per cent (Poppitt *et al.* 1994). This suggests that the adaptive strategies invoked to maintain energy balance have been very successful in producing a fetus which is appropriate to its mother's body size and will therefore have optimal survival prospects.

2 Possible detrimental effects

The fact that humans display these advanced mechanisms for protecting fetal growth does not necessarily mean that they can be achieved without a cost. For instance, a 10 per cent reduction in the metabolic rate of a mother and her fetus at 24 weeks' gestation (see Gambian curves in Fig. 2.8) implies an even greater reduction in the mother's own tissues since the gravid uterus is bound to have substantial requirements of its own by this stage. The biochemical and biophysical processes that are being down-regulated by the mother are unknown. They might have long-term consequences for her own health. The lower portion of Fig. 2.15 shows how frugal an under-nourished mother has to be during pregnancy. It shows fetal weight as a percentage of total pregnancy weight gain. In normal well-nourished women the fetus is only 25 per cent of total gain, whereas in the Gambian women the fetus represents almost half of the total pregnancy weight gain.

Adaptive strategies regulating energy balance in human pregnancy

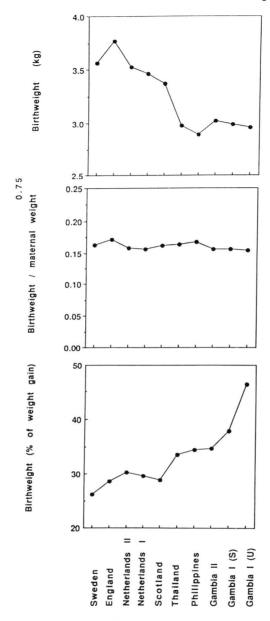

Fig. 2.15 Mean birthweights from 10 longitudinal studies ranked in order of decreasing cost of pregnancy. Upper section, absolute birthweight; middle section, birthweight as a proportion of maternal metabolic body size (weight$^{0.75}$); lower section, birthweight as a proportion of total pregnancy weight gain. (Reproduced with permission from Poppitt *et al.* 1994.)

This strongly suggests that the mother is struggling to achieve a satisfactory outcome.

It is also known that birthweight is an inadequate measure of the baby's overall condition at birth, and that subtle nutritional influences on the fetal environment might have long-term consequences for the offspring (CIBA 1991). It should not be automatically assumed that all is well just because fetal growth rate is relatively well preserved at different planes of nutrition.

Finally, the differences in fat storage may have important longer term implications. Fat deposition in pregnancy is often viewed as an intentional mechanism to provide an additional energy reserve for use in lactation. However, it appears that the women who are most likely to require an energy buffer in lactation are those who are least able to spare energy for fat deposition in pregnancy. Conversely the women who deposit large amounts of adipose tissue in pregnancy are least likely to mobilize it during lactation, and tend to retain a substantial amount of excess fat with implications in terms of the development of obesity (Lederman 1993; Sohlstrom 1993).

V CONCLUSIONS

We conclude that human pregnancy is associated with a high degree of metabolic plasticity which can modulate the energy cost of gestation over a wide range, and which is responsive to the mother's existing energy status in a direction that tends to normalize energy balance and protect the fetus. These adaptive strategies seem to be potentiated by pregnancy, and to be more pronounced in humans than in most other mammals. These conclusions are supported by comprehensive experimental evidence derived from a number of recent independent studies, and are reinforced by sound teleological arguments in their favour. The protective potential of these metabolic adjustments appears impressive, but it is of the utmost importance that they are not over-interpreted as indicating that pregnant women are not in need of special nutritional care.

REFERENCES

Allen, L. H., Backstrand, J. R., Chavez, A., and Pelto, G. H. (1992). *People cannot live by tortillas alone: the result of the Mexico Nutrition CRSP. Final report to the US Agency for International Development Cooperative Agreement*. University of Connecticut and Instituto Nacional de la Nutrition Salvador Zubiran, Washington.

Beal, V. A. (1971). Nutrition studies during pregnancy. 1. Changes in intakes of calories, carbohydrate, fat, protein and calcium. *Journal of the American Dietetic Association*, **58**, 312–20.

Black, A. E., Wiles, S. J., and Paul, A. A. (1986). The nutrient intakes of pregnant and lactating mothers of good socio-economic status in Cambridge, UK: some implications for recommended daily allowances of minor nutrients. *British Journal of Nutrition*, **56**, 59–72.

Blaxter, K. (1989). *Energy metabolism in animals and man.* Cambridge University Press.

Bronstein, M. N., Mak, R. P., and King, J. C. (1994). Basal and total energy expenditure in the obese pregnant woman. *American Journal of Clinical Nutrition*.

CIBA. (1991). *The childhood environment and adult disease*, CIBA Foundation Symposium No. 156. John Wiley, Chichester.

Cole, T. J. (1993). Seasonal effects on physical growth and development. In *Seasonality and human ecology* (ed. S. J. Ulijaszek and S. S. Strickland), pp. 89–106. Cambridge University Press.

Cowan, R. T., Robinson, J. J., McDonald, I., and Smart, R. (1980). Effects of body fatness at lambing and diet in lactation on body tissue loss, feed intake and milk yield of ewes in early lactation. *Journal of Agricultural Science, Cambridge*, **95**, 497–514.

de Groot, L. C. P. G. M., Boekholt, H. A., Spaaij, C. J. K., van Raaij, J. M. A., Drijvers, J. J. M. M., van Heide, L. J. M., *et al.* (1994). Energy balances of healthy Dutch women before and during pregnancy: limited scope for metabolic adaptations in pregnancy. *American Journal of Clinical Nutrition*, **59**, 827–32.

Doyle, W., Crawford, M. A., Laurence, B. M., and Drury, P. (1982). Dietary survey during pregnancy in a low socio-economic group. *Human Nutrition: Applied Nutrition*, **36A**, 95–106.

Durnin, J. V. G. A. (1992). Energy metabolism in pregnancy. In *Principles of perinatal–neonatal metabolism* (ed. R. Cowett), pp. 228–36. Springer-Verlag, New York.

Durnin, J. V. G. A., McKillop, F. M., Grant, S., and Fitzgerald, G. (1987). Energy requirements of pregnancy in Scotland. *Lancet*, **ii**, 897–900.

FAO/WHO/UNU (1985). Energy and protein requirements. *Technical Report Series* 724. WHO, Geneva.

Forsum, E., Sadurskis, A., and Wager, J. (1988). Resting metabolic rate and body composition of healthy Swedish women during pregnancy. *American Journal of Clinical Nutrition*, **47**, 942–7.

Forsum, E., Kabir, N., Sadurskis, A., and Westerterp, K. (1992). Total energy expenditure of healthy Swedish women during pregnancy and lactation. *American Journal of Clinical Nutrition*, **56**, 334–42.

Frisch, R. E. (1978). The effect of food intake on reproductive ability. In *Nutrition and human reproduction* (ed. W. H. Mosley), pp. 91–122. Plenum Press, New York.

Goldberg, G. R., Prentice, A. M., Coward, W. A., Davies, H. L., Murgatroyd, P. R., Sawyer, M. B., *et al.* (1991). Longitudinal assessment of the components of energy balance in well-nourished lactating women. *American Journal of Clinical Nutrition*, **54**, 788–98.

Goldberg, G. R., Prentice, A. M., Coward, W. A., Murgatroyd, P. R., Davies, H. L., Wensing, C., *et al.* (1993). Longitudinal assessment of energy expenditure in pregnancy by the doubly-labelled water method. *American Journal of Clinical Nutrition*, **57**, 494–505.

Heini, A., Schutz, Y., Diaz, E., Prentice, A. M., Whitehead, R. G., and Jequier, E. (1991). Free-living energy expenditure measured by two independent techniques in pregnant and non-pregnant Gambian women. *American Journal of Physiology*, **261**, E9–E17.

Hytten, F. and Chamberlain, G. (1980). *Clinical physiology in obstetrics*. Blackwell Scientific Publications, Oxford.

Hytten, F. and Leitch, I. (1971). *The physiology of human pregnancy*, (2nd edn). Blackwell Scientific Publications, Oxford.

Illingworth, P. J., Jung, R. T., Howie, P. W., and Isles, T. E. (1987). Reduction in postprandial energy expenditure during pregnancy. *British Medical Journal*, **294**, 1573–6.

Keys, A., Brozek, J., Henschel, A., Mickelsen, O., and Taylor, H. (1950). *The biology of human starvation*. University of Minnesota Press, Minneapolis.

Kihlstrom, J. E. (1972). Period of gestation and body weight in some placental mammals. *Comparative Biochemistry and Physiology*, **43A**, 673–9.

Kleiber, M., Cole, H. H., and Smith, A. H. (1943). Metabolic rate of rat fetuses *in vitro*. *Journal of Cellular and Comparative Physiology*, **22**, 170.

Lawrence, M., Lawrence, F., Coward, W. A., Cole, T. J., and Whitehead, R. G. (1987). Energy requirements of pregnancy in the Gambia. *Lancet*, **ii**, 1072–6.

Lederman, S. A. (1993). The effect of pregnancy weight gain on later obesity. *Obstetrics and Gynecology*, **82**, 148–55.

Leutenegger, W. (1973). Maternal–fetal weight relationships in primates. *Folia Primatologica*, **20**, 280–93.

Leutenegger, W. (1976). Allometry of neonatal size in eutherian mammals. *Nature*, **263**, 229–30.

Martin, R. D. (1980). Adaptation and body size in primates. *Zeitschrift für Morphologie und Anthropologie*, **71**, 115–24.

Martin, R. D. (1984). Scaling effects and adaptive strategies in mammalian lactation. *Symposium of the Zoological Society of London*, **51**, 87–117.

Millar, J. S. (1980). Pre-partum reproductive characteristics of eutherian mammals. *Evolution*, **35**, 1149–63.

Naismith, D. J. (1969). The foetus as a parasite. *Proceedings of the Nutrition Society*, **28**, 25–31.

Papoz, L., Eschwege, E., Pequignot, G., Barrat, J., and Schwartz, D. (1982). Maternal smoking and birth weight in relation to dietary habits. *American Journal of Obstetrics and Gynecology*, **142**, 870–6.

Payne, P. R. and Wheeler, E. F. (1967). Growth of the foetus. *Nature* **215**, 849–50.

Payne, P. R. and Wheeler, E. F. (1968). Comparative nutrition in pregnancy and lactation. *Proceedings of the Nutrition Society*, **27**, 129–38.

Pond, C. M. (1984). Physiological and ecological importance of energy storage in the evolution of lactation: evidence for a common pattern of anatomical organisation of adipose tissue in mammals. *Symposia of The Zoological Society of London*, **51**, 1–32.

Pond, C. M. (1991). Adipose tissue in human evolution. In *The aquatic ape: fact or fiction?* (ed. M. Roede, J. Wind, J. Patrick, and V. Reynolds), pp. 193–220. Souvenir Press, London.

Poppitt, S. D., Prentice, A. M., Jequier, E., Schutz, Y., and Whitehead, R. G. (1993). Evidence of energy-sparing in Gambian women during pregnancy: a longitudinal study using whole-body calorimetry. *American Journal of Clinical Nutrition*, **57**, 353–64.

Poppitt, S. D., Prentice, A. M., Goldberg, G. R., and Whitehead, R. G. (1994). Energy-sparing strategies to protect human fetal growth. *American Journal of Obstetrics and Gynecology*, **171**, 118–25.

Prentice, A. M. (1984). Adaptations to long-term low energy intake. In *Energy intake and activity: current topics in nutrition and disease* (ed. E. Pollitt and P. Amante), pp. 1–31. Alan R. Liss, New York.

Prentice, A. M. (ed.) (1990). The doubly-labelled water method for measuring energy expenditure: technical recommendations for use in humans, (ed A. M. Prentice), pp. 1–301. *International Dietary Energy Consultancy Group* IAEA, Vienna.

Prentice, A. M. and Whitehead, R. G. (1987). The energetics of human reproduction. *Symposia of the Zoological Society of London*, **57**, 275–304.

Prentice, A. M., Coward, W. A., Davies, H. L., Murgatroyd, P. R., Black, A. E., Goldberg, G. R., *et al.* (1985). Unexpectedly low levels of energy expenditure in healthy women. *Lancet*, **i**, 1419–22.

Prentice, A. M., Whitehead, R. G., Coward, W. A., Goldberg, G. R., Davies, H. L., and Murgatroyd, P. R. (1987). Reduction in post-prandial energy expenditure during pregnancy. *British Medical Journal*, **295**, 266–7.

Prentice, A. M., Goldberg, G. R., Davies, H. L., Murgatroyd, P. R., and Scott, W. (1989). Energy-sparing adaptations in human pregnancy assessed by whole-body calorimetry. *British Journal of Nutrition*, **62**, 5–22.

Prentice, A. M., Diaz, E., Goldberg, G. R., Jebb, S. A., Coward, W. A., and Whitehead, R. G. (1992). Famine and refeeding adaptations in energy metabolism. In *The biology of feast and famine: relevance to eating disorders* (ed. G. H. Anderson and S. H. Kennedy), pp. 245–67. Academic Press, San Diego.

Reiss, M. J. (1989). *The allometry of growth and reproduction.* Cambridge University Press.

Robbins, C. T. (1983). *Wildlife feeding and nutrition*, pp. 166–208. Academic Press, New York.

Schutz, Y. (1989). Energy metabolism in the pregnant woman. *Sozial und Praeventivmedizin*, **34**, 63–6.

Schutz, Y., Golay, A., and Jequier, E. (1988). 24 hour energy expenditure (24-EE) in pregnant women with a standardised activity level. *Experientia*, **44**, A31.

Searle, T. W., McGraham, N., and O'Callaghan, M. (1972). Growth in sheep. 1. The chemical composition of the body. *Journal of Agricultural Science, Cambridge*, **79**, 371–81.

Singh, J., Prentice, A. M., Diaz, E., Coward, W. A., Ashford, J., Sawyer, M., *et al.* (1989). Energy expenditure of Gambian women during peak agricultural activity measured by the doubly-labelled water method. *British Journal of Nutrition*, **62**, 315–29.

Smith, C. A. (1947). Effects of wartime starvation in Holland on Pregnancy and its products. *American Journal of Obstetrics and Gynecology*, **53**, 599–608.

Sohlstrom, A. (1993). *Body fat during reproduction in a nutritional perspective: studies in women and rats*, PhD Thesis. Karolinska Institute, Stockholm.

Spaaij, C. J. K. (1993). *The efficiency of energy metabolism during pregnancy and lactation in well-nourished Dutch women*, PhD Thesis. University of Wageningen, Netherlands.

Spray, C. M. (1950). A study of some aspects of reproduction by means of chemical analysis. *British Journal of Nutrition*, **4**, 354–60.

Stein, Z., Susser, M., Saenger, G., and Marolla, F. (1975). Famine and human development: the Dutch hunger winter of 1944–1945. Oxford University Press, New York.

Steingrimsdottir, L., Greenwood, M. R. C., and Brasel, J. A. (1980). Effect of pregnancy, lactation and a high-fat diet on adipose tissue in Osborne-Mendel rats. *Journal of Nutrition*, **110**, 600–9.

Thongprasert, K., Tanphaichitre, V., Valyasevi, A., Kittigool, J., and Durnin, J. V. G. A. (1987). Energy requirements of pregnancy in rural Thailand. *Lancet*, **ii**, 1010–2.

Truswell, A. S., Ash, S., and Allen, J. R. (1988). Energy intake during pregnancy. *Lancet*, **i**, 49.

Tuazon, M. A., van Raaij, J. M. A., Hautvast, J. G. A. V., and Barba, C. V. C. (1987). Energy requirements of pregnancy in the Philippines. *Lancet*, **ii**, 1129–30.

van Raaij, J. M. A., Vermaat-Miedema, S. H., Schonk, C. M., Peek, M. E. M., and Hautvast, J. G. A. J. (1987). Energy requirements of pregnancy in the Netherlands. *Lancet*, **ii**, 953–5.

van Raaij, J. M. A., Peek, M. E. M., Vermaat-Miedema, S. H., Schonk, C. M., and Hautvast, J. G. A. J. (1988). New equations for estimating body fat mass in pregnancy from body density or total body water. *American Journal of Clinical Nutrition*, **48**, 24–9.

van Raaij, J. M. A., Schonk, C. M., Vermaat-Miedema, S. H., Peek, M. E. M., and Hautvast, J. G. A. J. (1990). Energy cost of physical activity throughout pregnancy and the first year postpartum in Dutch women with sedentary lifestyles. *American Journal of Clinical Nutrition*, **52**, 234–9.

Widdowson, E. M. (1950). Chemical composition of newly born mammals. *Nature*, **166**, 626–8.

Zeigler, E. E., O'Donnell, A. M., Nelson, S. E., and Fomon, S. J. (1976). Body composition of the reference fetus. *Growth*, **40**, 329–41.

3 Angiogenic growth factor expression in the uterus

STEVEN K. SMITH

I   **Introduction**

II **Angiogenesis**

III **Angiogenic growth factors**
   1 Fibroblast growth factors
      i Structure and function
      ii FGF receptors
      iii Identification in the human reproductive tract
      iv Function in human reproduction
   2 Vascular endothelial growth factor
      i Structure and function
      ii VEGF receptors
      iii Expression of VEGF in human endometrium
      iv Expression of VEGF in the pregnant reproductive tract
      v Function in human reproduction

IV **Angiogenesis and its clinical relevance**
   1 Menstruation
   2 Implantation
   3 Endometriosis
   4 Endometrial bleeding with long-acting contraceptives and continuous combined hormone replacement therapy
   5 Placental development

V   **Other angiogenic agents**

I   INTRODUCTION

Angiogenesis is the process by which new blood vessels are formed (Folkman and Shing 1992) by the sprouting of capillaries from pre-existing blood vessels (Risau 1991). In most circumstances in adult life, angiogenesis occurs infrequently and the turnover of endothelial cells in blood vessels is of the order of years (Denekamp 1984). Normal angiogenesis is part of the body's repair system involved in wound healing and bone repair. In addition pathological angiogenesis arises in a wide range of diseases including the growth of solid tumours (Folkman 1990; Bicknell and Harris 1991), diabetic retinopathy, arthritis, and some autoimmune diseases (Folkman and Klagsburn 1987). However by far the most common exception to the rule that angiogenesis occurs infrequently is in the female reproductive tract. Here angiogenesis occurs monthly. In the ovary, angiogenesis occurs with the development of the follicle, its rupture, and the subsequent formation of the corpus luteum. In this review I wish to consider the other site of profound angiogenesis without which human reproduction would not occur, that is the endometrium and the placenta.

Primate reproduction has evolved a system of regular endometrial shedding in order to provide a receptive endometrium for implantation. In primate species there is an urgent need to repair the desquamated endometrium that is present at the end of menstruation. The subsequent growth and development of the endometrium, as with any other organ, is dependent on the development of an adequate blood supply. In this tissue it is of even greater importance because it is the endometrial vessels which will supply nutrition to the embryo at implantation. Finally, angiogenesis is required for the development of the placenta and of the embryonic cardiovascular system (Risau 1991).

Disturbances in the growth and development of the endometrial vasculature have profound consequences for women's health, ranging from menstrual dysfunction, endometriosis, failed implantation, and breakthrough bleeding with oral and injectable contraceptives, or combined continuous or single continuous hormone replacement therapies, to early pregnancy loss, intra-uterine growth retardation, and pre-eclampsia, the latter three conditions being characterized by disturbances of trophoblast–endothelial cell interactions. Considering the widespread importance of this subject to women's health it is surprising how little is known of the normal physiology of angiogenesis in the endometrium, decidua, and placenta.

II   ANGIOGENESIS

Angiogenesis is a closely regulated mechanism obligatory for reproduction and development. Endothelial cells undergo proliferation, migration, and morphogenic changes to form new tubular structures (Folkman and Shing 1992). In addition, the endothelium can regulate the development of vascular smooth muscle and is involved in the formation of larger blood vessels (De May and Schiffers 1993).

At menstruation, approximately the upper two-thirds of endometrium is shed, leaving the desquamated basal endometrium. Blood vessels are ruptured leaving open damaged vessels (Ludwig and Metzger 1976). Blood does not pass out of these vessels, partly because of the weak platelet fibrin thrombi which form in endometrium during the first 24 hours of menstruation, and partly because of the intense vasoconstriction which occurs in the basal parts of the vessels (Markee 1940). By the fifth day after menses, the blood vessels have been repaired and have re-established healthy tubular structures. The growth of the endometrium during the proliferative phase, including the increased thickening that arises from the action of oestradiol, is associated with growth of the spiral arterioles and the development of a sub-epithelial capillary plexus (Fig. 3.1). However, endometrial microvascular density does not alter during the cycle (Hourihan *et al.* 1990; Rogers *et al.* 1993).

The changes that occur at human implantation are sparsely described. In rodents, which share some features of human haemochorial placentation (Finn 1983), implantation is characterized by local oedema at the implantation site (Psychoyos and Martel 1985) and is associated with increased capillary dilatation (Tawia and Rogers 1992) and altered permeability. After attachment in the human interstitial implantation occurs with the formation of lacunae, but trophoblasts rapidly invade the endometrial blood vessels and replace the endothelial cells. Recent evidence from Doppler ultrasonography suggests that blood does not pass into the trophoblastic lacunae in the first trimester of pregnancy. At this stage trophoblast may act as a plug, invading the vasculature. Placental development requires the further growth of the maternal vasculature but also involves the formation and growth of fetal blood vessels within the villae.

Over the past 20 years an increasing number of factors have been indentified which exert angiogenic actions. These actions are assayed *in vitro* by investigating effects on endothelial cell proliferation, migration, differentiation, and tube formation in soft agar gels (Pepper *et al.* 1992), and *in vivo* in preparations like the chick chorio-allantoic membrane and cornea (Ausprunk *et al.* 1974; Gimbrone *et al.* 1974). Angiogenic factors may be broadly divided into peptides and non-peptides, the latter including 1-butyryl glycerol (Dobson *et al.* 1990), PGE_1 (Ziche and Gullino 1982), PGE_2 (Form and Auerbach 1983), nicotinamide (Morris *et al.* 1989), adenosine (Dusseau *et al.* 1986), and okadaic acid (Oikawa *et al.* 1992) (Table 3.1). This review will concentrate on the function of angiogenic peptides expressed in endometrium, decidua, and placenta.

III ANGIOGENIC GROWTH FACTORS

1 Fibroblast growth factors

i Structure and function

Fibroblast growth factors (FGFs) are a family of mitogenic growth factors with pleiotrophic actions, stimulating growth in endothelial cells, smooth

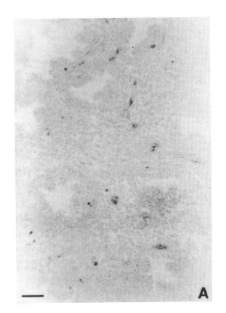
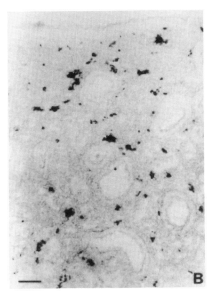
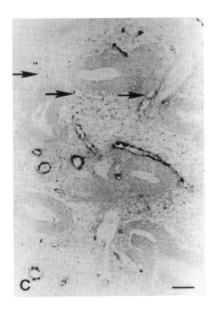

muscle cells, fibroblasts, and some epithelial cells (Burgess and Maciag 1989). Acidic and basic FGF are heparin-binding growth factors of about 18 kDa which share just over 50 per cent sequence homology (Esch *et al.* 1985). Longer forms of the peptides arise by transcription from alternative initiation sites (Florkiewicz and Sommer 1989). Other members of the family include several oncogenes which share about 40 per cent sequence homology with the FGFs (Delli-Bovi *et al.* 1987; Yoshida *et al.* 1988; Dickson *et al.* 1990). FGF-7 is also called keratinocyte growth factor and stimulates epithelial cell proliferation in addition to endothelial cell mitosis (Finch *et al.* 1989). FGF-8 is a uterine FGF isolated from the cow (Milner *et al.* 1989). Both acidic and basic FGFs are intracellular peptides and so lack a secretory signal sequence (Abraham *et al.* 1986; Vlodavsky *et al.* 1987). They bind heparin which protects them from heat, acid, and protease digestion (Gospodarowicz *et al.* 1986). The binding of FGF to heparin sulphate is needed for binding to the high-affinity cell surface receptor (Rapraeger *et al.* 1991; Yayon *et al.* 1991). In addition, the FGFs are bound to heparin and heparin sulphate proteoglycans in the extracellular matrix, providing a reservoir of growth factor for subsequent release by proteases. The FGFs also bind to a specific heparin sulphate proteoglycan in the cell membrane (Kiefer *et al.* 1990) which prolongs their biological activity.

Acidic and basic FGF are angiogenic in the chick chorio-allantoic membrane and corneal bioassays (Lobb *et al.* 1985; Abraham *et al.* 1986). They are mitogenic and morphogenic for endothelial cells (Gospodarowicz *et al.* 1987), stimulate plasminogen activator release from endothelial cells (Mignatti *et al.* 1992) and stimulate endothelial cells to form tubules in three-dimensional gels (Montesano *et al.* 1986).

ii FGF receptors

FGFs exert their actions through specific cell-surface receptors which have intrinsic tyrosine kinase activity (Huang *et al.* 1986; Friesel *et al.* 1989). Dionne *et al.* (1990) isolated clones from a human endothelial cell cDNA library which they have identified as the specific receptors for acidic and

Fig. 3.1 Immunohistochemical staining for von Willebrand factor in the normal menstrual cycle. (A) At the time of menstruation, there is very poor staining for von Willebrand factor. (B) Intense immunohistochemical staining is present in the microvasculature in late proliferative phase endometrium. (C) Immunohistochemical staining is still present in some of the blood vessels of the endometrium in the secretory phase of the cycle, though in other vessels indicated by the arrow there is less enhanced staining, demonstrating an increased heterogeneity of staining in this phase. These findings demonstrate the changes in microvascular density that arise during the menstrual cycle and changes in endothelial cell heterogeneity during the cycle. (Reproduced from by kind permission of the editor of *Human Reproduction.*)

Table 3.1 *Angiogenic growth factors*

Growth factor	*In vitro* endothelial actions	Angiogenesis	Reference
EGF	+	+	Connolly et al. 1989 b
aFGF	+	+	Shing et al. 1985
bFGF	+	+	Shing et al. 1985
Epidermal growth factor	+	+	Schreiber et al. 1986
TGFα	+	+	Schreiber et al. 1986
PD-ECGF	+	+	Miyazono et al. 1987
TNFα	±	+	Leibovich et al. 1987
TGFβ	±	+	Roberts et al. 1986
Angiotropin	±	+	Hockel et al. 1988
Angiogenin	±	+	Fett et al. 1985
Granulocyte–Macrophage colony-stimulating factor	+	±	Bussolino et al. 1989
Granulocyte colony-stimulating factor	+	±	Bussolino et al. 1989
Butyryl glycerol	±	+	Bussolino et al. 1989
PGE$_1$	±	+	Conn et al. 1990
PGE$_2$	±	+	Conn et al. 1990
Nicotinamide	±	+	Connolly et al. 1989 b
Adenosine	±	+	D'Amore and Klagsbrun 1989
Hyaluronic acid	±	+	Davidson et al. 1985
(12R)-hydroxyeicosatrienoic acid	±	+	Delli-Bovi et al. 1987
Okadaic acid	±	+	Dickson et al. 1990

A wide range of agents are now assumed to have angiogenic properties, some of which exert their effects directly and can be shown to have actions on endothelial cell proliferation and migration. Others do not have direct effects on endothelial cells *in vitro*. (For reviews see Klagsbrun and D'Amore 1991; Folkman and Shing 1992.)

basic FGF (a FGF and bFGF) previously designated *flg* (FGF-R1) and *bek* (FGF-R2). The extracellular domain has three immunoglobulin-like repeats and is similar to other receptors of the family. For example, receptors for colony stimulating factor (CSF), stem cell factor (SCF), and platelet-derived growth factor (PDGF) contain five, and the receptor for vascular endothelial growth factor (VEGF) contains seven immunoglobulin-like domains. Alternative splicing in the third immunoglobulin-like domain confers specificity in binding between acidic and basic FGF (Werner *et al.* 1992). The putative IIIb exon codes for a protein which will bind aFGF but not bFGF. Conversely, the peptide receptor containing the product of exon IIIc binds both aFGF and bFGF. Further alternative splicing results in a secreted receptor which differs from the membrane-spanning domain receptor, lacking a part of the third immunoglobulin-like domain and the transmembrane and cytoplasmic sequence (Reid *et al.* 1990). A secreted mouse protein is a functional receptor for acidic and basic FGF (Werner *et al.* 1992). These receptors confer cellular specificity for the FGFs but as yet their expression in human reproductive tissue remains unknown.

iii Identification in the human reproductive tract

Immunoreactive bFGF is present in endometrium at high concentration (Rusnati *et al.* 1990) and the concentrations do not alter significantly throughout the menstrual cycle. Certain endometrial carcinoma cell lines also release immunoreactive bFGF but in this case the release is stimulated by oestradiol and inhibited by progesterone (Presta 1988).

Immunohistochemistry indicates that the glandular and luminal epithelial cells of endometrium are the principal site of FGF immunoreactivity for both acidic and basic FGF (Ferriani *et al.* 1993) (Fig. 3.2). Whilst this is not quantitative, the intensity of staining did not alter throughout the menstrual cycle. There was no background staining in the stromal cells, but endothelial cells were positively stained. In early pregnancy, the staining for both growth factors was similar. Extravillous trophoblast in the decidua showed immunoreactive FGFs and stained with cytokeratin in serial sections (Ferriani *et al.* in press). In first trimester villae, FGF was present in and around the cytotrophoblast but not in the syncytiotrophoblast or in the intermediate trophoblast. By the third trimester of pregnancy, clear staining was present in the syncytiotrophoblast and the vascular smooth muscle of the fetal vessels within the tertiary villae.

iv Function in human reproduction

FGFs probably act in the endometrium to promote angiogenesis. However, they do not have a secretory signal sequence and are probably most active at menstruation. FGF can be released from the cell by trauma (Muthukrishnan *et al.* 1991) and may be released in large amounts at the time of menstruation. It may then be stored in the extracellular matrix or

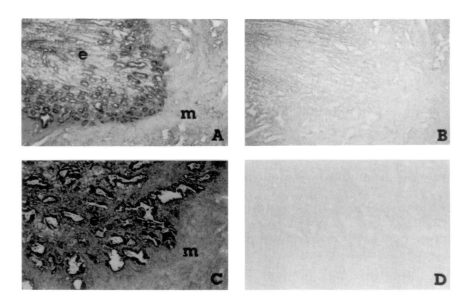

Fig. 3.2 Immunohistochemical identification of acidic fibroblast growth factors (aFGF) in normal endometrium. Tissue was obtained in the proliferative (A) and secretory (C) phases of the menstrual cycle. Control incubations (B and D) were performed in the presence of pre-immune serum. The most intense staining was demonstrated in the basal glands of the endometrium (e) close to the myometrium (m). Intensity of staining increased in the secretory phase of the menstrual cycle. This characteristic of staining was demonstrated for both acidic and basic fibroblast growth factor. (Reproduced from by kind permission of the editor of *Human Reproduction*.)

on the surface of epithelial cells by their association with heparin sulphate proteoglycans (Ruoslahti and Yamaguchi 1991). FGF could be released when proteolytic enzymes are secreted during apoptosis which occurs at menses.

In addition to their angiogenic activities FGFs may also mediate mesenchymal functions in endometrium. Isolated stromal cells from human endometrium, obtained in either the proliferative or secretory phases of the cycle, proliferate in the presence of bFGF, an action enhanced by the presence of progesterone, being greatest in cells obtained on day 21 of the luteal phase (Irwin *et al.* 1991). This action of FGF is mediated by enhanced phosphatidyinositol-phospholipase C and phospholipase D activity in the endometrium (Ahmed *et al.* 1994; Ferriani *et al.*, in press). It is not clear whether FGFs exert other actions on stromal cells or whether their actions depend on the lineage of the stromal cells. The function of FGFs in extravillous trophoblast is unclear but they do induce migration of endothelial cells and studies are needed to determine if they are involved in trophoblast migration. The presence of receptors on trophoblast has not been confirmed.

In view of the many permutations arising from the differential expression of acidic and basic FGF, their binding and release from heparin or heparan sulphate proteoglycans in the extracellular matrix and on the cell surface, and their potential for different receptor expression in target cells, not to mention the possibility of alternative splicing of the receptor in the cell membrane or the release of a secreted receptor, identification of the function of FGFs would seem daunting. In concert with VEGF, the FGFs represent a significant source of angiogenic growth factor in endometrium, capable of repairing and maintaining the endometrial vasculature throughout the menstrual cycle. Pathological changes in angiogenic growth factor expression in benign or malignant disease of the endometrium have yet to be determined but studies are now under way to investigate this possibility.

2 Vascular endothelial growth factor

i *Structure and function*
After the identification of FGFs as potent angiogenic growth factors, another group of heparin binding growth factors was identified whose specificity of action is less diffuse than that of FGFs. The VEGFs are a group of proteins of approximately 46 KDa consisting of two identical 23 KDa subunits (Connolly *et al.* 1989*b*; Ferrara and Henzel, 1989; Gospodarowicz *et al.* 1989; Tischer *et al.* 1991) which arise by alternative splicing of a single gene product. The human gene comprises eight exons (Tischer *et al.* 1991). At least five species of VEGF have been identified in the human arising from alternative splicing. The first four exons are transcribed in all species, the rest arising by alternative splicing of the exons encoding the carboxy-terminal of the peptides (Leung *et al.* 1989; Houck *et al.* 1991; Charnock-Jones *et al.* 1993). Exon 1 contains a 26 amino acid hydrophobic consensus secretory signal common to all peptides (Tischer *et al.* 1991). $VEGF_{189}$ is transcribed from all eight exons. $VEGF_{165}$ does not include exon 6 and contains a predicted 24 amino acid deletion, the splice site being at position 116, with a lysine to asparagine change at residue 115. $VEGF_{145}$ is predicted to include exon 6 but lacks exon 7. $VEGF_{121}$ lacks the products of exons 6 and 7. A further VEGF ($VEGF_{206}$ has been described which contains an additional 17 codons after the 24 amino acid insertion in $VEGF_{189}$ (Houck *et al.* 1991). This peptide has only been described in fetal liver and is not present in the reproductive tract.

The various species of VEGF have different patterns of secretion and significant differences in structure and isoelectric point. Initial studies suggested that $VEGF_{165}$ and $VEGF_{121}$ were secreted peptides, biological activity being present in the supernatant of cells transfected with the gene (Houck et al. 1991). More recent evidence suggests that $VEGF_{189}$ is released from the cell but is tightly bound to proteoglycans in the cell membrane or in the extracellular matrix, as indicated by the release of $VEGF_{189}$ into culture media with treatment by suramin and heparinases I and III (Houck et al. 1992). $VEGF_{121}$ is secreted and is freely soluble. The binding of VEGFs to heparin is similar to that found with the FGFs.

It is now clear that vascular permeability factor (Senger et al. 1987, 1990; Connolly et al. 1989a, b) is analogous to $VEGF_{189}$ (Keck et al. 1989) indicating that at least the larger molecular weight species of VEGF do not just promote angiogenesis but also increase vascular permeability. All species have direct effects on endothelial cells, stimulating proliferation of cloned bovine microvascular endothelial (BME) cells (Pepper et al. 1991) and human umbilical vein endothelial (HUVE) cells (Bikfalvi et al. 1991), and they are angiogenic in the chick chorio-allantotic membrane assay (Leung et al. 1989; Plouet et al. 1989). In the cases of $VEGF_{189}$ and $VEGF_{165}$, these actions are present or enhanced respectively only when the peptides are released from their bound states (Houck et al. 1991, 1992). $VEGF_{165}$ also induces plasminogen activator activity in BME cells (Pepper et al. 1991). Urokinase and tissue plasminogen activator mRNA levels in these endothelial cells are increased 7.5 and 8-fold, respectively, after 15 hours in culture and plasminogen activator inhibitor mRNA is increased 4.5 fold after 4 hours. VEGFs exert angiogenic actions *in vitro*, causing endothelial cells to invade three-dimensional collagen gels and forming tubules (Bikfalvi et al. 1991; Pepper et al. 1992). VEGF stimulates release of von Willebrand factor from endothelial cells (Brock et al. 1991) and release of tissue factor from HUVE cells (Clauss et al. 1990).

ii VEGF receptors

Two receptors for VEGF have been identified in humans. The *fms*-like tyrosine kinase receptor (*flt*) is a type III receptor tyrosine kinase which also includes the product of *c-kit* and the PDGF α and β chains (De Vreis et al. 1992). The predicted *flt* gene product has a molecular weight of about 150 KDa and consists of a 758 amino acid extracellular domain, a 22 amino acid transmembrane domain, and a 558 amino acid intracellular region which contains a tyrosine kinase domain (Shibuya et al. 1990). Because of partial gene duplication, the extracellular domain contains seven immunoglobulin-like domains, unlike the other members of the family which have five of these repeats (*fms*,, c-*kit*, and PDGF α β), and it is consequently 220 amino acids longer (Shibuya et al. 1990). The receptor has a 66 amino acid insertion in the middle of the tyrosine kinase domain, characteristic of type III receptor tyrosine kinases. The kinase domain contains a conserved lysine at the ATP binding residue and a putative autophosphorylation site (1053).

A further receptor was identified recently which binds VEGF with high affinity, is similar to *flt* in that it has seven immunoglobulin like domains in the extracellular region and a 70 amino acid insert in the tyrosine kinase domain, and has been identified as the kinase domain receptor (Terman et al. 1992).

iii Expression of VEGF in human endometrium

Four transcripts of VEGF are found in human endometrium at all stages of the menstrual cycle (Charnock-Jones et al. 1993). Three of these transcripts encode $VEGF_{189}$, $VEGF_{165}$, and $VEGF_{121}$. A fourth product, a splice variant encoding a peptide of 145 amino acids is translated from an mRNA lacking exon 7 but incorporating exon 6. This region of the VEGF gene encodes a part of the peptide which increases its binding to heparin and suggests that this 'new' peptide would have the same extracellular distribution as $VEGF_{189}$ and $VEGF_{165}$. Interestingly, in peripheral monocytes, there are two products of polymerase chain reaction consistent with $VEGF_{165}$ and $VEGF_{121}$. *In situ* hybridization can identify the specific cellular localization of the mRNAs for VEGF. In the proliferative phase of the menstrual cycle, hybridization is found both in glandular and luminal epithelial cells and in most cells of the stroma. These studies cannot discriminate between each splice variant nor do they identify the specific cells of the stroma. However, in the luteal phase of the cycle, hybridization does not occur in the stromal cells. The most intense hybridization occurs in the glandular cells of menstruating endometrium (Fig. 3.3). At this time, some stromal cells also contain VEGF mRNA. Northern blot and RNase protection assays show that oestradiol increases steady state levels of mRNA in endometrial carcinoma cell lines HEC-1a and HEC-1b, which contain mRNA encoding the oestradiol receptor. Greatest induction occurs at about 20 hours.

iv Expression of VEGF in the pregnant reproductive tract

VEGF is expressed in a variety of cells of the placenta and placental bed throughout human pregnancy (Sharkey et al. 1994). Glandular epithelium in the decidua continues to show hybridization. In addition, macrophages in the decidua show intense hybridization and in the first trimester, this appears to be the principal site of VEGF expression (Fig. 3.4). The signal is not present in all macrophages but in a subset of macrophages near the placenta. Weak hybridization is present in syncytio-and cytotrophoblast but is not present in extravillous trophoblast. Hofbauer cells, thought to be fetal macrophages, also contain the mRNA.

The pattern of hybridization in placentae and decidua of late pregnancy is quite different. Syncytiotrophoblast cells still express mRNA for VEGF

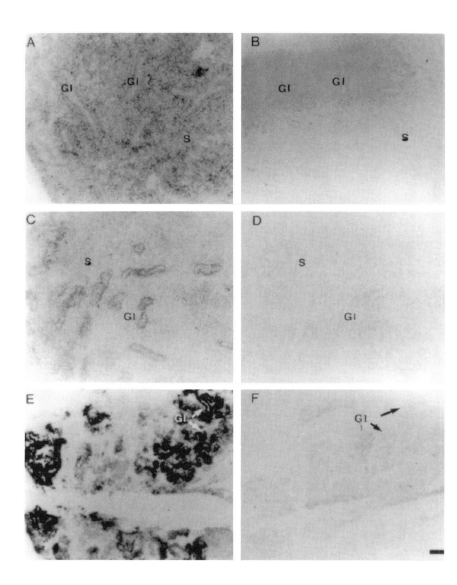

and intense hybridization is found in extravillous trophoblast cells in the decidua. Amnion demonstrates hybridization to the VEGF probe, and cells situated in the chorion, which also stain for CD14, showed strong hybridization. These are presumably macrophages. The fetal macrophages do not show hybridization in late pregnancy. The trophoblast cell lines BeWo, JEG, and JAR all demonstrate mRNA encoding all four splice variants of VEGF (Charnock-Jones et al. 1993; Sharkey et al. 1994).

v Function in human reproduction

Clearly, the most likely function of VEGF in non-pregnant endometrium is angiogenesis. At menstruation, glandular cells demonstrate intense hybridization suggesting, though not proving, increased expression of VEGF. In gliomas, hypoxia increases expression of VEGF and, in view of the ischaemia which preceeds menstruation due to vasoconstriction of spiral arterioles, it is possible that this mechanism increases VEGF synthesis, supplementing the FGFs in the extracellular matrix. In addition VEGF is also bound to heparin and is expressed in the luteal phase by glandular cells. It is not known whether secretion of VEGF by epithelial cells is preferentially from the basal or the apical cell membrane, Apical secretion would provide a reservoir of VEGF, as occurs with FGF. In addition to glandular secretion of VEGF and the extracellular matrix reservoir, macrophages present at the site of wound healing at menstruation may secrete VEGFs. The smallest transcript is not bound to heparin but is angiogenic, suggesting a more direct effect of macrophages on angiogenesis.

The continued growth of endometrium during the proliferative phase, is like all organ growth, dependent on the concomitant development of a blood supply. In the endometrium this arises from the growth of the end-arterioles, the spiral vessels. *In vivo* VEGF is angiogenic, though it is likely that this development involves other angiogenic growth factors.

In rodents, the implantation site is characterized by local oedema (Bell 1985). In view of the vascular permeability properties of VEGFs it is

Fig. 3.3 Localization of VEGF in endometrial tissue. The antisense VEGF RNA probe was used in sections A, C, and E; and the equivalent sections hybridised with sense probe are shown in B, D, and F. Tissue was obtained in (A) the proliferative phase, (C) the secretory phase, and (E) at the time of menstruation. In proliferative endometrium hybridization occurs in the glandular structures (Gl), and in the stromal compartment (S). In the luteal phase of the menstrual cycle there is more specific hybridization for VEGF mRNA in the gland, the hybridization in the stromal compartment being significantly reduced from that present in the proliferative phase of the cycle. At the time of menstruation, intense hybridization is found in the superficial glands of the endometrium with little or none in the stromal compartment. (Reproduced from by kind permission of the editor of *Biology of Reproduction*).

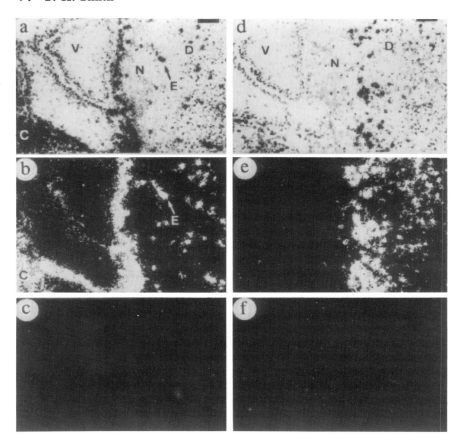

Fig. 3.4 Detection of human VEGF and *flt* mRNA by *in situ* hybridization in first trimester human placenta. (a) Brightfield view of section hybridized with an *flt* antisense probe. Villous trophoblast (V) is shown next to maternal decidua (D), separated by a necrotic area, the layer of Nitabuch (N). Extravillous trophoblast cells (E) are seen passing into the decidua. Highest levels of hybridization are seen in the extravillous trophoblast and the cytotrophoblast (C) cells passing away from the villous trophoblast. (b) Darkfield view of the same section, in which the antisense *flt* probe has been used. (c) The section in which the sense probe has been employed. (d) Brightfield view with the VEGF antisense probe. (e) Darkfield view of the same section. In these sections VEGF probe is seen to hybridize to individual cells within the decidua but there is only weak hybridization to trophoblast. Serial sections using immunohistochemical techniques confirm these cells as being maternal macrophages. (f) Darkfield presentation using the VEGF sense probe.

possible that they may contribute to this effect. Further studies are needed to determine if this oedema occurs at human implantation sites and to establish the role of VEGF in rodent pregnancy.

Until recently, VEGF has been assumed to have specific actions on endothelial cells because receptors have not been identified on other cell types. Present in situ hybridization studies in murine and human trophoblast now contradict this assumption (Charnock-Jones et al. 1994). The VEGF receptor, *flt*, is found in villous columns as they pass from the villae into the decidua and is present on extravillous trophoblast. VEGF is angiogenic because it promotes proliferation, migration, and tube formation of endothelial cells. It is possible that it has a similar effect on trophoblast which express the same receptor. A human choriocarainoma trophoblastic cell line (BeWo) expresses the receptor and responds to VEGF by phosphorylation of the signal transducer Mitosis-activated protein kinase and increased proliferation though only by 50 per cent, in cell numbers over four days (Charnock-Jones et al. 1994). Thus in the first trimester, macrophages expressing secreted VEGF may in some way regulate the migration of trophoblast into the decidua. In addition, villous VEGF may be needed for the development of the fetal vessels. Further studies are needed to clarify the function of VEGF in human implantation and placentation.

IV ANGIOGENESIS AND ITS CLINICAL RELEVANCE

1 Menstruation

The factors regulating the volume of menstrual blood loss are poorly understood. The contribution that vascular repair makes to the process is not known. At the time of menstruation a large number of angiogenic growth factors are presumably present in the endometrium. However, the process of wound repair that occurs at the endometrial surface is not the same as that which occurs at most sites in the body in that fibrous tissue is not deposited. In some ways the repair resembles that in fetal tissues. The contribution of angiogenic growth factors to the aetiology of menstrual dysfunction needs to be established.

2 Implantation

The finding of altered FGF expression at different sites in the endometrium of the rat at implantation raises the prospect of local regulation of angiogenic growth factor expression (Carlone and Rider 1993). The oedema at the implantation site suggests that growth factors like VEGF, which increase vascular permeability, could be involved in implantation. Developing this knowledge into treatments is likely to be extremely difficult.

3 Endometriosis

Recently, endometriotic explants were shown to contain immunoreactive aFGF and bFGF (Ferriani *et al.* 1994) and the same epithelial cells contain mRNA encoding bFGF (Ahmed, personal communication). McLaren *et al.* (unpublished) have shown mRNA and protein for VEGF co-existing in the same explants. The greatest level of expression, however, is not in the endometrial tissue itself but in the invading activated macrophages. These macrophages also express other angiogenic growth factors including TNFα and TGFβ. These agents may be responsible for the intense angiogenesis which surrounds explanted endometrium. Why and how these macrophages are activated remains a key unknown factor in the understanding of the aetiology of endometriosis.

4 Endometrial bleeding with long-acting contraceptives and continuous combined hormone replacement therapy

These treatment regimes are of increasing interest in the developing and developed worlds, respectively. In both circumstances, prolonged non-cyclical administration of steroids results in irregular unpredictable bleeding (Odlind and Fraser 1990). The regulation of the factors responsible for this bleeding is not known. Significant increases in microvascular density are found in the endometrium of women taking steroids continuously and it is likely that this reflects angiogenic growth factor expression. A significant effort is needed to determine the steroidal regulation of growth factor expression in the endometrium as it is probable that some of the effects of steroids on vascular development are mediated by these proteins. This would permit the development of steroidal treatment regimens that target the effects on endometrial vascular development.

5 Placental development

The importance of angiogenic growth factors in the preparation of the endometrium is clear. Similarly, the complex interactions that occur between the trophoblast and the maternal vessels in establishing the blood supply of the placenta may also require the full range of angiogenic growth factors present in endometrium and trophoblast. The exact function of angiogenic growth factor receptors on trophoblast remains unclear. The links for cross-talk between the maternal immune system and the trophoblast are in place and would seem to play a part in the mechanism of implantation. This is likely to continue into the latter stages of trophoblast invasion.

V OTHER ANGIOGENIC AGENTS

This review has described the structure and function of three well characterized angiogenic growth factors. This inevitably results in the exclusion of a range of other agents known to stimulate or antagonize new blood vessel development (see Table 3.1). Similarly, the function of oestradiol and progesterone in mediating these events, and the role of surface adhesion molecules in the growth, differentiation, and migration of endothelial cells have not been considered. All these factors are likely to be of crucial importance in the regulation of angiogenesis in the reproductive tract. The close association between the endometrium and the trophoblast further complicates the picture. In view of the importance of angiogenesis to women's and children's health, further research is urgently needed to understand this process and to develop novel means of regulating reproductive function.

REFERENCES

Abraham, J. A., Mergia, A., Whang, J. L., Tumolo, A., Friedman, J., Hjenild, K. A., *et al.* (1986). Nucleotide sequence of a bovine clone encoding the angiogenic protein, basic fibroblast growth factor. *Science*, **233**, 545–8.

Ahmed, A., Plevin, R., Shaobi, M., Fountain, S. A., Ferriani, R. A., and Smith, S. K. (1994). Basic fibroblast growth factor activates phospholipase D in human vascular endothelial cells in the absence of inositol-lipid hydrolysis. *American journal of Physiology*, **266**, C206–12.

Ausprunk, D. H., Knighton, D. R., and Folkman, J. (1974). Differentiation of vascular endothelium in the chick corioallantois; a structural and autoradiographic study. *Developmental Biology*, **38**, 237–48.

Bell, S. C. (1985). Comparative aspects of decidualization in rodents and humans: cell types, secreted products and associated function. In *Implantation of the human embryo* (ed. R. G. Edwards, J. Purdy, and P. C. Steptoe), pp. 71–122, Academic Press, London.

Bicknell, R. and Harris, A. L. (1991). Novel growth regulatory factors and tumour angiogenesis. *European Journal of Cancer*, **27**, 785–9.

Bikfalvi, A., Sanjeau, C., Maukadiri, H. M., Maclouf, J., Busso, N., Bryckaert, M., *et al.* (1991). Interaction of vasculotropin/vascular endothelial cell growth factor with human umbilical vein endothelial vessels: binding, internalization, degradation and biological effects. *Journal of Cellular Physiology*, **149**, 50–9.

Brock, T. A., Dvorak, H. F., and Senger, D. R. (1991). Tumor-secreted vascular permeability factor increases cytosolic Ca^{2+} and von Willebrand factor release in human endothelial cells. *American Journal of Pathology*, **138**, 213–21.

Burgess, W. H. and Maciag, T. (1989). Heparin binding (fibroblast) growth factor family of proteins. *Annual Review of Biochemistry*, **58**, 575–606.

Bussolino, F., Wang, J. M., Defilippi, P., Turrini, F., Sanavio, F., Edgell, C. J. S., *et al.* (1989). Granulocyte-and granulocyte–macrophage-colony stimulating factors induce human endothelial cells to migrate and proliferate. *Nature*, **337**, 471–3.

Carlone, D. L. and Rider, V. (1993). Embryonic modulation of basic fibroblast growth factor in the rat uterus. Biology of Reproduction, **49**, 653–65.

Charnock-Jones, D. S., Sharkey, A. M., Rajput-Williams, J., Burch, D., Schofield, J. P., Fountain, S. A., *et al.* (1993). Identification and localization of alternately spliced mRNAs for vascular endothelial growth factor in human uterus and steroid regulation in endometrial carcinoma cell lines. *Biology of Reproduction*, **48**, 1120–8.

Charnock-Jones, D. S., Sharkey, A. M., Fenwick, P., and Smith, S. K. (1994). LIF mRNA levels peak in human endometrium at the time of implantation and the blastocyst contains mRNA for the receptor at this time. *Journal of Reproduction and Fertility*, **101**, 421–6.

Clauss, M., Gerlach, M., Gerlach, H., Brett, J., Wang, F., Familletti, P. C., *et al.* (1990). Vascular permeability factor: a tumor-derived polypeptide that induces endothelial cell and monocyte procoagulant activity and promotoes monocyte migration. *Journal of Experimental Medicine*, **172**, 1535–45.

Conn, G., Bayne, M. L., Soderman, D. D., Kwok, P. W., Sullivan, K. A., Palisi, T. M., *et al.* (1990). Amino acid and cDNA sequences of a vascular endothelial cell mitogen that is homologous to platelet-derived growth factor. *Proceedings of the National Academy of Science of the USA*, **87**, 2628–32.

Connolly, D. T., Olander, J. V., Heuvelman, D., Nelson, R., Monsell, R., Siegel, N., *et al.* (1989a). Human vascular permeability factor. *Journal of Biological Chemistry*, **264**, 20017–24.

Connolly, D. T., Heuvelman, D. M., Nelson, R., Olander, J. V., Eppley, B. L., Delfino, J. J., *et al.* (1989b). Tumor vascular permeability factor stimulates endothelial cell growth and angiogenesis. *Journal of Clinical Investigation*, **84**, 1470–8.

D'Amore, P. A. and Klagsbrun, M. (1989). Angiogenesis: factors and mechanisms. In *The pathobiology of neoplasia* (ed. A. E. Sirica), pp. 513–531. Plenum Press, New York.

Davidson, J. M., Klagsbrun, M., Hill, K. E., Buckley, A., Sullivan, R., Brewer, P. S., *et al.* (1985). Accelerated wound repair, cell proliferation, and collagen accumulation are produced by a cartilage-derived growth factor. *Journal of Cell Biology*, **100**, 1219–27.

Delli-Bovi, P., Curatola, A. M., Kern, F. G., Greco, A., Ittman, M., and Basilico, C. (1987). An oncogene isolated by transfection of Kaposi's sarcoma DNA encodes a growth factor that is a member of the FGF family. *Cell*, **50**, 729–37.

De Mey, J. G. R. and Schiffers, P. M. (1993). Effects of the endothelium on growth responses in arteries. *Journal of Cardiovascular Pharmacology*, **21**, S22–5.

Denekamp, J. (1984). Vascular endothelium as the vulnerable element in tumours. *Acta Radiologica Oncologies*, **23**, 217–25.
De Vries, C., Escobedo, J. A., Ueno, H., Houck, K., Ferrara, N., and Williams, L. T. (1992). The *fms*-like tyrosine Kinase, a receptor for vascular endothelial growth factor. *Science*, **255**, 989–91.
Dickson, C., Smith, R., Brookes, S., and Peters, G. (1990). Proviral insertions within the *int*-2 can generate multiple anomalous transcripts but leave the protein-coding domain intact. *Journal of virology*, **64**, 784–93.
Dionne, C. A., Crumley, G., Bellot, F., Kaplow, J. M., Searfoss, G., Ruta, M *et al.* (1990). Cloning and expression of two distinct high-affinity receptors cross-reacting with acidic and basic fibroblast growth factors. *EMBO Journal*, **9**, 2685–92.
Dobson, D. E., Kambe, A., Block, E., Dion, T., Lu, H., Castellot, J. J., *et al.* (1990). 1-Butyryl-glycerol: a novel angiogenesis factor secreted by differentiating adipocytes. *Cell*, **61**, 223–30.
Dusseau, J. W., Hutchins, P. M., and Malbasa, D. S. (1986). Stimulation of angiogenesis by adenosine on the chick choriollantoic membrane. *Circulation Research*, **59**, 163–70.
Esch, F. A., Baird, N., Ling, N., Ueno, F., Hill, L., Denoroy, R., *et al.* (1985). Primary structure of bovine pituitary basic fibroblast growth factor (FGF) and comparison with the amino-acid terminal sequence of bovine brain acidic FGF. *Proceedings of the National Academy of Sciences of the USA*, **82**, 6507–11.
Ferrara, N. and Henzel, W. J. (1989). Pituitary follicular cells secrete a novel heparin-binding growth factor specific for vascular endothelial cells. *Biochemical and Biophysical Research Communications*, **161**, 851–8.
Ferriani, R. A., Charnock-Jones, D. S., Prentice, A., Thomas, E. J., Smith, S. K. (1993). Immunohistochemical localization of acidic and basic fibroblast growth factors in normal human endometrium and endometriosis and the detection of their mRNA by polymerase chain reaction. *Human Reproduction*, **8**, 11–16.
Ferriani, R. A., Ahmed, A., Sharkey, A., and Smith, S. K. (1994). Immunolocalisation of acidic and basic fibroblast growth factor (FGF) in human placenta and basic FGF-stimulated mitogenesis and phospholipase C activation in trophoblast cell line JEG-3. *Growth Factor*, **10**, 259–68.
Ferriani, R. A., Ahmed, A., and Smith, S. K. (1995). The regulation of phospholipase D activity in bradykinin stimulated human endometrial stromal cells. *Biology of Reproduction*. (In press.)
Fett, J. W., Strydom, D. J., Lobb, R. F., Alderman, E. M., Bethune, J. L., Riordan, J. F., *et al.* (1985). Isolation and characterisation of angiogenin, an angiogenic protein from human carcinoma cells. *Biochemistry*, **24**, 5480–6.
Finch, P. W., Rubin, J, S., Miki, T., Ron, D., and Aaronson, S. A. (1989). Human KGF is FGF-related with properties of a paracrine effector of epithelial cell growth. *Science*, **245**, 752–5.
Finn, C. A. (1983). Implantation of ova—assessment of the value of laboratory animals for the study of implantation in women. *Oxford Reviews of Reproductive Biology*, **5**, 272–89.
Florkiewicz, R. Z. and Sommer, A. (1989). Human basic fibroblast growth factor gene encodes four polypeptides: three initiate translation from non-AUG

codons. *Proceedings of the National Academy of Sciences of the USA*, **86**, 3978–81.

Folkman, J. (1990). What is the evidence that tumors are angiogenesis dependent? *Journal of the National Cancer Institute*, **82**, 4–6.

Folkman, J. and Klagsbrun, M. (1987). A family of angiogenic peptides. *Nature*, **329**, 671–2.

Folkman, J. and Shing, Y. (1992). Angiogenesis. *Journal of Biological Chemistry*, **267**, 10931–4.

Form, D. M. and Auerbach, R. (1983). PGE2 and angiogenesis. *Proceedings of the Society of Experimental Biology Medicine*, **172**, 214–18.

Friesel, R., Burgess, W. H., and Maciag, T. (1989). Heparin-binding growth factor 1 stimulates tyrosine phosphorylation in NIH 3T3 cells. *Molecular and Cellular Biology*, **9**, 1857–65.

Gimbrone, M. A., Cotran, R. S., and Folkman, J. (1974). Tumour growth neovascularization: an experimental model using rabbit cornea. *Journal of the National Cancer Institute*, **52**, 413–27.

Gospodarowicz, D., Baird, A., Cheng, J., Lui, G. M., Esch, F., and Bohlen, P. (1986). Isolation of fibroblast growth factor from bovine adrenal gland: physico-chemical and biological characterisation. *Endocrinology*, **118**, 82–90.

Gospodarowicz, D., Ferrara, N., Schweigerer, L., and Neufeld, G. (1987). Structural characterization and biological functions of fibroblast growth factor. *Endocrine Reviews*, **8**, 95–114.

Gospodarowicz, D., Abraham, J. A., and Schilling, J. (1989). Isolation and characterization of a vascular endothelial cell mitogen produced by pituitary-derived folliculo-stellate cells. *Proceedings of the National Academy of Sciences of the USA*, **86**, 7311–15.

Hockel, M., Jung, W., Vaupel, P., Rabes, H., Khaledpour, C., and Wissler, J. H. (1988). Purified monocyte-derived angiogenic substance (angiotropin) induces controlled angiogenesis associated with regulated tissue proliferation in rabbit skin. *Journal of Clinical Investigation*, **82**, 1075–90.

Houck, K. A., Ferrara, N., Winer, J., Cachianes, G., Li, B., and Leung, D. W. (1991). The vascular endothelial growth factor family: identification of a fourth molecular species and characterization of alternative splicing of RNA. *Molecular Endocrinology*, **5**, 1806–14.

Houck, K. A., Leung, D. W., Rowland, A. M., Winer, J., and Ferrara, N. (1992). Dual regulation of vascular endothelial growth factor bioavailability by genetic and proteolytic mechanisms. *Journal of Biochemical Chemistry*, **267**, 26031–7.

Hourihan, H. M., Sheppard, B. L., and Brosens, I. A. (1990). Endometrial haemostasis. In *Proceedings of symposium on contraception and mechanisms of endometrial bleeding* (ed. C. D'Arcangues, I. S. Fraser, J. R. Newton, and V. Odlind), pp. 95–116. Cambridge University Press.

Huang, J. S., Huang, S. S., and Kuo, M. D. (1986). Bovine brain-derived growth factor: purification and characterization of its interaction with responsive cells. *Journal of Biological Chemistry*, **261**, 11600–7.

Irwin, J. C., Utian, W. H., and Eckert, R. L. (1991). Sex steroids and growth factors differentially regulate the growth and differentiation of cultured human endometrial stromal cells. *Endocrinology*, **129**, 2385–92.

Keck, P. J., Hauser, S. D., Krivi, G., Sanzo, K., Warren, T., Feder, J., et al. (1989). Vascular permeability factor: an endothelial cell mitogen related to PDGF. *Science*, **246**, 1309–12.

Kiefer, M. C., Stephans, J. C., Crawford, K., Okino, K., and Barr, P. J. (1990). Ligand-affinity cloning and structure of a cell surface heparan sulfate proteoglycan that binds basic fibroblast growth factor. *Proceedings of the National Academy of Sciences of the USA*, **87**, 6985–9.

Klagsbrun, M. and D'Amore, P. A. (1991). Regulators of angiogenesis. *Annual Review of Physiology*, **53**, 217–39.

Leibovich, S. J., Polverini, P. J., Shepard, H. M., Wiseman, D. M., Shively, V., and Nuseir, N. (1987). Macrophage-induced angoigenesis is mediated by tumour necrosis factor-alpha. *Nature*, **329**, 630–2.

Leung, D. W., Cachianes, G., Kuang, W-J., Goeddel, D. V., and Ferrara, N. (1989). Vascular endothelial growth factor is a secreted angiogenic mitogen. *Science*, **246**, 1306–9.

Lobb, R. R., Alderman, E. M., and Fett, J. W. (1985). Induction of angiogenesis by bovine brain derived class 1 heparin-binding growth factor. *Biochemistry*, **24**, 4969–73.

Ludwig, H. and Metzger, H. (1976). The re-epithelialization of endometrium after menstrual desquamation. *Archives Gynaecologie*, **221**, 51–60.

Markee, J. E. (1940). Menstruation in intraocular endometrial transplants in the rhesus monkey. *Contributions to Embryology, Carnegie Institute*, **28**, 219–308.

Mignatti, P., Morimoto, T., and Rifkin, D. B. (1992). Basic fibroblast growth factor, a protein devoid of secretory signal sequence, is released by cells via a pathway independent of the endoplasmic reticulum-Golgi complex. *Journal of Cellular Physiology*, **151**, 81–93.

Milner, P. G., Li, Y-S., Hoffman, R. M., Kodner, C. M., Siegel, N. R., and Deuel, T. F. (1989). A novel 17 kD heparin-binding growth factor (HBGF-8) in bovine uterus: purification and N-terminal amino acid sequence. *Biochemical and Biophysical Research Communications*, **165**, 1096–103.

Miyazono, K., Okabe, T., Urabe, A., Takaku, F., and Heldin, C-H. (1987). Purification and properties of an endothelial cell growth factor from human platelets. *Journal of Biological Chemistry*, **262**, 4098–103.

Montesano, R., Vassali, J. D., Baird, A., Guillemin, R., and Orci, L. (1986). Basic fibroblast growth factor induces angiogenesis *in vitro*. *Proceedings of the National Academy of Sciences of the USA*, **83**, 7297–301.

Morris, P. B., Ellis, M. N., and Swain, J. L. (1989). Angiogenic potency of nucleotide metabolites: potential role in ischaemia-induced vascular growth. *Journal of Molecular and Cellular Cardiology*, **21**, 351–8.

Muthukrishnan, L., Warder, E., and McNeil, P. L. (1991). Basic fibroblast growth factor is efficiently released from a cytosolic storage site through plasma membrane disruptions of endothelial cells. *Journal of Cellular Physiology*, **148**, 1–16.

Odlind, V. and Fraser, I. S. (1990). Contraception and menstrual bleeding disturbances: a clinical overview. In *Contraception and mechanisms of endometrial bleeding* (ed. C. D'Arcangues, I. S. Fraser, J. R. Newton, and V. Odlind), pp. 5–32. Cambridge University Press.

Oikawa, T., Hirotani, K., Shimamura, M., Ashino-Fuse, H., and Iwaguchi, T.

(1992). Inhibition of angiogenesis by herbimycin. *Journal of Antibiotics*, **42**, 1202.

Pepper, M. S., Ferrara, N., Orci, L., and Montesano, R. (1991). Vascular endothelial growth factor (VEGF) induces plasminogen activators and plasminogen activator inhibitor-1 in microvascular endothelial cells. *Biochemical and Biophysical Research Communications*, **181**, 902–6.

Pepper, M. S., Ferrara, N., Orci, L., and Montesano, R. (1992). Potent synergism between vascular endothelial growth factor and basic fibroblast growth factor in the induction of angiogenesis *in vitro*. *Biochemical and Biophysical Research Communications*, **189**, 824–31.

Plouet, J., Schilling, J., and Gospodarowicz, D. (1989). Isolation and characterisation of a newly identified endothelial cell mitogen produced by AtT-20 cells. *EMBO Journal*, **8**, 3801–6.

Presta, M. (1988). Sex hormones modulate the synthesis of basic fibroblast growth factor in human endometrial adenocarcinoma cells: implications for the neovascularization of normal and neoplastic endometrium. *Journal of Cellular Physiology*, **137**, 593–7.

Psychoyos, A. and Martel, D. (1985). Embryo-endometrial interactions at implantation. In *Implantation of the human embryo* (ed. R. G. Edwards, J. M. Purdy, and P. C. Steptoe), pp 197–219. Academic Press, London.

Rapraeger, A. C., Krufka, A., and Olwin, B. B. (1991). Requirement of heparin sulfate for bFGF-mediated fibroblast growth and myoblast differentiation. *Science*, **252**, 1705–8.

Reid, H. H. Wilks, A. F., and Bernard, O. (1990). Two forms of the basic fibroblast growth factor receptor-like mRNA are expressed in the developing mouse brain. *Proceedings of the National Academy of Sciences of the USA*, **87**, 1596–600.

Risau, W. (1991). Vasculogenesis, angiogenesis and endothelial cell differentiation during embryonic development. In *Issues in biomedicine: the development of the vascular system* (ed. R. N. Feinberg, G. K. Sherer and R. Averbach), vol. 14, pp. 58–68. Karger, Basel.

Roberts., A. B., Sporn, M. B., Assoian, R. K., Smith, J. M., Roche, N. S., Wakefield, I. M. *et al.* (1986). Transforming growth factor type-beta: rapid induction of fibrosis and angiogenesis *in vivo* and stimulation of collagen formation *in vitro*. *Proceedings of the National Academy of Sciences of the USA*, **83**, 4167–71.

Rogers, P. A. W., Au, C. L., and Affandi, B. (1993). Endometrial microvascular density during the normal menstrual cycle and following exposure to long-term levonorgestrel. *Human Reproduction*, **8**, 1396–404.

Ruoslahti, E. and Yamaguchi, Y. (1991). Proteoglycans as modulators of growth factor activities. *Cell*, **64**, 867–9.

Rusnati, M., Casarotti, G., Pecorelli, S., Ragnotti, G., and Presta, M. (1990). Basic fibroblast growth factor in ovulatory cycle and postmenopausal human endometrium. *Growth Factors*, **3**, 299–307.

Schreiber, A. B., Winkler, M. E., and Derynck, R. (1986). Transforming growth factor-alpha: a more potent angiogenic mediator than epidermal growth factor. *Science*, **232**, 1250–3.

Senger, D. R., Connolly, D. T., Peruzzi, C. A., Alsup, D., Nelson, R.,

Leimgruber, R., et al. (1987). Purification of vascular permeability factor (VPF) from tumor cell conditioned medium. *Federation Proceedings*, **46**, 2102.

Senger, D. R., Connolly, D. T., Van De Water, L., Feder, J., and Dvorak, H. F. (1990). Purification and NH_2-terminal amino acid sequence of guinea pig tumor-secreted vascular permeability factor. *Cancer Research*, **50**, 1774–8.

Sharkey, A. M., Charnock-Jones, D. S., Boocock, C. A., Brown, K. D., and Smith, S. K. (1994). Expression of mRNA for vascular endothelial growth factor in human placenta. *Journal of Reproduction and Fertility*, **99**, 609–15.

Shibuya, M., Yamaguchi, S., Yamane, A., Ikeda, T., Tojo, A., Matsushime, H., et al. (1990). Nucleotide sequence and expression of a novel human receptor-type tyrosine kinase gene (*flt*) closely related to the (*fms*) family. *Oncogene*, **5**, 519–24.

Shing, Y., Folkman, J., Haudenschild, C., Lund, D., Crum, R., and Klagsbrun, M. (1985). Angiogenesis is stimulated by a tumor-derived endothelial cell growth factor. *Journal of Cellular Biochemistry*, **29**, 275–87.

Tawia, S. A. and Rogers, P. A. W. (1992). *In vivo* microscopy of the subepithelial capillary plexus of the endometrium of rats during embryo implantation. *Journal of Reproduction and Fertility*, **96**, 673–80.

Terman, B. I., Dougher-Vermazen, M., Carrion, M. E., Dimitrov, D., Armellino, D. C., Gospodarowicz, D., et al. (1992). Identification of the KDR tyrosine kinase as a receptor for vascular endothelial cell growth factor. *Biochemical and Biophysical Research Communications*, **187**, 1579–86.

Tischer, E., Mitchell, R., Hartman, T., Silva, M., Gospodarowicz, D., Fiddes, J. C., and Abraham, J. A. (1991). The human gene for vascular endothelial growth factor. *Journal of Biological Chemistry*, **266**, 11947–54.

Vlodavsky, I., Fridman, R., Sullivan, R., Sasse, J., and Klagsbrun, M. (1987). Aortic endothelial cells synthesize basic fibroblast growth factor which remains cell associated and platelet-derived growth factor which is secreted. *Journal of Cellular Physiology*, **131**, 402–8.

Werner, S., Dah-Shuhn, R. D., deVries, C., Peters, K. G., Johnson, D. E., and Williams, L. T. (1992). Differential splicing in the extracellular region of fibroblast growth factor receptor 1 generates receptor variants with different ligand-binding specificities. *Molecular and Cellular Biology*, **12**, 82–8.

Yayon, V., Klagsbrun, M., Esko, J. D., Leder, P., and Ornitz, D. M. (1991). Cell surface, heparin-like molecules are required for binding of basic fibroblast growth factor to its high affinity receptor. *Cell*, **64**, 841–8.

Yoshida, T., Muramatsu, H., Muramatsu, T., Sakamoto, H., and Katoh, O. (1988). Differential expression of two homologous and clustered oncogenes, *Hst1* and *Int-2* during differentiation of F9 cells. *Biochemical and Biophysical Research Communications*, **157**, 618–25.

Ziche, M., Jones, J., and Gullino, P. M. (1982). Role of prostaglandin E1 and copper in angiogenesis. *Journal of the National Cancer Institute*, **69**, 475–81.

4 Ovarian endocrine control of sperm progression in the Fallopian tubes
RONALD H. F. HUNTER

I   **Introduction**

II **Rate of functional sperm transport and pre-ovulatory distribution in isthmus**

III **The functional sperm reservoir and phase of sperm storage**
    1 Viscous secretions
    2 Chemical influences
    3 Temperature gradients
    4 Patency of lumen

IV **Peri-ovulatory activation and release from sperm reservoir**

V   **Ovarian endocrine regulation of sperm suppression and activation**

VI **Re-interpretation of capacitation of spermatozoa: chronological aspects**

VII **Regulation of spermatozoa ascending the isthmus**

VIII **Post-ovulatory increase in sperm numbers**

IX **Why are so many spermatozoa ejaculated?**

X   **Interpretation of competitive mating experiments**

XI **Fate of non-fertilizing spermatozoa**

I   INTRODUCTION

A vital component of fertility is the process of fertilization itself. However, a successful union of gametes represents the culmination of diverse physiological events, amongst which a timely arrival of male gametes at the site of fertilization in the Fallopian tubes cannot be taken for granted. This caution applies especially to domestic farm animals in which procedures of artificial insemination are widely employed, but it is also relevant to our own species where psychosomatic influences appear to play a significant role in gamete transport within the female duct system. This review focuses on the events of sperm transport and fertilization in sheep, cows, and pigs but, because of the reasonably comparable interval between the pre-ovulatory gonadotrophin surge and release of the oocyte at ovulation, it may shed light on events in the human Fallopian tube.

As an introduction to the topic, it is necessary to suggest why further studies of sperm transport by classical physiological approaches are still valid and important even though much current experimentation has moved to a molecular level. First, the rate of sperm transport to the site of fertilization has long been a controversial issue. Reported values for this process range from a few minutes to many hours (see Thibault 1973). Second, previous studies on the rate of sperm transport have seldom considered the competence of the population of cells in question, and yet the ability of spermatozoa to penetrate and activate secondary oocytes is the aspect of paramount importance. Measurements on the rate of sperm transport should therefore be concerned primarily with viable or functional spermatozoa that can promote formation of a zygote (Hunter 1980*a*). Third, the rate of transport may vary with the stage of oestrus and imminence of ovulation (Dauzier and Wintenberger 1952), and studies in the large domestic species with their relatively long pre-ovulatory intervals (the luteinizing hormone (LH) peak to ovulation interval is 26–42 hours, during which oestrous behaviour can be monitored) may reveal new aspects of Fallopian tube physiology, especially in relation to sperm storage and the process of capacitation.

Reviews on the topic of sperm transport in the female genital tract have been numerous during the last 30 years, but many of the observations and interpretations are compromised by the shortcomings outlined above, or by technical considerations. The main technical problems are: the artefacts imposed by procedures of slaughter, with massive contractions of the smooth musculature and a consequent displacement of unattached cells; and the fact that spermatozoa may remain in certain regions of the tract for prolonged periods—even for the duration of pregnancy (Martinet and Raynaud 1975; Mortimer 1983). Thus, sperm cells in the higher reaches of the female tract after an experimentally imposed mating are not necessarily derived from that particular mating. A further caution concerns the failure to distinguish rigorously between the results of natural mating and those of artificial insemination. Artificial insemination might have introduced a diluted and stored suspension of sperm cells deeper into the tract than would occur during coitus, and would have removed all behavioural interactions between the male and female. An additional point concerns the presumptive transport of dead spermatozoa: dead cells recovered from the tract at autopsy may not have been dead during the phase of sperm transport. Finally, there is a widespread and much repeated misunderstanding that the number of spermatozoa reaching the upper portions of the female tract is the same as the supposedly small number recorded during a single observation. This point is elaborated upon below.

Although the emphasis will be on findings in domestic animals, reference will be made to recent observations in laboratory species where these illuminate a particular component of the transport process. Notable essays and reviews that have appeared during the last twenty years include those of Blandau (1973), Hunter (1973a, 1975, 1980a, 1987a, 1989, 1990a, 1991a); Robinson (1973); Thibault (1973); Austin (1975); Thibault et al. (1975); Baker and Polge (1976); Overstreet and Katz (1977); Mortimer (1978, 1983); Polge (1978); Einarsson (1980); Yanagimachi (1981, 1988); Harper (1982, 1988); Overstreet (1983); Hawk (1987); Larsson (1988); Katz et al. (1989); Suarez and DeMott (1991); Drobnis and Overstreet (1992); Roldan and Gomendio (1992); Fournier-Delpech and Thibault (1993); and Croxatto (in press).

II RATE OF FUNCTIONAL SPERM TRANSPORT AND PRE-OVULATORY DISTRIBUTION IN ISTHMUS

The possibility of a very rapid phase of sperm transport into and through the uterus of ruminants, lasting 1–2 minutes or less, has been discussed previously (Hunter, 1980a, b; Hunter et al. 1980; Hunter and Nichol 1983; Hunter and Wilmut 1984). If this phenomenon is indeed a sequel to mating early in oestrus, it certainly does not apply to the spermatozoa involved directly in fertilization for, as in the rabbit, rapidly transported spermatozoa would be dead upon arrival in the Fallopian tubes (Overstreet and Katz 1977; Overstreet and Cooper 1978; Overstreet et al. 1978). However, it is conceivable that dead or dying spermatozoa release degradation products into the tubal lumen that interact with the epithelium and later arriving gametes. These may facilitate maturational changes, or perhaps sensitize the peritoneal phagocytosis system in preparation for the arrival of greater numbers of motile spermatozoa (Hunter 1980a). Similarly, Overstreet (1983) has suggested that rapidly transported cells may be local messengers acting to coordinate movement of the reproductive tract.

In the studies summarized below, the rate of functional sperm transport into the Fallopian tubes was assessed by post-coital ligation and section of the tract at the utero-tubal junction and subsequent recovery of the eggs. These were examined by phase-contrast microscopy for evidence of fertilization and to assess the number of accessory spermatozoa attached to or embedded within the zona pellucida. This approach differed from earlier experiments which simply recorded the number of spermatozoa in smears, flushings, or sections of the tubal contents or tissues—approaches susceptible to artefacts as described above. Fertilization would have occurred in the present studies only if viable spermatozoa were already established in the tubes at the time of transection. The number of spermatozoa associated with the egg membranes may provide some estimate of the size of the available sperm population(s) in the Fallopian tubes at different intervals after mating.

A minimum period of 6–8 hours is required after ejaculation of semen into the vagina for competent spermatozoa to be established in the Fallopian tubes of sheep and cows mated at the onset of oestrus (Hunter et al. 1980, 1982; Hunter and Wilmut 1982, 1984). Although highly motile spermatozoa are present in the uterine lumen, spermatozoa are arrested in the caudal 1–2 cm of the Fallopian tube isthmus for 17–18 hours or more, that is until just before the moment of ovulation (Table 4.1; Hunter and Nichol 1983; Hunter and Wilmut 1984). However, with impending ovulation, they are activated and a small number progress onwards the 6–7 cm to the site of fertilization at the isthmo–ampullary junction. Widely spaced episodes of multiple mating do not seem to alter the pre-ovulatory distribution of viable spermatozoa already established in the isthmus, since multiple mating has no detectable influence on the incidence of fertilization or on the number of accessory spermatozoa (Hunter and Nichol, 1983, 1986a). If spermatozoa are displaced from the caudal isthmus by multiple mating, they are no longer functional by the time of ovulation. Mating close to the time of ovulation results in accelerated and quantitatively enhanced transport of spermatozoa (Table 4.2), suggesting an endocrine influence on both these components of gamete passage (Hunter et al. 1982).

The timescale of events differs markedly in pigs, primarily as a result of the intra-uterine site of ejaculation, but also due to the 40–42 hour interval between the onset of oestrus and ovulation. Because the voluminous semen accumulates in the uterus and a dense sperm suspension ($1–2 \times 10^8$ cells per ml) bathes the utero–tubal junction at the end of the protracted period of mating, it is not surprising that sufficient spermatozoa have entered the Fallopian tubes within 15 minutes to fertilize at least a proportion of the eggs (Hunter and Hall 1974a). Sperm motility appears to be essential for traversing the oedematous polypoid processes of the utero–tubal junction and gaining access to the lumen of the isthmus throughout the pre-ovulatory phase of oestrus. During this passage, sperm cells are largely divested of seminal plasma and resuspended in female tract fluids. Boar seminal plasma could not be detected in the Fallopian tubes of oestrous animals by classical biochemical approaches (Mann et al. 1956), nor by a physiological approach (Hunter and Hall, 1974b), nor by tracing with radio-opaque fluids (Polge 1978). Even so, components

Table 4.1 *Timing of fertilization in ewes*

Interval from mating to transection (hours)	Condition of ovaries at transection	No. of ewes examined	No. of ewes with some fertilized eggs	No. of eggs recovered	No. of eggs fertilized	No. of accessory sperm per egg mean	No. of accessory sperm per egg range
10	Pre-ovulatory	8	0	9	0	0	—
12	Pre-ovulatory	8	0	8	0	0	—
14	Pre-ovulatory	8	0	8	0	0	—
18	Pre-ovulatory	8	0	8	0	0	—
20	Pre-ovulatory	8	0	11	0	0	—
21	Pre-ovulatory	8	0	8	0	0	—
22	Pre-ovulatory	8	1	8	1	0	—
23	Pre-ovulatory	10	1	11	1	0	—
24	Pre-ovulatory	11	0	14	0	0	—
25	Peri-ovulatory	12	3	14	3	0.7	0–2
26	Post-ovulatory	13	11	16	13	7.9	0–26
Total		102	16	115	18	7.4	0–26

The proportion of ewes yielding fertilized eggs and the incidence of fertilization, at various intervals from mating to transection of the isthmus 1.5–2.0 cm above the utero–tubal junction. All ewes operated on 26 hours after the onset of oestrus had recent ovulations. Animals were mated at the onset of oestrus. Note the restriction of viable spermatozoa to the caudal portion of the Fallopian tube until ovulation is imminent. Taken from Hunter and Nichol (1983).

Table 4.2 *Timing of establishment of a fertilizing population of sperm in ewes*

Time of mating	No. of animals		Egg recovery		Eggs fertilized		No. of accessory sperm per egg	
	examined	with fertilized eggs	no. of corpora lutea*	no. of eggs	no.	%	mean	range
Morning of 1st day of oestrus (pre-ovulatory)	15	4	20	20†	4	20.0	0.4	0–2
Morning of 2nd day of oestrus (peri-ovulatory)	15	14	18	18‡	17	94.4	5.1	0–37
Total	30	18	38	38	21	55.3	2.8	0–37

Results of an experiment designed to examine the rate of establishment of a fertilizing population of spermatozoa in the Fallopian tubes of ewes following mating on the first or second day of oestrus and transection at the utero–tubal junction approximately 8 hours later. (After Hunter et al. 1982.)

* The number of developing corpora lutea was equated with the number of eggs shed.
† Plus four fertilized eggs from control Fallopian tube on the contralateral side.
‡ Plus three fertilized eggs from control Fallopian tube on the contralateral side.

of the seminal plasma would be expected to remain in association with the sperm head membranes until the time of ovulation (see Einarsson *et al.* 1980).

Two studies based on the examination of large numbers of eggs indicated that sufficient boar spermatozoa have entered the tubes within one hour of mating to ensure subsequent fertilization of more than 90 per cent of the eggs ovulated (Hunter 1981), and that in pigs mated at the onset of oestrus, viable spermatozoa are arrested close to the utero–tubal junction for 36 hours or more. As in sheep, when multiple mating occurred, either it did not displace viable spermatozoa already established in the isthmus (Hunter 1984), or any displaced spermatozoa were no longer functional by the time of ovulation. In fact, it is now appreciated that live spermatozoa bind to the epithelium of the caudal isthmus during most of the pre-ovulatory interval (see below). Ad-ovarian release of spermatozoa is activated only shortly before ovulation (Table 4.3). Extremely limited numbers of spermatozoa progress to the site of fertilization at this time. These studies involved animals mated at the onset of oestrus. As initially suggested for domestic animals by Dauzier and Wintenberger (1952) and Du Mesnil du Buisson and Dauzier (1955), the rate of transport of competent spermatozoa is accelerated close to the time of ovulation.

III THE FUNCTIONAL SPERM RESERVOIR AND PHASE OF SPERM STORAGE

The above studies on the distribution of viable cells in sheep, cows, and pigs mated early in oestrus all indicated that the fertilizing population of spermatozoa spends most of the pre-ovulatory interval sequestered in the caudal portion of the Fallopian tube(s). This finding therefore emphasizes the fact that the caudal isthmus serves as the functional sperm reservoir, accumulating and liberating male gametes involved directly in the process of fertilization. A corresponding region, the intra-mural portion of the Fallopian tube, may act as the functional sperm reservoir in primates, including humans (Hunter 1987*b*; Mortimer 1991). This question has been touched on most recently by Williams *et al.* (1993) and Croxatto (in press) but, because of problems of experimental access, it has not yet been resolved. Apart from being responsive to local ovarian control mechanisms (discussed below), a site of pre-ovulatory storage in the isthmus would enable sperm cells to be maintained largely free of seminal plasma, beyond reach of the uterine population of polymorphonuclear leucocytes, and isolated from the metabolic stimulation afforded by uterine and ampullary fluids. A sufficient portion of the isthmus would remain before the site of fertilization to

Table 4.3 *Arrest of spermatozoa in the caudal portion of the isthmus before ovulation*

Interval from mating to transection (hours)	Condition of ovaries at transection	Transected isthmus			Control isthmus*		
		eggs recovered	no. of eggs fertilized	%	eggs recovered	no. of eggs fertilized	%
3	Pre-ovulatory	34	0	0	32	32	100
6	Pre-ovulatory	40	0	0	35	33	94
12	Pre-ovulatory	42	0	0	41	41	100
24	Pre-ovulatory	50	0	0	41	39	95
30	Pre-ovulatory	53	0	0	33	32	97
36	Pre-ovulatory	51	1	2	41	41	100
38	Pre-ovulatory	39	2	5.1	49	49	100
40	Peri-ovulatory	48	19	39.6	35	35	100
42–44	Post-ovulatory	46	46	100	34	34	100
Total		403	68	16.9	341	336	98.5

The influence of transecting the isthmus of pigs 1.5–2.0 cm proximal to the utero–tubal junction at increasing intervals after mating at the onset of oestrus on the proportion of eggs subsequently fertilized (six animals in each group). Viable spermatozoa are arrested in the caudal portion of the isthmus until shortly before ovulation. (Modified from Hunter 1984).

*Double ligatures were placed around the control Fallopian tube and then removed.

permit an incisive regulation of sperm numbers confronting the newly ovulated egg(s).

1 Viscous secretions

As first observed in rabbits (Overstreet and Cooper 1975), and as a sequel to the transport studies of Harper (1973a, b), sperm motility is specifically suppressed in the caudal portion of the isthmus during the pre-ovulatory interval, in marked contrast to sperm activity in the uterine lumen. Parallels have been drawn with sperm storage in the caudal portion of the epididymal duct (Hunter and Nichol 1983; Overstreet 1983). Suppression of motility may be primarily as a consequence of the viscous proteinaceous secretion that appears as a mucus-like substance in the isthmus. A comparable viscous secretion in the caudal portion of the epididymis is termed immobilin (Usselman and Cone 1983; Carr and Acott 1984). In scanning electron micrographs of this portion of the Fallopian tube in pigs (Hunter et al. 1987) and cows (Hunter et al. 1991), and as a consequence of the fixation procedures, the mucus-like secretion appears as relatively large spheres or globules together with a flocculent material displayed in a seemingly organized manner on the surface of the sperm head. This material probably corresponds to the mucus described in the mouse (Reinius 1970), rabbit (Overstreet and Cooper, 1978), and human Fallopian tube (Patek 1974; Jansen 1978).

Samples of mucus have been obtained at surgery by passing flame-polished glass micropipettes through the utero–tubal junction and into the lumen of the isthmus of oestrous pigs. Aspirating the contents indicated that it is whitish-cream in colour, completely occludes the caudal isthmus by acting as a plug, and is extremely viscous before ovulation (Hunter 1990a, 1991a). Indeed, the mucus is invariably pulled back into the isthmus during sampling due to its tenacious viscoelastic properties. Whether spermatozoa are physically arrested and their flagellar activity suppressed principally by the physical nature of such pre-ovulatory secretion (see Suarez and Dai 1992), or initially by epithelial interactions, cannot yet be stated with certainty. Our own observations in anaesthetized animals examined at surgery indicate that this mucus-like material becomes progressively less viscous after ovulation, in step with a reduction in tonicity of the myosalpinx. If this observation is representative of the physiological situation, then the physical condition of secretions in the caudal portion of the Fallopian tube must contribute in a major way to the phase of sperm storage and could act, together with epithelial binding, to prevent displacement of spermatozoa during episodes of multiple mating before ovulation. The post-ovulatory reduction in viscosity may in part represent an influence of enhanced fluid transudation into the tubal lumen and, together with an increased size of the lumen, would facilitate an accentuated flagellar beat and onwards progression of spermatozoa.

2 Chemical influences

Chemical, as distinct from physical, changes in composition of fluid in the lumen of the isthmus may also contribute to pre-ovulatory storage and then peri-ovulatory activation. Studies in rabbits suggested that relatively high extracellular concentrations of K^+ ions inhibited sperm motility in the isthmus whereas pyruvate stimulated such motility, there being some evidence for differences in the composition of isthmic and ampullary fluids before ovulation (Burkman et al. 1984). Our own studies on regional fluid environments within the Fallopian tubes of oestrous pigs indicated differences in lactate and glucose concentrations between the isthmus and ampulla, especially before ovulation, whereas such differences were reduced or eliminated after ovulation (Nichol et al. 1992). Accordingly, modified ionic and molecular concentrations in luminal fluid close to the time of ovulation may have a powerful influence on sperm arrest or progression, individual sperm cells varying in their sensitivity to such changes.

3 Temperature gradients

A further factor which could be critically involved in sperm arrest and storage is a temperature gradient in the lumen of the Fallopian tube. Before ovulation, the caudal isthmus has been noted to be approximately 0.75 °C cooler than the rostral ampulla in mated pigs, a temperature differential that effectively disappears at or just after ovulation (Hunter and Nichol, 1986b). Although relatively small, such changes in temperature could none the less influence the sperm cell directly—especially the potential activity and configuration of the flagellum—by acting on its structural proteins. Regional differences in pre-ovulatory temperatures had previously been recorded in the Fallopian tubes of rabbits (David et al. 1972), although a physiological interpretation in the context of gametes or embryos was not offered. Regional differences in temperature may reflect regional differences in the vascular and lymphatic beds. If this is the case, modified transudates and gas tensions within the duct might contribute to the phase of storage. Regional differences in myosalpingeal activity are also thought to contribute importantly to the temperature gradient, heat being generated by contractile activity of the ampullary portion.

4 Patency of lumen

Finally, in terms of pre-ovulatory sperm arrest within the isthmus, the smooth muscle layers are extremely tonic and contracted, and the mucosa is oedematous (Andersen 1928; Lee 1928), conditions enhanced under the influence of ovarian oestradiol secretion (reviewed by Hunter 1977).

Adrenergic receptors are relatively densely distributed in the circular muscle layers of the isthmus (Brundin 1965). These aspects impose a form of physiological stricture and together act very significantly to reduce the patency of the isthmus, thereby obstructing ad-ovarian passage of spermatozoa and reducing the scope for flagellar beat. Indeed, the contractile activity of the myosalpinx may be further accentuated in the porcine tube by an influence of seminal plasma spasmogens bathing the utero–tubal junction and perhaps entering the isthmus in strictly limited quantities (see Einarsson et al. 1980).

IV PERI-OVULATORY ACTIVATION AND RELEASE FROM SPERM RESERVOIR

After the relatively prolonged pre-ovulatory phase during which sperm motility is suppressed in the caudal portion of the isthmus, reactivation of such spermatozoa occurs with impending ovulation. Close coordination in the meeting of male and female gametes would be dictated by the maturity of a Graafian follicle on the verge of ovulation. This statement can be made with confidence on the basis of data from extensive studies in domestic farm animals (Tables 4.1 and 4.3). Not only is there a gradual and tightly controlled ad-ovarian release and onwards progression of spermatozoa in limited numbers, but there are also changes in the sperm head membranes that culminate morphologically in the acrosome reaction. Whilst there is overall agreement on these observations in both laboratory and farm species (see Overstreet 1983; Drobnis and Overstreet 1992), precision is lacking. This is because studies look at a population of spermatozoa, and descriptions will refer to only a proportion of sperm cells at any one time. There also remains the question as to exactly where motility and morphological changes are initiated, and whether most cells respond synchronously to regulatory cues or are 'staggered' in terms of the timing of hyperactivation and the acrosome reaction (Hunter 1990 a, b, 1991a).

On the basis of careful analysis in sheep, cows, and pigs, it is important to stress that sperm activation and gradual release from the isthmus reservoir commence shortly before ovulation, rather than as a consequence of it (Tables 4.1 and 4.3). Events associated with ovulation may enhance these processes of activation and release but physiological mechanisms must occur which precede, and are distinct from, any influence of the products of ovulation entering the Fallopian tubes (see Harper 1973a, b). This changing physiology clearly underlies the accelerated sperm transport noted with mating at the time of ovulation (Table 4.2). A working model of the pre-and peri-ovulatory events is presented in Fig. 4.1. The model also takes account of observations in cows by Herz et al. (1985), in which there were 85 per cent of acrosome-reacted spermatozoa in the ampulla of the ipsilateral tube close to ovulation, compared with less than 20 per cent in the contralateral tube or before ovulation.

The left-hand side of the model (Fig. 4.1a) portrays the situation in mated animals soon after the onset of oestrus. The pre-ovulatory Graafian follicles are not yet of maximal diameter, the wall of the isthmus is oedematous, the myosalpinx is highly tonic and, as a consequence, the lumen is extremely narrow. A majority of viable sperm cells at this early stage of oestrus have intact acrosomes, and the flagellum as seen in scanning electron micrographs is almost straight or only slightly undulating. The cell is judged to be stabilized and motility suppressed.

The situation for follicles on the verge of ovulation is depicted in Fig. 4.1b. Associated with increasing pre-ovulatory synthesis and secretion of progesterone, the extent of oedema in the wall of the isthmus is already diminishing and tension in the myosalpinx is reduced (Hunter 1972, 1977). As a consequence, the patency of the duct lumen is gradually increasing. This influence of progesterone has been demonstrated experimentally using the surgical approach of instilling microdroplets of a solution of

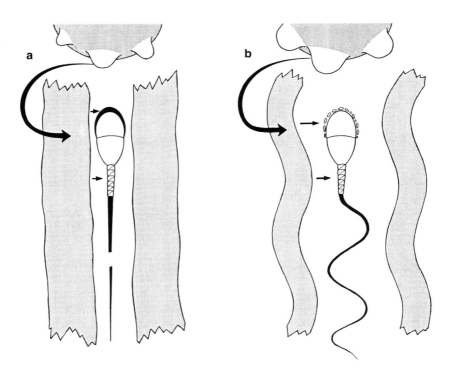

progesterone in oil under the serosal layer of the tube shortly before ovulation (Hunter 1972). Acrosome-reacted spermatozoa are conspicuous, and the form of the flagellum in scanning electron micrographs indicates activation.

Whether true hyperactivation of spermatozoa (Katz et al. 1978; Yanagimachi 1981; Suarez et al. 1983) can and does occur in the caudal isthmus of farm animals remains to be clarified. The size of the lumen and presence of mucus shortly before ovulation may preclude full expression of the whiplash pattern of flagellar beat until the ampullary–isthmic junction is reached. In line with this point of view, Suarez et al. (1992) removed the Fallopian tubes surgically from 13 gilts close to the time of ovulation and found a total of only five hyperactivated spermatozoa from these mated animals, all from the ampulla. Although an acrosome reaction can be distinguished in cells still resident in the isthmus, such spermatozoa may lose their functional potential before they reach the site of fertilization. Debate continues as to whether the eggs or their investments provide a specific—if not unique—stimulus for induction of the acrosome reaction and a concomitant hyperactivation of the fertilizing spermatozoon (Szollosi and Hunter 1978; Yanagimachi 1981, 1988; Crozet and Dumont 1984). Certainly, the zona

Fig. 4.1 Model to illustrate the way in which the endocrine activity of pre- or peri-ovulatory Graafian follicles acts locally to programme the membrane configuration and motility of spermatozoa in the lumen of the Fallopian tube isthmus. Gonadal hormones (from the follicles) act on the tubal epithelium whose transudates and secretions in turn influence the nature of the luminal fluids. Completion of capacitation is reasoned to be a peri-ovulatory event, at least in the large farm species with a protracted interval between the gonadotrophin surge and ovulation. (a) Intact, relatively quiescent spermatozoon under the overall influence of pre-ovulatory follicles. Membrane vesiculation on the anterior part of the sperm head is suppressed, as is the development of whiplash activity in the flagellum, presumably due to local molecular control mechanisms. The lumen of the isthmus is extremely narrow and contains viscous secretions, and myosalpingeal contractions are reduced. In reality, the heads of viable spermatozoa are bound to the epithelium. (b) An acrosome-reacted, hyperactive spermatozoon under the influence of Graafian follicles on the point of ovulation. The patency of the isthmus has begun to increase, allowing a more powerful pattern of flagellar beat. Progression of such spermatozoa to the site of fertilization may also be aided by enhanced contractile activity of the myosalpinx, and yet involves a strict numerical regulation. In fact, hyperactivation and an acrosome reaction in a fertilizing spermatozoon may only develop in the vicinity of the egg. A premature acrosome reaction would risk displacement of the vesiculated membranes before the site of fertilization. This model accords with the observation that high proportions of spermatozoa undergoing the acrosome reaction in ruminants are found predominantly in the ampulla adjoining the ovulatory ovary and only at or following ovulation.

pellucida can initiate an acrosome reaction in various *in vitro* models (see Yanagimachi 1988), but simple extrapolation from such models may not be warranted since subtle components of the physiological control mechanisms may be overlooked.

As a separate point, none of the preceding description and argument is intended to deny an influence of the ovulated cumulus (granulosa cell) mass or its associated follicular fluid and hormonal constituents on the subsequent phases of sperm transport. However, such an influence needs to be interpreted against the controlled ascent of spermatozoa in the isthmus observed initially in very limited numbers.

V OVARIAN ENDOCRINE REGULATION OF SPERM SUPPRESSION AND ACTIVATION

In the light of the preceding observations on pre-ovulatory storage of spermatozoa in the caudal portion of the isthmus, the specific means of activation of sperm motility and progression to the site of fertilization close to the time of ovulation requires explanation. Although the contributing mechanisms are not fully understood, there is some information on the manner of ovarian control and on the changing composition of tubal luminal fluids during the stage of oestrus (Hunter 1990a; Nichol *et al.* 1992). Overall control of sperm motility and progression resides in the endocrine function of the Graafian follicle(s) close to ovulation (Hunter 1977). In other words, gonadal activity regulates and coordinates a final maturation and meeting of the gametes.

The mechanism appears to operate primarily through a local transfer of hormonal information rather than via a systemic route, so that both Fallopian tube and sperm function are reprogrammed in an incisive and sensitive manner at a time just preceding release of the egg(s). Evidence for a counter-current transfer in the ovarian pedicle of relatively high concentrations of follicular hormones such as steroids and prostaglandins to the vasculature supplying the wall of the isthmus has been presented elsewhere (Table 4.4; Hunter *et al.* 1983). The studies compared the concentrations of ovarian hormones in the systemic blood with those in the arterioles supplying the isthmus of the Fallopian tube. Findings in 12 out of 12 oestrous animals supported the existence of a counter-current transfer of hormones from the ovarian vein into the tubal branch of the ovarian artery (Fig. 4.2; reviewed by Einer-Jensen 1988). The experimental evidence did not preclude a contribution from lymphatic transfer of hormones, for the caudal portion of the Fallopian tube is particularly rich in lymphatic sinuses (Andersen 1927). Nor should it be overlooked that follicular hormones would also be reaching the wall of the isthmus through the systemic circulation, although at a much reduced concentration (Table 4.4).

How could an endocrine influence of mature Graafian follicles act to

Table 4.4 *Evidence for a counter-current mechanism within the ovarian pedicle of oestrous animals for relatively high concentrations of gonadal hormones*

Animal No.	Time of collection*	Sample details		Oestradiol (pmol/l)	Testo-sterone	Androste-nedione	Proges-terone†	PGF$_{2\alpha}$	PGE$_2$
		site of collection	adjacent ovary		(nmol/l)				
125	1st day of oestrus (12–16 h)	Right isthmus	11 follicles of 8 mm	>1836	8.7	132	36.2	17.5	0.5
		Left isthmus	4 follicles of 8 mm	460	1.5	13	2.7	24.3	1.3
		Peripheral		109	<0.7	1	1.2	3.1	0.4
126	2nd day of oestrus (20–24 h)	Right isthmus	5 follicles of 10 mm	229	0.7	5.3	9.7	7.3	0.7
		Left isthmus	7 follicles of 10 mm	1691	1.2	17.4	10.1	15.5	1.1
		Peripheral		136	<0.7	<0.7	3.3	2.3	0.4
127	2nd day of oestrus (24–30 h)	Right isthmus	11 follicles of 9 mm	410	<0.7	3.5	10.3	7.3	0.3
		Left isthmus	5 follicles of 9 mm	69	<0.7	<0.5	5.4	7.6	0.5
		Peripheral		45	<0.7	0.7	2.1	3.4	0.4
132	2nd day of oestrus (30–34 h)	Right isthmus	7 follicles of 9 mm	<15	<0.7	1.1	5.0	9.3	0.4
		Left isthmus	12 follicles of 9 mm	<15	<0.7	1.3	6.0	8.8	0.9
		Peripheral		<15	<0.7	<0.7	4.2	1.2	0.3
124	2nd day of oestrus (36–40 h)	Right isthmus	9 recent ovulations	74	<0.7	1.4	267	9.0	0.9
		Left isthmus	5 recent ovulations	47	<0.7	1.2	149	6.5	0.2
		Peripheral		30	<0.7	<1.0	9.1	2.5	0.7

The measurements are of steroids and prostaglandins in blood plasma sampled from the arterial arcade bordering the isthmus of each Fallopian tube (see Fig. 4.2) and from the systemic circulation in five pigs at different stages of oestrus. (Modified from Hunter *et al.* 1983.)
The symbol < indicates below figure given, which is the sensitivity of the assay.
* Approximate time after onset of oestrus. † The sensitivity of this assay was 1 nmol/l.

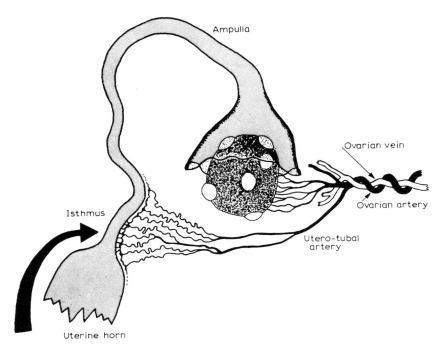

Fig. 4.2. The arterial blood supply to the ovary and isthmus of the pig Fallopian tube. A portion of the ovarian vein is also shown. A counter-current transfer of follicular hormones occurs from the ovarian vein to the corresponding artery and thus into the utero–tubal branch. This transfer mechanism is thought to provide the endocrine basis for local gonadal regulation of gamete activity in the lumen of the Fallopian tubes. Blood was sampled from three to five arterioles entering the caudal region of the isthmus (*arrow*). (Adapted from Hunter *et al.* 1983.)

coordinate a meeting of the gametes? Progressively changing hormonal information could be transduced through the tissues of the Fallopian tube (Fig. 4.1), and then expressed in a modified composition of the secretion and transudates (Hunter 1977, 1988), and perhaps also in modified gas tensions in the lumen of the isthmus. To achieve specific changes in sperm motility and release from the storage site close to the time of ovulation, thresholds of follicular hormone secretion would need to be invoked, the pattern of hormone secretion being modified critically as ovulation approaches. As noted, increasing synthesis of progesterone rather than of oestradiol is a principal feature (Table 4.4), arising as a sequel to the pre-ovulatory gonadotrophin surge and loss of follicular aromatase activity. A particular role for progesterone interactions with the wall of the Fallopian tube in facilitating sperm transport has been demonstrated (Hunter 1972). Ovarian prostaglandins and peptide hormones such as

Table 4.5 *Mean concentration of prostaglandins in ovarian follicular fluid*

Interval from onset of oestrus or HCG (hours)	No. of animals sampled	Mean concentration in ng/ml		
		PGE_2	$PGF_{2\alpha}$	6-oxo-$PGF_{1\alpha}$
8–12	4	1.4	2.3	2.1
20–30	8	4.0	1.6	1.4
31–39	10	6.2	1.8	1.8
40–42	8	166.1	24.6	11.2
Ovulation begun	3	1.1	0.8	1.5

Fluid was aspirated at laparotomy or autopsy from five different groups of pigs at different times from the onset of standing oestrus or a pro-oestrous injection of human chorionic gonadotrophin. Note the very high concentration of prostaglandins in follicular fluid immediately before ovulation, and their rapid diminution once ovulation had begun. An influence of prostaglandins upon sperm motility in the lumen of the Fallopian tube could be exerted via the counter-current mechanism, or more directly upon release at ovulation. (Adapted from Hunter and Poyser 1985.) Ovulation in this breeding herd took place 41–42 hours after a single pro-oestrous injection of 500 I.U. HCG.

relaxin and oxytocin have also been implicated (Hunter *et al.* 1983; Hunter and Poyser 1985), and receptor molecules for each of these hormone types have been demonstrated in the isthmus (Ayad *et al.* 1990). Concentrations of prostaglandins in follicular fluid achieve impressive values just before ovulation (Table 4.5). Interactions between the ovarian endocrine profile and the adrenergic nervous system to alter the pattern of contractile activity and reduce tonicity must not be overlooked when modifications to the myosalpinx are under consideration (Brundin 1964, 1969). In addition, as discussed for Fig. 4.1, a progressively reduced oedema of the mucosa and increased patency of the lumen would also be involved in regulating the potential activity of spermatozoa in the isthmus of the tube.

In the context of sperm transport, the influence of a counter-current transfer of ovarian follicular hormones on the Fallopian tube may be reinforced at ovulation by release of follicular fluid into the tubal ostium. Whilst there is no evidence to indicate that neat follicular fluid passes down the tube to reach the region of the isthmus in which spermatozoa are stored, it would none the less enter the proximal portion of the ampulla owing to the manner in which the fimbriated infundibulum tightly envelopes the ovary at ovulation (see Hunter 1977, 1988). A subsequent reflux into the peritoneal cavity represents the fate of most follicular fluid since bulk flow is from the tubes into the abdomen shortly after ovulation. In line with this proposition, only approximately 0.5 per cent of the anticipated volume of follicular fluid was detectable in the Fallopian tubes of pigs close to the time of ovulation, a proportion that decreased 10–12 fold within four hours of ovulation (Hansen *et al.* 1991). On the other hand, liberated granulosa cells pass down the ampulla with the oocyte and may provide a further, internal route for influencing and reprogramming tubal physiology. Since granulosa cells can be successfully cultured *in vitro* and actively synthesize steroid hormones when provided with appropriate substrates (Channing 1966), a similar role should not be overlooked *in vivo*. Not only would these cells synthesize steroids but also they might retain an ability to secrete diverse peptides, thus influencing the preliminaries to fertilization in a local manner (Hunter 1988).

VI RE-INTERPRETATION OF CAPACITATION OF SPERMATOZOA: CHRONOLOGICAL ASPECTS

On the basis of the preceding analysis, and in the light of straightforward biological reasoning, classical views on the time-course of mammalian capacitation may not be strictly correct. These views have been modified significantly since discovery of the phenomenon more than 40 years ago (Austin 1951; Chang 1951). The summary that follows offers an alternative perspective, presented in greater detail elsewhere (Hunter and Nichol 1983; Hunter 1986, 1987*a*, *c*).

Ever since the reports of Austin (1951, 1952), Chang (1951), and Noyes (1953), the process of capacitation has been viewed as the final phase of maturation of spermatozoa in the female genital tract, conferring upon a proportion of the cells an ability to penetrate the egg investments and thereby to achieve fertilization. Ejaculated or cauda epididymal spermatozoa deposited at the site of fertilization cannot penetrate and activate the egg(s) for a period usually measured in hours. Capacitation has been thought primarily to involve escape from the seminal plasma, resuspension of spermatozoa in uterine and then tubal fluids, and an altered metabolic activity. More subtle cellular changes were always suspected, but remained elusive until the 1970s (Bedford 1970; Chang and Hunter, 1975). Overall, capacitation was considered to require an interval characteristic for each species, a timing which may be closely mimicked in systems for *in vitro* fertilization. However, the finding of a relatively prolonged period of sperm storage in the caudal isthmus with suppressed motility and a stabilized acrosome has led to a reinterpretation.

More recent studies have focused on events that are now interpreted as a consequence of the capacitation process. These include:

(1) membrane vesiculation due to point fusions on the anterior portion of the sperm head, termed the acrosome reaction, which facilitates the release and action of proteolytic enzymes (Barros *et al.* 1967; Bedford 1967, 1968);

(2) the dramatically modified flagellar beat—the so-called whiplash or hyperactivated motility—that gives an incisive force of penetration (Katz et al. 1978; Yanagimachi 1981; Suarez and DeMott 1991; Suarez et al. 1991a; 1992).

These changes are closely associated with Ca^{2+} influx into the sperm cell and increased levels of cAMP.

Because of these changes in membrane configuration and flagellar beat, and bearing in mind their negligible reserves of cytoplasm, capacitated spermatozoa are accepted to be fragile, unstable, and short-lived (Bedford 1970; Chang and Hunter 1975). This situation would seem to pose major problems in females undergoing a single mating at the onset of receptivity, especially in species in which the period of oestrus extends for one or two days or more before the oocyte is released. If sperm cells were to achieve full capacitation soon after mating, they would be moribund or dead by the time of ovulation, owing to membranous modifications and metabolic exhaustion. Explanations for the prolonged availability of capacitated spermatozoa have therefore invoked the heterogeneous nature of the cell population in the ejaculate, with flexibility arising from a progressive ripening of spermatozoa in different states of maturity at deposition: in other words, a series of curves of ripening and decay within the millions of cells in an ejaculate (see Dziuk 1970; Bedford 1982). However, studies in farm animals (see above) indicate that this interpretation may need to be modified, at least in species with a protracted interval between the pre-ovulatory gonadotrophin surge and ovulation. Two examples of the dilemma will suffice. Ram spermatozoa have been reported to require 1.0–1.5 hours for capacitation *in vivo* (Mattner 1963), yet sheep ovulate approximately 26 hours after the onset of oestrus. Boar spermatozoa require 2–3 hours for capacitation *in vivo* (Hunter and Dziuk 1968), whereas pigs ovulate 40–42 hours after the onset of oestrus. Even invoking the heterogeneous condition of the ejaculate and the subsequent curves of ripening, the time relationships suggest that these do not furnish an adequate explanation for fertility. This argument becomes even more persuasive for dog and stallion spermatozoa, which may remain in the female tract for four to seven days before ovulation and still give rise to pregnancies.

The new understanding of Fallopian tube physiology derived from observations in domestic animals indicates that traditional views on the process of capacitation do require modification. Rather than this terminal maturation needing a given period of time in the female tract (i.e. 1–5 hours according to species), completion of capacitation—at least in a proportion of the stored spermatozoa—appears to be closely coordinated with the events of ovulation (Hunter 1981, 1984; Hunter and Nichol 1983). This has been judged by activation of the previously stored spermatozoa to a hyperactivated condition and appearance of a distinct acrosome reaction. Because both hyperactivation and the acrosome reaction are processes that follow very soon or immediately upon completion of capacitation, they therefore serve as valuable monitors of the stage of sperm maturation. Clearly, there would be little value in achieving full capacitation within a few hours of semen deposition if such capacitated cells were to be non-functional by the time of ovulation. In fact, glycoproteins and lipids secreted by the Fallopian tubes may suppress the completion of capacitation or the ensuing processes of the acrosome reaction and whiplash motility until shortly before ovulation. Although the endocrine activity of the pre-ovulatory Graafian follicle(s) is considered essential for programming the final maturation of spermatozoa, the contribution of the autonomic nervous system to the events of capacitation still requires further examination. Sperm cells are known to possess adrenergic receptors and to respond to various preparations of catecholamines (Bavister *et al.* 1979; Cornett *et al.* 1979). Interactions between the autonomic nervous system and ovarian activity might therefore be anticipated. None the less, the overall message remains that completion of capacitation is an ovulation-related event (Hunter 1986, 1987*a*, *c*).

As a final remark on this topic, it may be instructive to consider why chronological aspects of the process of capacitation were initially misinterpreted. A principal reason must be that classical studies of capacitation, both *in vivo* and *in vitro*, were invariably performed in the presence of eggs and their investments. This would have obscured the pre-ovulatory physiology discussed above. Another factor is that most of the early studies were performed in laboratory species such as rats, rabbits, or hamsters, which have relatively short pre-ovulatory intervals of approximately 10 hours. However, following the reinterpretation based on farm animals, capacitation in species such as the golden hamster has also been accepted to be a peri-ovulatory event (Smith and Yanagimachi 1989).

As to sperm motility, an oversimplified but useful working perspective may be to regard spermatozoa as being an essentially immotile, highly concentrated suspension in the storage region of the male duct system, the cauda epididymidis, only becoming activated at the time of ejaculation upon dilution by the voluminous seminal plasma. Such dilution and activation would be accentuated during resuspension of a proportion of the ejaculated cells in uterine fluid. Spermatozoa entering the caudal region of the isthmus in the pre-ovulatory interval would then have their motility specifically suppressed by viscous fluids in an extremely narrow lumen coupled with an influence of epithelial binding. Finally, a proportion of such spermatozoa would become reactivated with imminent ovulation and would further reach the state of hyperactivated motility essential for penetration of the egg investments. Failure to contact an egg promptly following hyperactivation would precipitate metabolic exhaustion and thereby an irreversible loss of fertilizing ability.

VII REGULATION OF SPERMATOZOA ASCENDING THE ISTHMUS

Concerning progression of spermatozoa from the caudal region of the isthmus to the site of fertilization, a number of difficult questions remain to be answered. Despite the original views of Hammond (1934) on a putative swarm or wave of spermatozoa ascending the Fallopian tubes, an appreciation has been growing since the observations of Moricard and Bossu (1951) in rats, and those of Stefanini *et al.* (1969) and Reinius (1970) in mice, that sperm:egg ratios in the ampulla may be close to unity at the time of initial penetration and activation—that is shortly after ovulation. At this time, there are seldom any other spermatozoa in the immediate vicinity of the egg(s), as noted in a variety of laboratory and farm animals (Zamboni 1972; Yanagimachi and Mahi 1976; Shalgi and Kraicer, 1978; Overstreet *et al.* 1978; Bavister 1979; Cummins 1982; Cummins and Yanagimachi 1982). In contrast to the initial tight regulation, the number of spermatozoa may increase slowly and progressively with time elapsing after ovulation (see next section). However, if several hundred or several thousand viable spermatozoa (depending upon species) are stored in the pre-ovulatory reservoir in the isthmus, and activation of these reserves occurs with imminent ovulation, then the manner of generating low sperm:egg ratios needs careful consideration (Hunter 1993). Failure of such control would lead to the pathological condition of polyspermy when the egg membranes are initially confronted by too large a number of competent spermatozoa. This has been demonstrated by various experimental approaches that over-ride the regulatory functions of the isthmus (Hunter, 1990*b*, 1991*b*).

As suggested above, a partial answer to this problem may be that full hyperactivation of the flagellum can occur only in the larger dimensions of the tube close to the ampullary–isthmic junction and in the ampulla. An alternative but not exclusive argument might be that capacitation and activation among individual sperm cells in the isthmus are not achieved synchronously but rather are staggered over a period of time. Even so, any explanation for the regulation of sperm ascent must also take account of the peri-ovulatory waves of myosalpingeal contraction that are sufficiently powerful in the golden hamster, for example, to displace droplets of oil in the isthmus towards the ampulla (Battalia and Yanagimachi 1979). How does the isthmus impose such a formidable and effective sperm gradient at the time of ovulation? Recent observations from three independent laboratories shed some light on this apparent paradox of peri-ovulatory sperm activation together with distinct myosalpingeal contractions without a consequent swarm of competent spermatozoa ascending the tract.

106 R. H. F. Hunter

Interactions between spermatozoa and the endosalpinx at stages close to ovulation have been monitored directly in mice and hamsters using trans-illumination of tissue mounted between slide and coverslip, a technique not applicable to the thicker tissues of farm animals. Spermatozoa in the caudal isthmus that become activated at the time of ovulation seem to proceed towards the site of fertilization by phases of free swimming alternating with phases of specific contact adhesion to the epithelium by means of the rostral tip of the head (Suarez 1987; Smith *et al.* 1987; Pollard *et al.* 1990, 1991; Smith and Yanagimachi 1990, 1991). The time-course of these alternating phases of binding and migration has not yet been reported for an individual

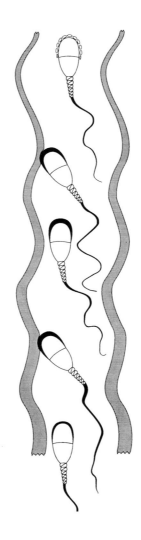

spermatozoon nor has the precise nature of the sperm–epithelial contacts (Fig. 4.3).

Adhesion to the epithelium must be remarkably avid to bring about sperm arrest (Fig. 4.4), and may well involve a class of molecules such as cadherins. These are Ca^{2+}-dependent glycoproteins that form an integral part of the cell membrane. Distribution and exposure of glycoproteins on the sperm head are thought to be in a dynamic state close to the time of capacitation, rendering the sperm cell susceptible to phases of contact interaction and adhesion with the tips of cilia in pigs (Hunter *et al.* 1987; Suarez and DeMott 1991; Suarez *et al.* 1991*b*) or microvillous engagement of the sperm head in cows (Hunter *et al.* 1991). The molecular nature of the changes leading to modification of surface charge and termination of an adhesive phase with release from the endosalpinx has yet to be determined. The alternate phases probably reflect progressive exposure of binding sites on the anterior portion of the sperm head. Loss of surface antigens and migration of protein molecules through different domains of the plasmalemma and acrosome could offer an underlying explanation, and be associated with renewed energetic inputs expressed through the sperm

Fig. 4.3 The peri-ovulatory progression of spermatozoa in the isthmus of the mammalian Fallopian tube, based on *in situ* observations in mice and hamsters. Spermatozoa arrested in the caudal isthmus during a relatively prolonged pre-ovulatory interval become activated at the time of ovulation and proceed towards the site of fertilization at the ampullary–isthmic junction by phases of highly active free swimming alternating with phases of contact adhesion to the epithelium by the rostral tip of the head. Such phases of adhesion may offer one means of reducing the risk of polyspermic fertilization by regulating the number of competent spermatozoa confronting the newly ovulated eggs. Even so, it remains uncertain whether a fertilizing spermatozoon reattaches after initial detachment from the epithelium. Three critical facts remain to be clarified in this model:

(1) Whether vesiculation of membranes on the anterior portion of the sperm head—the acrosome reaction—starts in the isthmus or is induced physiologically near the eggs at the site of fertilization. Scanning electron micrographs of both boar and bull spermatozoa in peri-ovulatory animals have recorded either a full acrosome reaction or an incipient reaction in the isthmus of the Fallopian tube. However, such spermatozoa are unlikely to be capable of fertilization.
(2) Whether the dramatic increase in beat of the flagellum that assumes a whiplash form—the so-called hyperactivation response—can be fully expressed in the isthmus or if the spermatozoon needs first to progress to the ampullary–isthmic junction. In rodents in particular, the size of the duct lumen may be a factor limiting full expression of hyperactivation.
(3) Precisely how sperm: egg ratios of close to unity at the time of initial penetration of the egg(s) are obtained. Interactions between the sperm head and microvilli and/or cilia on the endosalpinx are thought to be vital.

108 R. H. F. Hunter

flagellum. In other words, a reduced adhesion of the sperm cell could be linked to increased propulsion. The studies of DeMott and Suarez (1992) in mice have indicated that achievement of hyperactivation is a prime cause of spermatozoa breaking free from the epithelium.

Differential timing of the release from epithelial contact may account for the initial sperm:egg ratios. Phases of adhesion might also provide a temporary stabilization of the sperm membranes in situations of an incipient acrosome reaction in order to prevent agglutination of spermatozoa before they reach the egg surface. The number of spermatozoa released initially from the reservoir and passing along the isthmus could be adjusted by a form of ovarian programming, for the number of pre-ovulatory Graafian

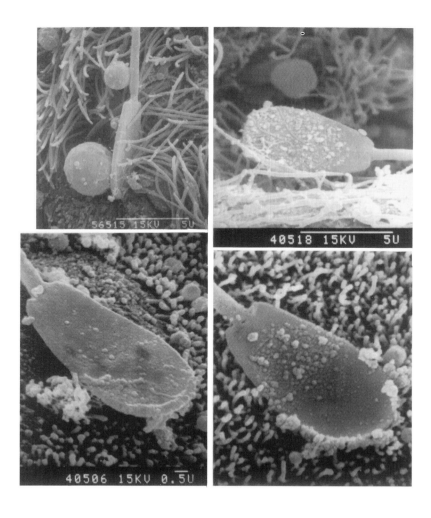

follicles would of course represent the number of eggs to be shed at ovulation. Sperm:egg ratios in polyovular mammals might also be adjusted by means of molecular gradients that divert spermatozoa penetrating a cumulus matrix away from eggs already activated and towards eggs as yet unfertilized. These points have been examined in some detail (Hunter 1993), and concern molecular cues released from an egg or its investments at the time of initial penetration. Such a means of redirecting incoming spermatozoa would increase the chances of fertilization of other eggs in the cumulus mass and reduce the chances of polyspermic fertilization in the first egg to be penetrated.

Two outstanding questions concern release of spermatozoa from their binding interactions with the epithelium of the caudal isthmus. How does release of spermatozoa occur progressively and in strictly limited numbers? And is only a particular sub-population of spermatozoa so involved? The best hypothesis may be that under the influence of the peri-ovulatory endocrine changes:

(1) there are regional differences in ionic fluxes across the duct epithelium, thereby creating a series of microenvironments along the length of the isthmus;
(2) the extent and degree (intensity) of sperm head–epithelial binding also vary progressively along the isthmus;
(3) sperm cells detaching from the epithelium release molecular signals (possibly metabolites) that retard the activation and release of further competent cells in the vicinity.

Fig. 4.4 Scanning electron micrographs of the Fallopian tube epithelium and luminal contents of direct relevance to the present review. The specimens were prepared from animals shortly before or after the time of ovulation. (a) Boar spermatozoon with intact acrosome in the caudal portion of the isthmus viewed shortly before ovulation. Secretory material and cilial tips appear to be in close apposition with the surface of the sperm head, and threads of mucus-like material can also be distinguished. (From Hunter *et al.* 1987.) (b) Bull spermatozoon viewed before ovulation in the caudal portion of the isthmus close to the utero-tubal junction. The sperm head is still intact, and the extensive surface coating of secretions is distributed predominantly over the anterior portion of the sperm head. (From Hunter *et al.* 1991.) (c) More detailed view of another bull spermatozoon in caudal portion of the isthmus. Note especially the intact acrosome, and distribution of fine secretory material essentially over the anterior portion of the head. Note also the apparently swollen microvillous tips, and the suggestion of an incipient microvillous interaction with the acrosomal ridge. (From Hunter *et al.* 1991.) (d) Bull spermatozoon viewed shortly after ovulation, illustrating important functional interactions. Larger elements of the secretory material are now disposed primarily over the surface of the post-nuclear cap; the microvillous tips appear less swollen than in (c); and there is evidence of a specific transfer of secretion, perhaps programmed in part by the sperm cell itself. (From Hunter *et al.* 1991.)

Only after attenuation of such signals might further spermatozoa be released. As to modifications of the sperm surface, one can reasonably infer that if spermatozoa bind to the epithelium of the tubal isthmus before completion of the capacitation process and to the surface of the zona pellucida after achievement of full capacitation, then different classes or layers of receptor molecule must be exposed during this terminal phase of sperm migration.

VIII POST-OVULATORY INCREASE IN SPERM NUMBERS

In marked contrast to the extremely tightly regulated peri-ovulatory ascent of spermatozoa to the site of fertilization, the wall of the isthmus undergoes progressive relaxation under the influence of increasing titres of plasma progesterone in the hours that follow ovulation. Indeed, this physical change can be detected during *in vivo* perfusion of fluid in procedures of egg recovery. As a consequence also of a progressive reduction in oedema of the mucosa, the lumen increases in patency enabling increased numbers of spermatozoa to pass from the isthmus into the ampulla (Hunter 1972, 1977). Although the size of the isthmic lumen is clearly a major factor, facilitation of sperm progression must also involve a reduced viscosity of the mucous secretion, a reduction in the incidence or degree of sperm head–epithelial engagement, and perhaps a greater freedom of flagellar activity. Underlying these remarks is the inference that reserves of spermatozoa in the isthmus have not all been activated simultaneously at the time of ovulation.

An increase in sperm numbers on the egg surface has been noted within one hour of ovulation in sheep (Hunter and Nichol 1983, 1986*a*) and one to two hours of ovulation in pigs (Hunter 1984), in due course achieving mean figures per egg of 7.9 and 53.8, respectively. In terms of the physiology of fertilization, the essential point is that this augmentation in the number of spermatozoa reaching the ampullary–isthmic junction occurs only after the events of fertilization have been completed and a fully functional block to polyspermy has been established. Thus, the arrival of relatively large numbers of competent spermatozoa at the surface of the zona pellucida should not compromise the ploidy of the egg. The situation is well illustrated in normally fertilized pig eggs in which the heads of 200 or more accessory spermatozoa may become attached to or embedded in the zona pellucida (Hancock 1961), although only a single spermatozoon has traversed the zona to fertilize the egg. Indeed, even surgical insemination of large numbers of boar spermatozoa directly into the ampulla shortly after completion of fertilization does not overcome the stability of the primary block to polyspermy in the zona pellucida, suggesting that it is irreversible

Ovarian control of sperm progression in the Fallopian tubes

at least until embryos enter the uterus (Hunter 1974, 1991b). Evidence for a secondary defence mechanism at the level of the vitelline membrane has been presented elsewhere (Hunter and Nichol 1988). Failure of the block to polyspermy occurs under conditions of post-ovulatory ageing of the egg before the time of mating or insemination (Hunter 1967). In this situation, the twin hazards of enhanced numbers of ascending spermatozoa and a retarded or defective establishment of the egg's defence mechanism cause an increasing incidence of polyspermic fertilization. The consequences of such multiple penetration are invariably lethal in mammals (Beatty 1957, 1961; Piko 1961; Austin 1963, 1965).

IX WHY ARE SO MANY SPERMATOZOA EJACULATED?

In the light of the quantitative control of sperm progression in the female tract, it is appropriate to consider the significance of the vast number of spermatozoa ejaculated in mammals. This can be assessed from diverse points of view (Parker 1984; Roldan and Gomendio 1992). However, for students of fertilization and in the context of the preliminaries to sperm–egg interactions, sperm numbers can be considered principally in terms of events within the Fallopian tubes. As mentioned in the Introduction, a long-standing misconception needs to be addressed. Many of the established texts and more recent reviews on sperm transport (for example, Drobnis and Overstreet 1992; Roldan and Gomendio 1992) stress the exceptionally small number of cells found in the upper portions of the Fallopian tubes after mating, frequently being cited as less than 100 spermatozoa (for example, Austin 1964). This is not strictly correct, for such figures do not indicate the total number of spermatozoa reaching the site of fertilization but simply those counted during a single observation. In fact, relatively large numbers of spermatozoa may pass through the Fallopian tubes into the peritoneal cavity before embryos have entered the uterus, and an initial transport of significant numbers of spermatozoa may have occurred during a phase of very rapid transport (Overstreet and Cooper 1978).

An overwhelming reason for the vast numbers of spermatozoa introduced into the female tract would be to ensure establishment of adequate pre-ovulatory reservoirs of spermatozoa in the caudal portion of the isthmus after a single mating early in oestrus. It needs to be emphasized that sperm progression to the Fallopian tubes is not an efficient process when viewed quantitatively. Not only do most spermatozoa fail to reach and enter the tubes, but the hazards of sperm loss associated with internal fertilization in mammals may not be appreciably less than those found under conditions of external (aquatic) fertilization in many lower species.

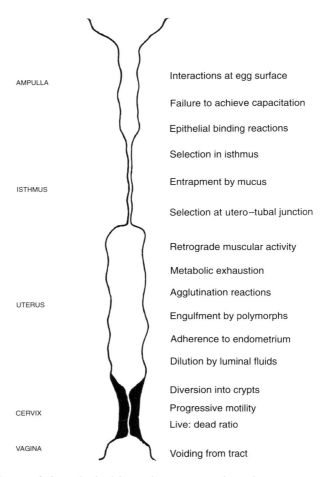

Fig. 4.5 Some of the principal hazards to progression of spermatozoa that may prevent potential gamete interactions at the ampullary–isthmic junction of the Fallopian tube.

This would be for two major reasons. First, the time of mating in mammals can be quite widely separated from the time of egg release at ovulation (for example, 1–2 days in sheep, cows, and pigs; 4–6 days in dogs; 5–7 days in horses), whereas, in aquatic species, spermatozoa are usually emitted in the vicinity of the freshly shed eggs. Second, passage of spermatozoa through the lower reaches of the mammalian genital tract is subject to many constraints (Fig. 4.5), resulting in an extraordinary wastage. Indeed, as demonstrated by the procedure of surgical insemination, the closer a sperm suspension is deposited to the site of fertilization, the smaller the number of spermatozoa required for fertilization (Hunter, 1973*b*).

Two examples from domestic farm animals are instructive. Bulls commonly ejaculate 4–8 ml of semen into the anterior vagina of oestrous cows close to the external cervical os. The density of spermatozoa in the ejaculate is of the order of $1-2 \times 10^9$ cells per ml. However, when a single ejaculate is diluted 400–800 or even 1000 times, and stored frozen, a thawed aliquot of 0.5 ml can lead to an incidence of fertilization of 75–85 per cent or more when introduced through the cervix into the uterus with an insemination straw. Such success is, of course, the basis of the artificial insemination industry. Thus, by overcoming the barrier presented by the cervical canal and its content of mucus, it is possible to reduce the number of spermatozoa by a factor of at least 1000 and still maintain acceptable levels of fertility. Deeper inseminations, that is intra-cornual insemination, enable sperm numbers to be reduced even further without compromising fertility (Senger *et al.* 1988).

By contrast, mature boars ejaculate 300–500 ml of semen via the cervix almost directly into the lumen of the uterus. The mean concentration of spermatozoa in the overall ejaculate would be approximately $1-2 \times 10^8$ per ml, and this may therefore represent the concentration bathing the utero–tubal junction by the completion of mating. Preferential transport of the sperm-rich fraction would present an even higher concentration at the utero–tubal junction (Du Mesnil du Buisson and Dauzier 1955). However, by means of surgical insemination during mid-ventral laparotomy, as little as 0.01–0.02 ml of whole ejaculate can be introduced into the base of each Fallopian tube at the onset of oestrus with a round-tipped 21- or 22-gauge needle (Polge *et al.* 1970). Not only will this tiny aliquot promote fertilization some 40 hours later, but polyspermic penetration of the eggs may be a frequent sequel (Hunter 1973*b*, 1976). Thus, a reduction of sperm numbers by at least a factor of 10 000 is possible if the barrier at the utero–tubal junction is overcome.

Together, these two examples serve to underline the fact that progression towards and into the Fallopian tubes after mating is a formidable undertaking for the majority of spermatozoa deposited. Moreover, even for those spermatozoa safely sequestered in the isthmus, only a small proportion may be competent to undergo specific sperm–egg interactions. In the golden hamster isthmus, a majority of spermatozoa are dead by the time of ovulation (Smith and Yanagimachi 1990). In the domestic pig, by contrast, although recently penetrated eggs usually have one or two spermatozoa in the zona pellucida, the number of sperm heads actually embedded in the zona may increase to 400 per egg by the time that four-celled embryos are passing to the uterus (Hunter 1974), indicating survival of substantial numbers of spermatozoa in the isthmus.

X INTERPRETATION OF COMPETITIVE MATING EXPERIMENTS

The physiology of spermatozoa established in the isthmus early in oestrus may enable the results of competitive mating experiments to be interpreted in a new light. This would be especially so for circumstances in which mating by a second male closer to the time of ovulation is successful in generating a majority of recorded pregnancies (Dziuk, 1970; Jewell et al. 1986). In brief, an interpretation that appears feasible focuses not so much on the size of the respective populations of spermatozoa as on the ready availability and physiological competence of the two populations. Spermatozoa reaching the isthmus from a first mating performed early in oestrus would be arrested in the caudal portion of the duct, and subjected to the full influence of storage mechanisms. Thus, they would undergo binding to the epithelium, suppression of flagellar activity and stabilization of the plasma membrane. By contrast, spermatozoa introduced from a second mating closer to the time of ovulation and gaining the isthmus might—depending on the time of ovulation—escape the full impact of the adhesion, suppression, and stabilization events. Even though spermatozoa from the first mating would be already established in the isthmus, those from a second, peri-ovulatory mating could achieve fertilization due to a progressive and accelerated completion of the processes of transport and capacitation. In fact, this very situation may be exploited by mature, experienced males. For example, in herds or flocks of ruminants also containing younger, less experienced males, experienced males may choose to mate closer to the time of ovulation (Jewell et al. 1986; Jewell 1989). Precisely what happens to spermatozoa from the first or earlier matings in this situation is unclear. The myosalpinx in sheep reaches a peak of sensitivity to oxytocin during oestrus (Gilbert et al. 1992), and displacement from the isthmus reservoir by oxytocin-promoted contractile activity at a later mating certainly needs to be considered. However, such displacement seems questionable in the light of our own observations mentioned above, perhaps because of a down-grading of oxytocin receptors, and in the light of conditions in the isthmus such as the viscous secretions in a narrow lumen and epithelial binding of spermatozoa.

XI FATE OF NON-FERTILIZING SPERMATOZOA

This chapter would not be complete without reference to routes of sperm disposal. In species not possessing an ovarian bursa, most male gametes are lost from the two extremities of the female tract. The great majority of spermatozoa are voided by passive leakage out of the female tract in the hours following mating. Of those gaining the uterus, a high proportion will be engulfed by polymorphonuclear leucocytes. These white cells are mobilized into the uterine lumen as a response to entry of seminal products (Lovell and Getty 1968). As to spermatozoa within the Fallopian tubes that reach the site of fertilization, substantial numbers become associated with the eggs or their investments. As already noted, as many as 200–400 spermatozoa may be trapped by the zona pellucida of pig eggs (Hancock 1961; Hunter and Léglise 1971), and considerable numbers may also be incorporated by associated or liberated cumulus (granulosa) cells (Szollosi and Hunter 1973). Spermatozoa not suffering these fates or passing into the peritoneal cavity (Croxatto *et al.* 1975; Settlage *et al.* 1975; Templeton and Mortimer 1980) may succumb to at least one further possibility. Those remaining in the isthmus as moribund or dead cells may be displaced in a retrograde manner into the uterine lumen at the time of embryo passage, there to be engulfed by cells of the proliferating trophoblast after its emergence from the zona pellucida (Hunter 1978, 1980*a*). Although Austin (1959) noted incorporation of sperm cells into the tubal mucosa of rodents, this has remained a controversial topic and is unlikely to represent a major quantitative avenue for disposal of surplus tubal spermatozoa—perhaps due to a protective role of specific glycoprotein secretions (Hunter 1994).

ACKNOWLEDGEMENTS

Most of the author's studies cited in this review were supported by UK (ARC) and French (INRA) research councils, for which grateful acknowledgement is made. Many friends and colleagues contributed to these studies, especially Drs B. Cook, P. J. Dziuk, B. and J. E. Fléchon, H. J. Leese, R. Nichol, N. L. Poyser, D. G. Szollosi, and I. Wilmut. Professor E. J. C. Polge, FRS, in Cambridge and Professors C. Thibault and F. Du Mesnil du Buisson in Paris were sources of inspiration throughout. Drs J. M. Bedford, H. B. Croxatto, M. J. K. Harper, J. W. Overstreet, E. R. S. Roldan, and S. S. Suarez generously forwarded preprints and reprints, and Mrs Frances Anderson kindly prepared the typescript.

REFERENCES

Andersen, D. H. (1927). Lymphatics of the Fallopian tube of the sow. *Contributions to Embryology of the Carnegie Institution*, **19**, 135–48.

Andersen, D. H. (1928). Comparative anatomy of the tubo–uterine junction. Histology and physiology in the sow. *American Journal of Anatomy*, **42**, 255–305.

Austin, C. R. (1951). Observations on the penetration of the sperm into the mammalian egg. *Australian Journal of Scientific Research B*, **4**, 581–96.

Austin, C. R. (1952). The 'capacitation' of the mammalian sperm. *Nature (London)*, **170**, 326.

Austin, C. R. (1959). Entry of spermatozoa into the Fallopian tube mucosa. *Nature (London)*, **183**, 908–9.

Austin, C. R. (1963). Fertilisation and transport of the ovum. In *Mechanisms concerned with conception* (ed. C. G. Hartman), pp. 285–320. Pergamon Press, Oxford.

Austin, C. R. (1964). Behaviour of spermatozoa in the female genital tract and in fertilisation. *Proceedings of the 5th International Congress on Animal Reproduction (Trento)*, **3**, 7–22.

Austin, C. R. (1965). *Fertilisation*. Prentice-Hall, New Jersey.

Austin, C. R. (1975). Sperm fertility, viability and persistence in the female tract. *Journal of Reproduction and Fertility Supplement*, **22**, 75–89.

Ayad, V. J., McGoff, S. A., and Wathes, D. C. (1990). Oxytocin receptors in the oviduct during the oestrous cycle of the ewe. *Journal of Endocrinology*, **124**, 353–9.

Baker, R. D. and Polge, C. (1976). Fertilisation in swine and cattle. *Canadian Journal of Animal Science*, **56**, 105–19.

Barros, C., Bedford, J. M., Franklin, L. E., and Austin, C. R. (1967). Membrane vesiculation as a feature of the mammalian acrosome reaction. *Journal of Cell Biology*, **34**, C1–5.

Battalia, D. E. and Yanagimachi, R. (1979). Enhanced and coordinated movement of the hamster oviduct during the periovulatory period. *Journal of Reproduction and Fertility*, **56**, 515–20.

Bavister, B. D. (1979). Fertilisation of hamster eggs in vitro at sperm: egg ratios close to unity. *Journal of Experimental Zoology*, **210**, 259–64.

Bavister, B. D., Chen, A. F., and Fu, P. C. (1979). Catecholamine requirement for hamster sperm motility *in vitro*. *Journal of Reproduction and Fertility*, **56**, 507–13.

Beatty, R. A. (1957). *Parthenogenesis and polyploidy in mammalian development*. Cambridge University Press.

Beatty, R. A. (1961). Genetics of mammalian gametes. *Animal Breeding Abstracts*, **29**, 243–56.

Bedford, J. M. (1967). Experimental requirement for capacitation and observations on ultrastructural changes in rabbit spermatozoa during fertilisation. *Journal of Reproduction and Fertility*, Supplement, **2**, 35–48.

Bedford, J. M. (1968). Ultrastructural changes in the sperm head during fertilisation in the rabbit. *American Journal of Anatomy*, **123**, 329–58.

Bedford, J. M. (1970). Sperm capacitation and fertilisation in mammals. *Biology of Reproduction*, Supplement, **2**, 128–58.

Bedford, J. M. (1982). Fertilisation. In *Reproduction in mammals* Vol. 1, (ed. C. R. Austin and R. V. Short), pp. 128–63. Cambridge University Press.

Blandau, R. J. (1973). Gamete transport in the female mammal. In *Handbook of physiology*, Section 7, Vol. 2 (ed. R. O. Greep and E. B. Astwood), pp. 153–63. American Physiological Society, Washington.

Brundin, J. (1964). The distribution of noradrenaline and adrenaline in the Fallopian tube of the rabbit. *Acta Physiologica Scandinavica*, **62**, 156–9.

Brundin, J. (1965). Distribution and function of adrenergic nerves in the rabbit Fallopian tube. *Acta Physiologica Scandinavica*, **66**, Supplement 259, 1–57.

Brundin, J. (1969). Pharmacology of the oviduct. In *The mammalian oviduct* (ed. E. S. E. Hafez and R. J. Blandau), pp. 251–69. University of Chicago Press.

Burkman, L. J., Overstreet, J. W., and Katz, D. F. (1984). A possible role for potassium and pyruvate in the modulation of sperm motility in the rabbit oviducal isthmus. *Journal of Reproduction and Fertility*, **71**, 367–76.

Carr, D. W. and Acott, E. S. (1984). Inhibition of bovine spermatozoa by cauda epididymal fluid. I. Studies of a sperm motility quiescence factor. *Biology of Reproduction*, **30**, 913–25.

Chang, M. C. (1951). Fertilising capacity of spermatozoa deposited into the Fallopian tubes. *Nature (London)*, **168**, 697–8.

Chang, M. C. and Hunter, R. H. F. (1975). Capacitation of mammalian sperm: biological and experimental aspects. In *Handbook of physiology*, (ed. D. W. Hamilton and R. O. Greep), Section 7, Vol. 5, pp. 339–51. American Physiological Society, Washington.

Channing, C. P. (1966). Progesterone biosynthesis by equine granulosa cells growing in tissue culture. *Nature (London)*, **210**, 1266.

Cornett, L. E., Bavister, B. D., and Meizel, S. (1979). Adrenergic stimulation of fertilizing ability in hamster spermatozoa. *Biology of Reproduction*, **20**, 925–9.

Croxatto, H. B. Gamete transport. In *Reproductive endocrinology, surgery, and technology* (ed. E. Y. Adashi, J. A. Rock, and Z. Rosenwaks). Raven Press, New York. (In press.)

Croxatto, H. B., Faundes, A., Medel, M., Avendano, S., Croxatto, H. D., Vera, C., *et al.* (1975). Studies on sperm migration in the human female genital tract. In *The biology of spermatozoa* (ed. E. S. E. Hafez and C. Thibault), pp. 56–62. Karger, Basel.

Crozet, N. and Dumont, M. (1984). The site of the acrosome reaction during *in vivo* penetration of the sheep oocyte. *Gamete Research*, **10**, 97–105.

Cummins, J. M. (1982). Hyperactivated motility patterns of ram spermatozoa recovered from the oviducts of mated ewes. *Gamete Research*, **6**, 53–63.

Cummins, J. M. and Yanagimachi, R. (1982). Sperm–egg ratios and the site of the acrosome reaction during *in vivo* fertilisation in the hamster. *Gamete Research*, **5**, 239–56.

Dauzier, L. and Wintenberger, S. (1952). Recherches sur la fécondation chez les mammifères: la remontée des spermatozoides dans le tractus génital de la brebis. *Comptes Rendus des Seances de la Société de Biologie*, **146**, 67–70.

David, A., Vilensky, A., and Nathan, H. (1972). Temperature changes in the different parts of the rabbit's oviduct. *International Journal of Gynecology and Obstetrics*, **10**, 52–6.

DeMott, R. P. and Suarez, S. S. (1992). Hyperactivated sperm progress in the mouse oviduct. *Biology of Reproduction*, **46**, 779–85.

Drobnis, E. Z. and Overstreet, J. W. (1992). Natural history of mammalian spermatozoa in the female reproductive tract. *Oxford Reviews of Reproductive Biology*, **14**, 1–45.

Du Mesnil du Buisson, F. and Dauzier, L. (1955). Distribution et résorption du sperme dans le tractus génital de la truie: survie des spermatozoides. *Annales d'Endocrinologie*, **16**, 413–22.

Dziuk, P. J. (1970). Estimation of optimum time for insemination of gilts and ewes by double-mating at certain times relative to ovulation. *Journal of Reproduction and Fertility*, **22**, 277–82.

Einarsson, S. (1980). Site, transport and fate of inseminated semen. *Proceedings of 9th International Congress on Animal Reproduction (Madrid)*, **1**, 147–58.

Einarsson, S., Jones, B., Larsson, K., and Viring, S. (1980). Distribution of small-and medium-sized molecules within the genital tract of artificially inseminated gilts. *Journal of Reproduction and Fertility*, **59**, 453–7.

Einer-Jensen, N. (1988). Countercurrent transfer in the ovarian pedicle and its physiological implications. *Oxford Reviews of Reproductive Biology*, **10**, 348–81.

Fournier-Delpech, S. and Thibault, C. (1993). Acquisition of sperm fertilizing ability: epididymal maturation, accessory glands and capacitation. In *Reproduction in mammals and man* (ed. C. Thibault, M. C. Levasseur, and R. H. F. Hunter), pp. 257–78. Ellipses, Paris.

Gilbert, C. L., Cripps, P. J., and Wathes, D. C. (1992). Effect of oxytocin on the pattern of electromyographic activity in the oviduct and uterus of the ewe around oestrus. *Reproduction, Fertility and Development*, **4**, 193–203.

Hammond, J. (1934). The fertilization of rabbit ova in relation to time. A method of controlling the litter size, the duration of pregnancy and the weight of the young at birth. *Journal of Experimental Biology*, **11**, 140–61.

Hancock, J. L. (1961). Fertilisation in the pig. *Journal of Reproduction and Fertility*, **2**, 307–31.

Hansen, C., Srikandakumar, A., and Downey, B. R. (1991). Presence of follicular fluid in the porcine oviduct and its contribution to the acrosome reaction. *Molecular Reproduction and Development*, **30**, 148–53.

Harper, M. J. K. (1973a). Stimulation of sperm movement from the isthmus to the site of fertilisation in the rabbit oviduct. *Biology of Reproduction*, **8**, 369–77.

Harper, M. J. K. (1973b). Relationship between sperm transport and penetration of eggs in the rabbit oviduct. *Biology of Reproduction*, **8**, 441–50.

Harper, M. J. K. (1982). Sperm and egg transport. In *Reproduction in mammals* Vol. 1, (ed. C. R. Austin and R. V. Short), pp. 102–27. Cambridge University Press.

Harper, M. J. K. (1988). Gamete and zygote transport. In *The physiology of reproduction* (ed. E. Knobil, J. Neill, L. L. Ewing, G. S. Greenwald, C. L. Markert, and D. W. Pfaff), pp. 103–34. Raven Press, New York.

Hawk, H. W. (1987). Transport and fate of spermatozoa after insemination of cattle. *Journal of Dairy Science*, **70**, 1487–503.

Herz, Z., Northey, D., Lawyer, M., and First, N. L. (1985). Acrosome reaction of bovine spermatozoa *in vivo*: sites and effects of stages of the oestrous cycle. *Biology of Reproduction*, **32**, 1163–8.

Hunter, R. H. F. (1967). The effects of delayed insemination on fertilisation and early cleavage in the pig. *Journal of Reproduction and Fertility*, **13**, 133–47.

Hunter, R. H. F. (1972). Local action of progesterone leading to polyspermic fertilisation in pigs. *Journal of Reproduction and Fertility*, **31**, 433–44.

Hunter, R. H. F. (1973a). Transport, migration and survival of spermatozoa in the female genital tract: species with intra-uterine deposition of semen. *INSERM (Paris)*, **26**, 309–42.

Hunter, R. H. F. (1973b). Polyspermic fertilisation in pigs after tubal deposition of excessive numbers of spermatozoa. *Journal of Experimental Zoology*, **183**, 57–64.

Hunter, R. H. F. (1974). Chronological and cytological details of fertilisation and early development in the domestic pig, Sus scrofa. *Anatomical Record*, **178**, 169–86.

Hunter, R. H. F. (1975). Transport, migration and survival of spermatozoa in the female genital tract. In *The biology of spermatozoa* (ed. E. S. E. Hafez and C. Thibault), pp. 145–55. Karger, Basel.

Hunter, R. H. F. (1976). Sperm-egg interactions in the pig: monospermy, extensive polyspermy, and the formation of chromatin aggregates. *Journal of Anatomy*, **122**, 43–59.

Hunter, R. H. F. (1977). Function and malfunction of the Fallopian tubes in relation to gametes, embryos and hormones. *European Journal of Obstetrics, Gynaecology and Reproductive Biology*, **7**, 267–83.

Hunter, R. H. F. (1978). Intraperitoneal insemination, sperm transport and capacitation in the pig. *Animal Reproduction Science*, **1**, 167–79.

Hunter, R. H. F. (1980a). Transport and storage of spermatozoa in the female tract. *Proceedings of 9th International Congress on Animal Reproduction*, (Madrid), **2**, 227–33.

Hunter, R. H. F. (1980b). *Physiology and technology of reproduction in female domestic animals*. Academic Press, London.

Hunter, R. H. F. (1981). Sperm transport and reservoirs in the pig oviduct in relation to the time of ovulation. *Journal of Reproduction and Fertility*, **63**, 109–17.

Hunter, R. H. F. (1984). Pre-ovulatory arrest and peri-ovulatory redistribution of competent spermatozoa in the isthmus of the pig oviduct. *Journal of Reproduction and Fertility*, **72**, 203–11.

Hunter, R. H. F. (1986). Peri-ovulatory physiology of the oviduct, with special reference to sperm transport, storage and capacitation. *Development, Growth and Differentiation*, **28**, Suppl., 5–7.

Hunter, R. H. F. (1987a). Peri-ovulatory physiology of the oviduct, with special reference to progression, storage and capacitation of spermatozoa. In *New horizons in sperm cell research* (ed. H. Mohri), pp. 31–45. Japan Science Society Press, Tokyo.

Hunter, R. H. F. (1987b). Human fertilisation *in vivo*, with special reference to progression, storage and release of competent spermatozoa. *Human Reproduction*, **2**, 329–32.

Hunter, R. H. F. (1987c). The timing of capacitation in mammalian spermatozoa—a reinterpretation. *Research in Reproduction*, **19**, 3–4.

Hunter, R. H. F. (1988). *The Fallopian tubes: their rôle in fertility and infertility*. Springer-Verlag, Berlin.

Hunter, R. H. F. (1989). Ovarian programming of gamete progression and maturation in the female genital tract. *Zoological Journal of the Linnean Society*, **95**, 117–24.

Hunter, R. H. F. (1990*a*). Physiology of the Fallopian tubes, with special reference to gametes, embryos and microenvironments. In *Proceedings of 7th Reinier de Graaf Symposium: From ovulation to implantation*, pp. 101–19, (International Congress Series No. 917). Excerpta Medica, Amsterdam.

Hunter, R. H. F. (1990*b*). Fertilisation *in vivo* and *in vitro*. *Journal of Reproduction and Fertility Supplement*, **40**, 211–26.

Hunter, R. H. F. (1991*a*). Behaviour of spermatozoa in the oviduct of farm animals. (Proceedings of international symposium on the biology of the oviduct.) *Archivos de Biologia y Medecina Experimentales (Chile)*, **24**, 349–59.

Hunter, R. H. F. (1991*b*). Oviduct function in pigs, with particular reference to the pathological condition of polyspermy. *Molecular Reproduction and Development*, **29**, 385–91.

Hunter, R. H. F. (1993). Sperm: egg ratios and putative molecular signals to modulate gamete interactions in polytocous mammals. *Molecular Reproduction and Development*, **35**, 324–7.

Hunter, R. H. F. (1994). Modulation of gamete and embryonic microenvironments by oviduct glycoproteins. *Molecular Reproduction and Development*, **39**, 176–81

Hunter, R. H. F. and Dziuk, P. J. (1968). Sperm penetration of pig eggs in relation to the timing of ovulation and insemination. *Journal of Reproduction and Fertility*, **15**, 199–208.

Hunter, R. H. F. and Hall, J. P. (1974*a*). Capacitation of boar spermatozoa: the influence of post-coital separation of the uterus and Fallopian tubes. *Anatomical Record*, **180**, 597–604.

Hunter, R. H. F. and Hall, J. P. (1974*b*). Capacitation of boar spermatozoa: synergism between uterine and tubal environments. *Journal of Experimental Zoology*, **188**, 203–14.

Hunter, R. H. F. and Léglise, P. C. (1971). Polyspermic fertilisation following tubal surgery in pigs, with particular reference to the rôle of the isthmus. *Journal of Reproduction and Fertility*, **24**, 233–46.

Hunter, R. H. F. and Nichol, R. (1983). Transport of spermatozoa in the sheep oviduct: preovulatory sequestering of cells in the caudal isthmus. *Journal of Experimental Zoology*, **228**, 121–8.

Hunter, R. H. F. and Nichol, R. (1986*a*). Post-ovulatory progession of viable spermatozoa in the sheep oviduct, and the influence of multiple mating on their pre-ovulatory distribution. *British Veterinary Journal*, **142**, 52–8.

Hunter, R. H. F. and Nichol, R. (1986*b*). A preovulatory temperature gradient between the isthmus and ampulla of pig oviducts during the phase of sperm storage. *Journal of Reproduction and Fertility*, **77**, 599–606.

Hunter, R. H. F. and Nichol, R. (1988). Capacitation potential of the Fallopian tube: a study involving surgical insemination and the subsequent incidence of polyspermy. *Gamete Research*, **21**, 255–66.

Hunter, R. H. F. and Poyser, N. L. (1985). Ovarian follicular fluid concentrations of prostaglandins E_2, $F_{2\alpha}$ and I_2 during the pre-ovulatory period in pigs. *Reproduction, Nutrition, Développement*, **25**, 909–17.

Hunter, R. H. F. and Wilmut, I. (1982). The rate of functional sperm transport into the oviducts of mated cows. *Animal Reproduction Science*, **5**, 167–73.

Hunter, R. H. F. and Wilmut, I. (1984). Sperm transport in the cow: peri-ovulatory redistribution of viable cells within the oviduct. *Reproduction, Nutrition, Développement*, **24**, 597–608.

Hunter, R. H. F., Nichol, R., and Crabtree, S. M. (1980). Transport of spermatozoa in the ewe: timing of the establishment of a functional population in the oviduct. *Reproduction Nutrition Développement*, **20**, 1869–75.

Hunter, R. H. F., Barwise, L., and King, R. (1982). Sperm transport, storage and release in the sheep oviduct in relation to the time of ovulation. *British Veterinary Journal*, **138**, 225–32.

Hunter, R. H. F., Cook, B., and Poyser, N. L. (1983). Regulation of oviduct function in pigs by local transfer of ovarian steroids and prostaglandins: a mechanism to influence sperm transport. *European Journal of Obstetrics, Gynaecology and Reproductive Biology*, **14**, 225–32.

Hunter, R. H. F., Fléchon, B., and Fléchon, J. E. (1987). Pre-and peri-ovulatory distribution of viable spermatozoa in the pig oviduct: a scanning electron microscope study. *Tissue and Cell*, **19**, 423–36.

Hunter, R. H. F., Fléchon, B., and Fléchon, J. E. (1991). Distribution, morphology and epithelial interactions of bovine spermatozoa in the oviduct before and after ovulation: a scanning electron microscope study. *Tissue and Cell*, **23**, 641–56.

Jansen, R. P. S. (1978). Fallopian tube isthmic mucus and ovum transport. *Science*, **201**, 349–51.

Jewell, P. A. (1989). Factors that affect fertility in a feral population of sheep. *Zoological Journal of Linnean Society*, **95**, 163–74.

Jewell, P. A., Hall, S. J. G., and Rosenberg, M. M. (1986). Multiple mating and siring success during natural oestrus in the ewe. *Journal of Reproduction and Fertility*, **77**, 81–9.

Katz, D. F., Yanagimachi, R., and Dresdner, R. D. (1978). Movement characteristics and power output of guinea-pig and hamster spermatozoa in relation to activation. *Journal of Reproduction and Fertility*, **52**, 167–72.

Katz, D. F., Drobnis, E. Z., and Overstreet, J. W. (1989). Factors regulating mammalian sperm migration through the female reproductive tract and oocyte vestments. *Gamete Research*, **22**, 443–69.

Larsson, B. (1988). *Distribution of spermatozoa in the bovine genital tract after artificial insemination*. PhD thesis. Swedish University of Agricultural Sciences, Uppsala.

Lee, F. C. (1928). The tubo–uterine junction in various animals. *Bulletin of the Johns Hopkins Hospital*, **42**, 335–57.

Lovell, J. E. and Getty, R. (1968). Fate of semen in the uterus of the sow: histologic study of endometrium during the 27 hours after natural service. *American Journal of Veterinary Research*, **29**, 609–25.

Mann, T., Polge, C., and Rowson, L. E. A. (1956). Participation of seminal plasma during the passage of spermatozoa in the female reproductive tract of the pig and horse. *Journal of Endocrinology*, **13**, 133–40.

Martinet, L. and Raynaud, F. (1975). Prolonged spermatozoan survival in the female hare uterus: explanation of superfetation. In *The biology of spermatozoa* (ed. E.S.E. Hafez and C. G. Thibault), pp. 134–44. Karger, Basel.

Mattner, P. E. (1963). Capacitation of ram spermatozoa and penetration of the ovine egg. *Nature (London)*, **199**, 772–3.

Moricard, R. and Bossu, J. (1951). Arrival of fertilising sperm at the follicular cell of the secondary oocyte: a study of the rat. *Fertility and Sterility*, **2**, 260–6.

Mortimer, D. (1978). Selectivity of sperm transport in the female genital tract. In *Spermatozoa, antibodies and infertility* (ed. J. Cohen and W. F. Hendry), pp. 37–53. Blackwells, Oxford.

Mortimer, D. (1983). Sperm transport in the human female reproductive tract. *Oxford Reviews of Reproductive Biology*, **5**, 30–61.

Mortimer, D. (1991). Behaviour of spermatozoa in the human oviduct. *Archivos de Biologia y Medicina Experimentales (Chile)*, **24**, 339–48.

Nichol, R., Hunter, R. H. F., Leese, H. J., and Cooke, G. M. (1992). Concentrations of energy substrates in porcine oviduct fluid and blood plasma during the peri-ovulatory period. *Journal of Reproduction and Fertility*, **96**, 699–707.

Noyes, R. W. (1953). The fertilizing capacity of spermatozoa. *Western Journal of Surgery, Obstetrics and Gynecology*, **61**, 342–9.

Overstreet, J. W. (1983). Transport of gametes in the reproductive tract of the female mammal. In *Mechanism and control of animal fertilisation* (ed. J. F. Hartmann), pp. 499–543. Academic Press, New York.

Overstreet, J. W. and Cooper, G. W. (1975). Reduced sperm motility in the isthmus of the rabbit oviduct. *Nature (London)*, **258**, 718–9.

Overstreet, J. W. and Cooper, G. W. (1978). Sperm transport in the reproductive tract of the female rabbit. I. The rapid transit phase of transport. *Biology of Reproduction*, **19**, 101–14.

Overstreet, J. W. and Katz, D. F. (1977). Sperm transport and selection in the female genital tract. In *Development in mammals* (ed. M. H. Johnson), Vol. 2, pp. 31–65. Elsevier, Amsterdam.

Overstreet, J. W., Cooper, G. W., and Katz, D. F. (1978). Sperm transport in the reproductive tract of the female rabbit. II. The sustained phase of transport. *Biology of Reproduction*, **19**, 115–32.

Parker, G.A. (1984). Sperm competition and the evolution of animal mating strategies. In *Sperm competition and the evolution of animal mating systems* (ed. R. L. Smith), pp. 1–60. Academic Press, Orlando.

Patek, E. (1974). The epithelium of the human Fallopian tube. *Acta Obstetrica Gynecologica Scandinavica*, **53**, Supplement **31**, 1–28.

Piko, L. (1961). La polyspermie chez les animaux. *Annales de Biologie Animale, Biochimie et Biophysique*, **1**, 323–83.

Polge, C. (1978). Fertilisation in the pig and the horse. *Journal of Reproduction and Fertility*, **54**, 461–70.

Polge, C., Salamon, S., and Wilmut, I. (1970). Fertilising capacity of frozen boar semen following surgical insemination. *Veterinary Record*, **87**, 424–8.

Pollard, J. W., Plante, C., King, W. A., Hansen, P. J., Suarez, S. S., and Betteridge, K. J. (1990). Sperm fertilizing capacity is maintained by binding to oviductal epithelial cells. In *Symposium on fertilisation in mammals* (ed. B. D. Bavister, J. Cummins, and E. R. S. Roldan), p. 61. (Abstract) Serono, Boston.

Pollard, J. W., Plante, C., King, W. A., Hansen, P. J., Betteridge, K. J., and Suarez, S. S. (1991). Fertilizing capacity of bovine sperm may be maintained by binding to oviductal epithelial cells. *Biology of Reproduction*, **44**, 102–7.

Reinius, S. (1970). Morphology of oviduct, gametes and zygotes as a basis of oviductal function in the mouse. I. Secretory activity of oviductal epithelium. *International Journal of Fertility*, **15**, 191–209.

Robinson, T. J. (1973). Factors involved in the failure of sperm transport and survival in the female reproductive tract. *Journal of Reproduction and Fertility*, Supplement **18**, 103–9.

Roldan, E. R. S. and Gomendio, M. (1992). Morphological, functional and biochemical changes underlying the preparation and selection of fertilising spermatozoa *in vivo*. *Animal Reproduction Science*, **28**, 69–78.

Senger, P. L., Becker, W. C., Davidge, S. T., Hillers, J. K., and Reeves, J. J. (1988). Influence of cornual insemination on conception in dairy cattle. *Journal of Animal Science*, **66**, 3010–6.

Settlage, D. S. F., Motoshima, M., and Tredway, D. (1975). Sperm transport from the vagina to the Fallopian tubes in women. In *The biology of spermatozoa* (ed. E. S. E. Hafez and C. Thibault), pp. 74–82. Karger, Basel.

Shalgi, R. and Kraicer, P. F. (1978). Timing of sperm transport, sperm penetration and cleavage in the rat. *Journal of Experimental Zoology*, **204**, 353–60.

Smith, T. T. and Yanagimachi, R. (1989). Capacitation status of hamster spermatozoa in the oviduct at various times after mating. *Journal of Reproduction and Fertility*, **86**, 255–61.

Smith, T. T. and Yanagimachi, R. (1990). The viability of hamster spermatozoa stored in the isthmus of the oviduct: the importance of sperm-epithelium contact for sperm survival. *Biology of Reproduction*, **42**, 450–7.

Smith, T. T. and Yanagimachi, R. (1991). Attachment and release of spermatozoa from the caudal isthmus of the hamster oviduct. *Journal of Reproduction and Fertility*, **91**, 567–73.

Smith, T. T., Koyanagi, F., and Yanagimachi, R. (1987). Distribution and number of spermatozoa in the oviduct of the golden hamster after natural mating and artificial insemination. *Biology of Reproduction*, **37**, 225–34.

Stefanini, M., Oura, C., and Zamboni, L. (1969). Ultrastructure of fertilisation in the mouse. II. Penetration of sperm into the ovum. *Journal of Submicroscopical Cytology*, **1**, 1–23.

Suarez, S. S. (1987). Sperm transport and motility in the mouse oviduct: observations *in situ*. *Biology of Reproduction*, **36**, 203–10.

Suarez, S. S. and Dai, X. (1992). Hyperactivation enhances mouse sperm capacity for penetrating viscoelastic media. *Biology of Reproduction*, **46**, 686–91.

Suarez, S. S. and DeMott, R. P. (1991). Functions of hyperactivated motility of sperm in the oviduct. *Archivos de Biologia y Medicina Experimentales (Chile)*, **24**, 331–7.

Suarez, S. S., Katz, D. F., and Overstreet, J. W. (1983). Movement characteristics and acrosomal status of rabbit spermatozoa recovered at the site and time of fertilisation. *Biology of Reproduction*, **29**, 1277–87.

Suarez, S. S., Katz, D. F., Owen, D. H., Andrew, J. B., and Powell, R. L. (1991*a*). Evidence for the function of hyperactivated motility in sperm. *Biology of Reproduction*, **44**, 375–81.

Suarez, S. S., Redfern, K., Raynor, P., Martin, F., and Philips, D. M. (1991*b*).

Attachment of boar sperm to mucosal explants of oviducts *in vitro*: possible role in formation of a sperm reservoir. *Biology of Reproduction*, **44**, 998–1004.

Suarez, S. S., Dai, X. B., DeMott, R. P., Redfern, K., and Mirando, M. A. (1992). Movement characteristics of boar sperm obtained from the oviduct or hyperactivated *in vitro*. *Journal of Andrology*, **13**, 75–80.

Szollosi, D. and Hunter, R. H. F. (1973). Ultrastructural aspects of fertilisation in the domestic pig: sperm penetration and pronucleus formation. *Journal of Anatomy*, **116**, 181–206.

Szollosi, D. and Hunter, R. H. F. (1978). The nature and occurrence of the acrosome reaction in spermatozoa of the domestic pig, *Sus scrofa*. *Journal of Anatomy*, **127**, 33–41.

Templeton, A. A. and Mortimer, D. (1980). Laparoscopic sperm recovery in infertile women. *British Journal of Obstetrics and Gynaecology*, **87**, 1128–31.

Thibault, C. (1973). Sperm transport and storage in vertebrates. *Journal of Reproduction and Fertility*, Supplement **18**, 39–53.

Thibault, C., Gerard, M., and Heyman, Y. (1975). Transport and survival of spermatozoa in cattle. In *The biology of spermatozoa* (ed. E. S. E. Hafez and C. G. Thibault), pp. 156–65. Karger, Basel.

Usselman, M. C. and Cone, R. A. (1983). Rat sperm are mechanically immobilized in the cauda epididymidis by 'immobilin', a high molecular weight glycoprotein. *Biology of Reproduction*, **29**, 1241–53.

Williams, M., Hill, C. J., Scudamore, I., Dunphy, B., Cooke, I. D., and Barratt, C. L. R. (1993). Sperm numbers and distribution within the human Fallopian tube around ovulation. *Human Reproduction*, **8**, 2019–26.

Yanagimachi, R. (1981). Mechanisms of fertilisation in mammals. In *Fertilisation and embryonic development in vitro* (ed. L. Mastroianni and J. D. Biggers), pp. 81–182. Plenum Press, New York.

Yanagimachi, R. (1988). Mammalian fertilisation. In *The physiology of reproduction* (ed. E. Knobil and J. Neill), pp. 135–85. Raven Press, New York.

Yanagimachi, R. and Mahi, C. A. (1976). The sperm acrosome reaction and fertilisation in the guinea pig: a study *in vivo*. Journal of Reproduction and Fertility, **46**, 49–54.

Zamboni, L. (1972). Fertilisation in the mouse. In *Biology of mammalian fertilisation and implantation* (ed. K. S. Moghissi and E. S. E. Hafez), pp. 213–62. Thomas, Springfield.

5 The origin of genetic defects in the human and their detection in the pre-implantation embryo

JOY D. A. DELHANTY and ALAN H. HANDYSIDE

I   Introduction

II Origin of chromosome abnormalities
    1 Developmental stage of origin
    2 Karyotype analysis in clinically recognized pregnancy
    3 Karyotype analysis of human gametes
    4 Karyotype analysis of human embryos

III Molecular cytogenetic investigations
    1 Spermatozoa
    2 Preimplantation embryos
    3 Mosaicism and cell-cycle checkpoints

IV Origin of single-gene defects

V   Preimplantation diagnosis of genetic defects
    1 Early approaches
    2 The application of dual FISH for sex determination
    3 Specific diagnosis of single-gene defects
    4 The potential for misdiagnosis
    5 Future developments

I   INTRODUCTION

At birth, approximately 2 per cent of humans have a genetic defect. Half of these defects involve chromosome imbalance; the other half, a combination of genetic contributions to congenital abnormalities and common diseases, and single-gene defects, of which approaching 5000 have now been described (McKusick 1992). Important though these surviving cases are, in terms of clinical, economic, and social problems, they represent a small fraction of those present in early developmental stages. Some indication of the high levels of fertilization failure, gametic abnormalities or errors in embryogenesis that result in inviability prior to implantation is given by the observation that in humans the fecundity rate (probability of achieving a clinically recognized pregnancy within a given cycle), is about 25 per cent (Wilcox *et al.* 1988). This figure was arrived at by studying a group of 220 women attempting to conceive, 95 per cent of whom were under 35 years of age and fertile. Although in this group of relatively young women, the rate of clinically recognized spontaneous abortions was only 9 per cent, pregnancy loss before this stage was more than double this figure.

II ORIGIN OF CHROMOSOME ABNORMALITIES

For all age groups, clinically recognized pregnancy loss is usually quoted as 15–20 per cent (Hassold 1986). It is this fraction of failed pregnancies that have been extensively studied cytogenetically and in which a chromosome abnormality rate of at least 50 per cent has been found (Hassold 1986). This contrasts with a figure of 5 per cent for this type of abnormality in stillbirths, illustrating clearly the *in utero* selection process that eliminates 95 per cent of chromosomally unbalanced conceptions. Clinical prenatal diagnosis can thus be seen as an extension of this natural process.

1 Developmental stage of origin

There are essentially three developmental stages when chromosomal defects may arise: gametogenesis, fertilization, and embryogenesis. Errors of gametogenesis are usually considered to be meiotic in origin. DNA polymorphisms at loci close to the centromere provide information not only on the parental origin of additional chromosomes in trisomic zygotes but also on the meiotic stage when the error occurred (Table 5.1). In maternally derived trisomy, meiosis I errors predominate for chromosomes 16, 21, and X, but almost all cases of 18 trisomy have their origin in maternal meiosis II (Antonarakis *et al.* 1992; Jacobs, personal communication). The 5 per cent trisomy 21 cases that are paternally derived are predominantly due to meiosis II error (Petersen *et al.* 1993), but in this trisomy as in trisomy 18 and triple X, DNA analysis has provided evidence for postzygotic mitotic error in 5–20 per cent of cases (Antonarakis *et al.* 1993; Fisher *et al.* 1993; Jacobs, personal communication). Segregational analysis of DNA markers cannot, however, detect premeiotic mitotic errors or low-level somatic mosaicism that leads to gonadal mosaicism in a parent. Additional chromosomes present in a zygote due to gonadal mosaicism would masquerade as being of meiotic origin, leading to an underestimate of mitotic errors (Sensi and Ricci 1993). Mosaicism detected by embryonic karyotype analysis has usually been seen in cleavage stage embryos; the fate

Table 5.1 Parental origin of aneuploidy

Aneuploidy		Maternal (%)	Paternal (%)	Postzygotic mitosis (%)	Reference
Trisomies					
21	M I	64	1		Antonarakis et al. 1992
	M II	19	3.5		
	M I/M II	11			
	M I	73	2	4.6	Antonarakis et al. 1993
	M II	18	2.5	(6 maternal, 5 paternal)	
18	M II	87	0	13	P. Jacobs, personal communication
				(2/7 paternal)	
16	M I	100			
13	M I/M II	80			
XXX	M I	59		18	Macdonald et al. 1994
	M II	16		(7 maternal, 2 paternal)	
	M I/M II		6		
XXY	M I	31	46	2	
	M II	11		(3/3 maternal)	
	M I/M II	9			
Monosomies					
X0		Maternal X	Paternal X		Hassold et al. 1988
		83	17		

of the abnormal cell line will depend on the partition of the different cell lineages to the inner cell mass or the trophectoderm. If the aneuploid line is in the minority, it may well not contribute to the embryo proper, but the fact that a substantial proportion of trisomic fetuses are due to a postzygotic mitotic error shows that aneuploid cells are not necessarily diverted to the trophectoderm.

2 Karyotype analysis in clinically recognized pregnancy

The early cytogenetic studies of spontaneous abortion material were carried out by culture of products of conception. To circumvent bias due to culture failure, a more recent approach has been the direct preparation methods using chorionic villi. One large study, interestingly, still gave an abnormality rate of 50.1 per cent, although the excess of females in chromosomally normal cases (male : female sex ratio 0.71) and the fact that the material was passed from the uterus raises the possibility of maternal cell contamination in some cases (Eiben *et al.* 1990). In contrast, in a much smaller survey, transcervical or transabdominal chorionic villus sampling was offered to all women presenting with either a spontaneous abortion or a blighted ovum (Strom *et al.* 1992). In this group, 83 per cent of karyotypes were unbalanced, and in the normal cases there was no excess of females. In the much larger German study, the frequency of the different types of anomalies mirrored the earlier findings from culture, namely predominantly trisomy (62 per cent), followed by triploidy (12.4 per cent), monosomy X (10.5 per cent), tetraploidy (9.2 per cent) and structural anomalies (4.7 per cent), (Eiben *et al.* 1990). Among trisomies, extra copies of chromosomes 16 (21.8 per cent), 22 (17.9 per cent), and 21 (10.0 per cent were prevalent, with conspicuous absence of involvement of chromosomes 1, 5, 17, and 19. Although the overall chromosomal abnormality rate was 50 per cent, this ranged from 40 per cent for women aged up to 29 years to 82.2 per cent for those aged 40 or above, but the increase was due solely to the increase in trisomies (range 14.3–80.0 per cent of abnormals); monosomy X and polyploidy decreased significantly with age. Similarly interesting data on the effect of maternal age was gathered by Hassold and Chiu (1985) on studying 2264 spontaneous abortions karyotyped from 1976–83 in a large maternity hospital in Honolulu. In women 40 years and older, trisomy was estimated to occur in approximately 30 per cent of recognized human conceptions. Assuming that the corresponding monosomies occur with equal frequency, it is estimated that in women over the age of 40 the majority of oocytes do not have the correct haploid number.

While it is clear that, for the population as a whole, rising maternal age is the main factor predisposing to the conception of a chromosomally unbalanced zygote and structural anomalies play a relatively small part, for couples where one partner carries a balanced rearrangement this can convey a high risk of reproductive failure or of producing a handicapped child. Individually the most frequent rearrangements are the Robertsonian or centric fusions, involving the acrocentric chromosomes, numbers 13 to 15, 21, and 22. The main risk of handicap is Down syndrome, as trisomies of all the other acrocentrics are essentially lethal. Risks associated with reciprocal translocations and inversions are specific not only to the individual chromosomal segments involved, but also to the sex of the carrier and the particular family, possibly depending on genetic background.

The overiding influence of maternal age in the production of trisomic conceptions is in agreement with data from restriction fragment length polymorphism (RFLP) analysis indicating a maternal origin of the additional chromosome for at least 80 per cent of cases for the autosomal trisomies 13, 18 and 21 and for triple X (Hassold *et al.* 1987; May *et al.* 1990; Antonarakis *et al.* 1991; Fisher *et al.* 1993). For Klinefelter's syndrome, caused by the sex chromosome constitution XXY, approximately half have a maternal origin, and in these cases there is a maternal age effect (Jacobs *et al.* 1988). This contrasts with fetuses that have a single X chromosome and females with Turner's syndrome in which 80 per cent lack the paternal sex chromosome (Hassold *et al.* 1988). Similarly, point mutations and structural chromosomal rearrangements seem to arise much more commonly in males (Chandley 1991).

3 Karyotype analysis of human gametes

Data on cytogenetic anomalies derived from spontaneous abortions is obviously limited to recognized pregnancies but the development of techniques for the investigation and treatment of infertility has provided the opportunity to study gametes and embryos before implantation. Studies with gametes have confirmed the origins of chromosomal anomalies. The labour-intensive and difficult technique of hamster egg penetration has been used to study the chromosomes of human sperm over the last decade, providing information on around 16 000 karyotypes (Jacobs 1992). Fewer than 200 hyperhaploid sperm were identified, whereas the figure for structural anomalies was tenfold higher, but these findings are of course limited to sperm capable of fertilization. The figure for structural change is so extraordinarily high that it has been considered to be a product of the technique (Jacobs 1992).

In vitro fertilization (IVF) clinics provide a ready source of oocytes remaining unfertilized after insemination that may be used for cytogenetic analysis. As these are conveniently arrested at metaphase II of meiosis at ovulation, the cytogeneticist merely has to spread the cells and stain the chromosomes. However, in the spreading process chromosomes are easily lost and the contracted morphology of the chromosomes makes them unsuitable for precise banding analysis, limiting the specificity of the information available. Additionally, oocytes remaining unfertilized after exposure to sperm are inevitably aged by at least 40 hours since retrieval and 10–15 per cent, when analysed, turn out to be fertilized. The combination of these factors results in low numbers giving a conclusive answer in each study series, with consequently variable proportions of hyperhaploidy, ranging between 2 and 14.5 per cent in eleven studies (reviewed by Zenzes and Casper 1992). Because it is not possible to distinguish true hypohaploidy from artefactual chromosome loss, total aneuploidy is usually estimated by doubling the figure for hyperhaploidy. Based on the 1120 oocytes in the eleven studies, the reviewers estimated the weighted mean percentage of aneuploidy as about 13 per cent; as expected, this figure is considerably higher than that found in sperm.

It is usually assumed that gametic hyperhaploidy is caused by failure of paired homologous chromosomes to disjoin at anaphase of meiosis I (non-disjunction). Studies made possible by the availability of polymorphic DNA markers located at the centromere of chromosomes are showing that trisomy is associated with anomalies of crossing over during prophase I of meiosis (Hassold *et al.* 1991; Sherman *et al.* 1991, Lorda Sanchez *et al.* 1992). Failure of (or reduced) chiasma formation, particularly in small bivalents such as that for chromosome 21, can result in premature separation of homologues, with consequent random arrangement of univalents on the metaphase spindle and equally random separation at anaphase I. This could also provide a predisposing mechanism for precocious division of univalents at first anaphase leading to oocytes with separated, and sometimes missing or extra, chromatids as has been observed at metaphase II (Angell *et al.* 1994). Fertilization of such oocytes could lead to a trisomic zygote. However, premature separation of chromatids at anaphase I is unlikely to apply to those bivalents where an excess of recombination has occurred, as in a proportion of triple X cases (Jacobs, personal communication).

4 Karyotype analysis of human embryos

In addition to anomalies of gametogenesis, abnormal zygotes can arise by errors of fertilization. A common abnormality in IVF is fertilization by more than one sperm—polyspermy—but the cytogenetic data show that this is also a common cause of spontaneous miscarriage *in vivo*, possibly as high as 1 per cent of conceptions, and again mainly due to dispermy (Jacobs *et al.* 1978). Polyspermy and apparent parthenogenetic activation of oocytes at IVF can be detected by examining the pronuclei 12–18 hours after insemination. However, the detection of precise chromosomal anomalies in the preimplantation human embryo is extremely difficult by conventional means. The usual approach has been to incubate the whole embryo overnight with colchicine to arrest dividing blastomeres at metaphase, followed by attempts to spread the intact embryo. Not surprisingly, this produces poor quality chromosomes, which are either contracted and difficult to group, or more elongated but overlapping. In these circumstances, complete analysis of a single metaphase is counted as a success and there are few studies with analysis of several cells from individual embryos. Most studies are carried out on 'spare' embryos that are surplus to requirements after those suitable for transfer to the mother have been chosen. A few studies have used donated oocytes that are then fertilized specifically for research. In a review of four studies in which a reasonable number of embryos were analysed (30–50) abnormality rates ranging between 23 and 40 per cent were found (Zenzes and Casper 1992). Where sufficient detail was given, mosaicism with normal and aneuploid or polyploid cell lines appeared to be the most common abnormality. In an interesting comparison of the chromosome status of untransferred embryos between two groups of women undergoing IVF treatment—those who became pregnant and those that did not—Zenzes and co-workers analysed one to four mitoses per embryo; overall, 13 per cent of embryos only were normal diploid, 28 per cent were aneuploid, and 36 per cent were mosaic (Zenzes *et al.* 1992). They concluded that the proportion of spare embryos that are chromosomally normal is significantly greater in pregnant than in age-matched IVF patients who did not become pregnant and also that detection of chromosomally normal embryos for transfer should improve the success rate in IVF.

To examine the feasibility of karyotype analysis in single cells biopsied from human embryos with a view to preimplantation diagnosis of chromosome status, zona-free cleavage stage embryos were disaggregated and individual cells cultured overnight in colcemid (Delhanty and Penketh, unpublished). In total, 164 cells from 33 embryos were processed. From these, 22 single-celled preparations of chromosomes were obtained, including two sets from each of two embryos. Chromosomes from two embryos were in the triploid range, one was haploid, and the remainder appeared to be diploid. Exact counts were possible in eight of the diploid cells, with chromosome numbers from 44 to 46; no structurally abnormal chromosomes were observed. Complete analysis of two revealed a normal female karyotype. The haploid metaphase may have been from a parthenogenetic embryo since there was no evidence of a Y chromosome, but since there were only 22 chromosomes either a sex chromosome or a member of the C group had been lost in preparation. One of the triploid cells was of interest because it also contained a *de novo* reciprocal translocation, 69, XXY, −D−F, +t(DqF). Both parents were chromosomally normal but analysis of Q-band polymorphisms revealed that the triploid was of digynic origin due to failure of the second meiotic division and retention of the second polar body. The second cell with an approximately triploid karyotype (more than 70 chromosomes) was derived from an embryo which also contained a diploid cell, providing evidence of mosaicism. An interesting and perhaps pertinent observation was that chromosome breakage was seen in all four cells that could be fully analysed, an exceptionally high frequency compared with cultured somatic cells.

Although it is possible to achieve moderate success in arresting single blastomeres in metaphase after overnight incubation it can be seen from the results described that, because of the difficulty in spreading chromosomes, the quality of the results would be unlikely to be adequate to obtain precise information on the karyotype of an embryo from single cell analysis. It is also apparent that the high frequency of observed chromosomal mosaicism in the human preimplantation embryo would make single-cell diagnosis unreliable.

III MOLECULAR CYTOGENETIC INVESTIGATIONS

Molecular cytogenetics essentially consists of techniques to investigate the hybridization *in situ* of labelled DNA probes to metaphase chromosome or interphase nuclei. Subsets of alphoid satellite DNA have been isolated that are chromosome specific and the use of these as DNA probes allows chromosome identification in metaphase and additionally enables chromosome copy number to be determined in interphase, so-called interphase cytogenetics. This approach is very useful in situations where chromosomes are not easily obtainable, as in sperm nuclei and preimplantation embryos. Although initially enzymatic methods were used for non-radioactive *in situ* hybridization, these have largely been superseded by fluorescent methods that combine advantages of speed, specificity, and sensitivity. Probes for flourescent *in situ* hybridization (FISH) are labelled by nick translation, most usually with biotin, and detected by fluorescein isothiocyanate−avidin (green fluorescence). For dual labelling, digoxigenin is used as a second label; detection is then by anti-digoxigenin antibody and tetramethylrhodamine isothiocyanate (red fluorescence). Chromosomal or nuclear DNA has first to be denatured and it is important that every trace of cytoplasm is removed before hybridization to the labelled probes is attempted.

1 Spermatozoa

FISH techniques are very appropriate for the study of sperm chromosomes. In contrast to the hamster egg penetration technique, thousands of decondensed sperm nuclei can be scored for chromosomal signals by FISH in a single experiment. In addition, the parameters are different because the total ejaculate is sampled, rather than analysis being limited to sperm capable of interspecies egg penetration. The first studies to employ FISH in the analysis of the chromosome complement of sperm used single DNA probes, commonly specific for chromosome 1 or the Y chromosome, for technical reasons. Despite the analysis of over 10 000 sperm in the majority of studies, a wide variation in the frequency of disomy was observed, from 0.06 to 0.8 per cent for chromosome 1, and 0 to 2.7 per cent for the Y chromosome (Martin et al. 1994). Technical reasons for this wide variability were suspected and indeed Martin and co-workers were able to show that differing results were obtained with fresh and stored sperm nuclei. Another reason for inaccuracy might be the production of appreciable numbers of diploid sperm; with the use of a single probe these would be scored as disomic. Hassold and colleagues have recently published results of an elegant study using two and three probes for the analysis of non-disjunction in sperm (Williams et al. 1993). The importance of using this approach was demonstrated by the fact that most donors had levels of diploidy higher than the disomy levels of individual chromosomes. A range of 0.2–1 per cent diploidy was found. It is interesting to note that not a single diploid human sperm was recorded in the hamster egg penetration analysis, suggesting that although frequently formed, diploid sperm may not be capable of fertilization. Simultaneous use of X and Y probes with an autosomal probe in three-colour FISH allowed an estimate of meiosis I sex chromosome disomy to be made, at 0.09 per cent. Adding in the results of meiosis II errors, the total frequency of sex chromosome disomy came to 0.2 per cent, 1.5 times as common as disomy 16 and twice as common as disomy 18. Studies of DNA from 47, XXY individuals and their parents indicate that nearly half are attributable to paternal meiosis I error (Hassold et al. 1991), suggesting that the XY bivalent is particularly likely to fail to disjoin, a property that is not shared by the autosomes at male meiosis. This is shown by studies of parental origin in cases of trisomy 21 and 18 which suggest that the small proportion of paternal errors are due to meiosis II or postzygotic non-disjunction (Table 5.1) (Antonarakis et al. 1991; Fisher et al. 1993).

In view of the overwhelmingly maternal origin of trisomy 16, and to a lesser extent, trisomy 18, the relatively high frequency of sperm disomic for these chromosomes is surprising and warrants further investigation. Similarly, the estimated frequency of sperm disomic for the X chromosome far exceeds the incidence of paternally derived 47, XXX conceptions in clinically recognized pregnancies. The possible reasons for these discrepancies include differential fertilizing ability of sperm carrying the extra chromosome and earlier lethality of paternally derived cases of these trisomies.

2 Preimplantation embryos

There are few published studies employing multicolour FISH for the study of the chromosome constitution of preimplantation human embryos. In part, this reflects the paucity of material for research purposes. Most IVF units freeze normally fertilized spare embryos for future transfer, leaving only abnormal embryos available for study. Nevertheless, interesting data can be obtained from such material. In one recent investigation employing multicolour FISH with DNA probes for chromosomes X, Y, 18, and 13/21 combined, 20 abnormally developing monospermic embryos, up to the 12-cell stage were studied; these were derived from patients undergoing IVF treatment for infertility (Munné *et al.* 1993). Most embryos were arrested in development and showed morphological abnormalities, the remainder were fragmenting or other wise showing abnormal development (Table 5.2). Between one and ten blastomeres were analysed per embryo. For the summary, the following criteria were employed: embryos were considered to be chromosomally abnormal if all nuclei from the same embryo showed identical anomalies of chromosome number when results were available from at least two blastomeres. Embryos were classed as mosaic if one or more blastomeres had additional copies of chromosomes with the remainder showing the normal copy number. In the case of embryos with fewer than the normal number of signals in some nuclei, these were included as mosaic only if identical abnormalities were seen in more than one blastomere or if more than one chromosome was involved in a single abnormal cell. Owing to the known high frequency in control lymphocytes of nuclei with one less than the expected number of FISH signals (11.5 per cent overall in this study) the observation of a blastomere nucleus with a single missing signal was not counted as abnormal. An exception to this rule was made when evidence for mitotic non-disjunction was obtained by the reciprocal observation of one and three signals for a particular chromosome in sibling blastomeres. Unclassified abnormals were those embryos where signal anomalies were limited to a single nucleus and no results were available for sibling blastomers; in these cases it was impossible to determine whether the observed anomaly was representative of the embryo as a whole. A final category was created of 'uncontrolled division' for embryos where apparently random abnormalities were observed with great variation from cell to cell. Using this conservative classification system, it is apparent from Table 5.2 that developmental abnormality in these monospermic embryos is correlated with a high degree of abnormality of chromosome number. Of the 20 embryos, only six were classed as normal, five were entirely abnormal (four due to aneuploidy), three were mosaic (of which one was basically tetraploid initially), and the remainder were unclassified abnormals or showed uncontrolled division. Assuming that the tetraploid embryo arose by doubling of a diploid chromosome set before the first cleavage division, then four of the 13 classified abnormalities are likely to be due to meiotic errors and the remainder appear to be postfertilization. It has to be remembered that only five chromosomes were investigated; investigation of the remainder

Table 5.2 *Chromosome abnormalities detected by FISH with probes for chromosomes X, Y, 18, and 13/21 in 20 abnormally developing human embryos*

Normal	Abnormal			Mosaic		Uncontrolled division	Unclassified abnormals (abnormal ploidy or mosaic)
	trisomy	monosomy	tetraploid	Diploid/tetraploid	Aneuploid		
6	2	2	1*	1	2	5	2

* Tetraploid embryo also aneuploid mosaic due to mitotic non-disjunction. Data from Munné et al. 1993.

may well have shown that the six embryos classed as normal were in fact chromosomally abnormal.

In the same study, Munné and colleagues (1993) also investigated 10 normally developing monospermic embryos donated by a group of patients in whom IVF had repeatedly failed. The mean maternal age of this group was 40 years. The same chromosomal probes were employed as previously and using the same classification system, the results for this group of patients are shown in Table 5.3. Although apparently developing normally up to about the eight-cell stage, of these 10 embryos only two showed the normal number of signals for the chromosomes investigated. Five were uniformaly aneuploid, two were mosaic aneuploid and one showed uncontrolled division; this is an abnormality rate considerably higher than in the arrested embryos from typical IVF patients, and includes a much higher proportion of pre-fertilization errors (five out of eight).

An interesting contrast is shown by consideration of our own data on donated embryos from an average group of IVF patients with a mean maternal age of 35 years (Harper et al. 1995). Thirty-four normally fertilized and normally developing spare cleavage-stage embryos from 19 donors were investigated using dual FISH for X and Y chromosomes. Between three and 12 nuclei were analysed per embryo, except in the case of a two-cell embryo in which hybridization occurred in one cell only. In total, 202 nuclei were spread for analysis and hybridization signals were seen in 197 nuclei, an efficiency of 97.5 per cent, close to that for lymphocyte controls which was 99 per cent. Using the classification system described previously, results of this study are summarized in Table 5.4. In 30 embryos, all nuclei had the normal number of X or X and Y signals; 17 were XY and 13 XX. The remaining four embryos were mosaic: diploid–tetraploid, or diploid–triploid (or aneuploid, XXY), and two diploid–haploid (or X monosomy). An additional 25 normally developing spare embryos from 13 IVF donors (mean age 35 years) were investigated with dual FISH, using probes for chromosomes 1 and 7 (Harper et al., 1995). In total, 137 nuclei were spread and 132 gave hybridization signals. Thirteen of the 25 embryos had the normal number of signals in all nuclei examined (Table 5.5). Only one embryo was completely abnormal, and that was considered most likely to be triploid. Using the criteria as before, eight embryos were classed as mosaic: three diploid–tetraploid, one diploid–triploid (or double trisomy), and three diploid–haploid (or double monosomy). A single embryo was considered to be an aneuploid mosaic due to mitotic non-disjunction. Considering these two series of normally fertilized and developing embryos together gives a total of 59 embryos, of which 43 gave normal results in dual FISH. Only one was completely abnormal, and that was probably triploid, whereas 12 (20 per cent) were mosaic, and one showed uncontrolled division; virtually all were likely to be errors during cleavage. This gives a completely different picture from that shown

Table 5.3 *Chromosome abnormalities detected by FISH with probes for chromosomes X, Y, 18, and 13/21 in 10 normally developing human embryos*

	Abnormal			Mosaic	
Normal	trisomy	monosomy	double aneuploid	diploid/aneuploid	Uncontrolled division
2	2	1	2	2	1

Donors were of mean age 40 years, and had repeated failures at IVF. Data from Munné *et al.* 1993.

Table 5.4 *Chromosome abnormalities detected by FISH with probes for X and Y chromosomes in 34 normally developing embryos*

Normal	Mosaic		
	diploid/tetraploid	diploid/triploid or aneuploid	diploid/haploid or X monosomy
30	1	1	2

Embryos were from 19 donors, mean age 35 years. From Harper *et al.* 1995.

Table 5.5 *Chromosome abnormalities detected by FISH with probes for chromosomes 1 and 17 in 25 normally developing embryos*

Normal	Abnormal	Mosaic			
	triploid	diploid/ tetraploid	diploid/triploid or double trisomy	diploid/haploid or double monosomy	diploid/ aneuploid	Uncontrolled division
13	1	3	1	3	1	1

Embryos were from 13 donors, mean age 35 years. From Harper *et al.* 1995.

by the analysis of either arrested embryos or normally developing embryos from older women with repetitive IVF failure (Munné *et al.* 1993). In the latter two groups there was a high proportion of abnormality due to total aneuploidy and mosaicism played a relatively minor role. Further studies with multiprobe FISH are needed to clarify the situation, but several interesting and important conclusions can already be drawn.

1. Arrested development in human preimplantation embryos is highly correlated with the presence of chromosome abnormality.
2. Most apparently normally developing embryos from older women with repetitive IVF failure are chromosomally abnormal. Transfer of such embryos would undoubtedly result in implantation failure or spontaneous abortion.
3. Normally developing spare embryos from average IVF patients may rarely be totally abnormal, but they do have a high level of mosaicism for chromosome abnormalities of various types.

The data from FISH studies thus confirm the rather sparse karyotyping data on human embryos, in that mosaicism is a common finding and the production of a subpopulation of tetraploid cells is also common (Zenzes and Casper 1992).

3 Mosaicism and cell-cycle checkpoints

The discovery, arising from FISH analysis of the majority of interphase nuclei from individual embryos with multiple chromosome-specific probes, that human embryos *in vitro* are often mosaics for cells with aneuploidies involving several chromosomes ('uncontrolled division') (Munné *et al.* 1993; Harper *et al.* 1995) is intriguing and raises questions about the mechanism that causes these abnormalities and the fate of these cells. Clearly, these chromosomal abnormalities cannot have arisen during gametogenesis or at fertilization but must arise postzygotically during early cleavage. As complementary abnormalities in other cells are not evident, the mechanism is unlikely to involve those normally associated with aneuploidy such as non-disjunction. The chaotic nature of the abnormalities observed is reminiscent of those which occur with transformed cells in culture. Cell-cycle checkpoints, first identified in yeast, would normally protect cells from genetic damage by ensuring that successive phases of the cell cycle are completed before the next is initiated (Hartwell and Weinert 1989; Murray 1992). For example, there is a checkpoint that prevents mitosis proceeding before completion of chromosomal segregation. In transformed cells, it is believed that these checkpoints are defective, resulting in the sporadic occurrence of secondary chromosomal and other genetic defects.

Cell-cycle checkpoints are also not operative during the early cleavage divisions of invertebrates and lower vertebrates and may not operate at this stage in mammalian embryos. In *Drosophila* and *Xenopus*, this is thought to be an adaptation to allow exceptionally rapid cycles during a period when there is no cellular growth and the G_1 phase of the cycle is absent (O'Farrell *et al.* 1989). Co-ordination of the cell cycle in this situation is thought to be dependent simply on the appropriate relative timing of the different processes occuring in parallel. With mammalian embryos, there is a short G_1 phase in the early cleavage divisions and the cycles are long. The absence of cell-cycle checkpoints may therefore relate more generally to the absence of embryonic gene transcription and dependence on maternal products inherited in the oocyte. However, many events that drive the cell cycle through the cyclical accumulation and degradation of cyclins are post-transcriptional.

The possibility that cell-cycle checkpoints do not operate during cleavage of the human embryo may also explain the relatively high incidence of various nuclear abnormalities which have been observed (Winston *et al.* 1991, Hardy *et al.* 1993). For example, binucleate blastomeres are common and probably arise from failure of cytokinesis during mitosis (Hardy *et al.* 1993). The presence of cells with nuclear abnormalities or highly abnormal chromosome complements may, therefore, reflect both a lack of co-ordination of the different processes of the cell cycle, possibly exacerbated by sub-optimal culture conditions or deficiencies in oocyte maturation, and the absence of checkpoint control. Some apparently abnormal cells may have a role in embryogenesis. For example, binucleate cells may give rise to tetraploid cells and ultimately contribute polyploid cells to the trophectoderm of the placenta. It is more likely, however, that most of these cells become arrested and contribute to the low developmental potential of human embryos after transfer to the uterus (Hardy 1993). Nevertheless, this mechanism does provide a basis for the aetiology of more minor genetic defects arising postzygotically which, if they arose in early cleavage and were compatible with viability and further division, could theoretically contribute substantially to the developing embryo.

IV ORIGIN OF SINGLE-GENE DEFECTS

Compared with chromosomal abnormalities, there is very little information on the origin and incidence of single-gene defects (Chandley 1991). Approaching 5000 single-gene defects have been identified through analysis of syndromes with Mendelian inheritance patterns (McKusick 1992) but there is no reason to suppose that mutations do not occur with similar frequencies in the vast majority of other genes, most of which have yet to be mapped and identified. A proportion of these mutations would be expected to affect the function of essential housekeeping genes and may therefore be lethal either in preimplantation development following the initiation of embryonic gene transcription and loss of maternally inherited products or in postimplantation development.

Most *de novo* mutations causing inherited disease are paternal in origin. This is thought to be related to differences between spermatogenesis and oogenesis. Before gametogenesis, the numbers of primordial germ cells increase by mitosis as they migrate to the gonad. In the female, this mitotic expansion continues until late in gestation when oogonia enter meiosis and arrest at the dictyate stage of meiosis I until menstrual cycles are initiated at puberty. In the male, by contrast, mitotic division of spermatogonia continues throughout life. Thus the number of mitotic divisions preceding gametogenesis is much greater in the male, increases with age, and may increase the risk of replication errors.

A newly discovered class of mutations involving the unstable expansion of triplet repeats is now known to cause a number of mainly dominant neurodegenerative diseases. These include fragile X syndrome (the most common from of inherited mental retardation), myotonic dystrophy, and Huntington's disease. Each of these diseases is characterized by increasing severity and earlier onset in successive generations, a phenomenon known as anticipation. The molecular basis for this appears to be related to the length of the triplet repeats involved in each case. Normal individuals have stable repeats with alleles falling within a certain range. In rare cases, however, it appears that the repeat expands to the premutation range. Repeats in the premutation range are unstable and carriers are at high risk of transmitting the full mutation, in which the number of repeats is further increased sometimes dramatically, to their offspring. Fragile X syndrome, for example, is caused by expansion of an unstable CGG repeat close to a CpG island in the 5' untranslated of the *FMR1* gene on the X chromosome associated with the fragile site at Xp27.3 (Verkerk *et al.* 1991). Expansion of the repeat causes methylation of the CpG island and transcriptional inactivation.

The mechanism responsible for expansion of these repeats is not known but could involve the processes of DNA recombination, replication, or repair. Interestingly, there is now considerable indirect evidence that the expansion occurs either during meiosis or at a postzygotic stage during mitosis in early development possible after segregation of the germline. This evidence includes mosaicism of repeat length in the tissues of carriers and affected individuals, differences between sibs but similarity in identical twins, absence of the full mutation in sperm (Reyniers *et al.* 1993), and apparent mitotic stability of adult cells (Wöhrle *et al.* 1993). In the mouse, a tetranucleotide repeat analogous to the trinucleotide repeat causing myotonic dystrophy is susceptible to expansion during early cleavage (Gibbs *et al.* 1993).

V PREIMPLANTATION DIAGNOSIS OF GENETIC DEFECTS

1 Early approaches

It might be thought that, with the availability of prenatal diagnosis in the first or second trimester of pregnancy, there was no need to develop techniques for diagnosis of genetic defects at the preimplantation stage. There are, however, several groups of patients for whom preimplantation diagnosis is preferable to that carried out prenatally, when induced abortion is implicit if the diagnosis is unfavourable. The first group comprises couples at risk of passing on X-linked disorders for which there is, as yet, no specific molecular diagnosis. At present, the only alternative procedure on offer is prenatal sex determination and selective abortion of all male fetuses, of which half will be unaffected. Secondly, there are the couples who have had to suffer repeated terminations of pregnancy, with or without a specific diagnosis (Delhanty et al. 1993). Finally, there are those who are morally opposed to termination of an established pregnancy but are prepared to consider selection of embryos at the preimplantation stage. The wish to help all three groups of families led to the initiation of research towards preimplantation genetic diagnosis in the mid-1980s in the UK. The approaches available are first polar body analysis and biopsy at the blastocyst or cleavage stage; the latter has found the most favour (Handyside 1994). Work at the Hammersmith Hospital in London showed that biopsy of cleavage-stage embryos at about the eight-cell stage on day three and removal of one or two cells does not adversely affect development to the blastocyst stage (Hardy et al. 1990). The approach decided upon therefore was to biopsy normally developing embryos early on day three after insemination, by which time most would have reached the eight-cell stage. The aim was to carry out genetic diagnosis within one working day to allow selective transfer of embryos to take place in the evening of day three, to avoid the need for cryopreservation of the embryo.

The earliest application was in the avoidance of X-linked disease in the offspring of carrier mothers by establishing the sex of *in vitro* fertilized embryos and selecting females for subsequent transfer to the mother. Technically, there were two possible strategies for determining the sex: polymerase chain reaction (PCR) amplification of DNA from the biopsied blastomeres, using primers specific for a sequence derived from the Y chromosome, or fixing the nuclei and spreading on slides for non-radioactive hybridization *in situ* again using a Y-specific DNA probe. Although non-isotopic detection of a biotinylated Y-specific probe using a streptavidin-linked alkaline phosphatase-based detection system had been achieved in less than a day, the efficiency of hybridization with human embryonic nuclei was relatively low (Penketh *et al.* 1989). It was therefore decided to proceed with the PCR-based approach to embryo sexing in the first clinical application.

PCR-based embryo sexing was achieved using a Y-specific centromeric alphoid and long arm sequences, pregnancies were established and several normal girls were born (Handyside *et al.* 1990). Following counselling about the risks of misdiagnosis, couples were urged to allow confirmation of the diagnosis by prenatal testing. Of seven fetuses tested following sexing by PCR amplification, one singleton was male and the couple elected to terminate the pregnancy (Handyside and Delhanty 1993). The reason for the misdiagnosis was presumed to be amplification failure.

By 1988, FISH had become established (Pinkel *et al.* 1986) and it was decided to adapt the technique for use with single human blastomeres. Initial results with Y-or X-specific repetitive DNA probes were encouraging (Griffin *et al.*1991). However, each probe used alone had disadvantages. Hybridization failure with a Y-derived probe could lead to misdiagnosis as with PCR, but with the X-probe the inadvertent sampling of tetraploid cells could again lead to the classification of a male, XXYY cell as female. To ensure reliability, a dual FISH technique was developed, with simultaneous hybridization of a biotinylated X probe and two digoxigenin labelled Y probes, to guard against Y hybridization failure (Fig. 5.1) (Griffin *et al.* 1992).

2 The application of dual FISH for sex determination

The dual FISH technique for embryo sexing has been successfully applied clinically. In all, at the Hammersmith Hospital, 18 couples at risk for X-linked disease have undergone treatment in a total of 27 IVF cycles. Seven clinical pregnancies were established, of which two miscarried. Seven normal babies have resulted from five births and all babies were female, as predicted. In all cases, no more than two embryos were transferred, so the two sets of twins each resulted from the transfer of two biopsied embryos (Griffin *et al.* 1994).

To control for amplification failure in the use of PCR for embryo sexing, strategies have been developed that employ sequences present on both X and Y, such as the homologous ZFX and ZFY, to guard against independent failure of amplification from different primer pairs (Chong *et al.* 1993). This approach is highly efficient at correctly assigning the sex of single-cell samples from a variety of somatic cells and from blastomeres. However, in addition to amplification failure, there are two drawbacks when amplifying DNA from single nuclei for the purposes of sex determination. The first is a general one, that the risk of amplifying contaminating DNA is high. The second is that the detection of an amplified band only confirms the presence

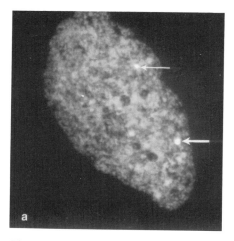 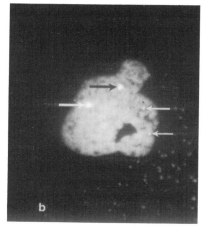

Fig. 5.1 (a) Blastomere nucleus after dual FISH with X and Y chromosome-specific probes. *Broad arrow*, large Y signal; *small arrow*, X signal. (b) Presumed tetraploid blastomere nucleus after dual FISH, showing two Y signals (*broad arrows*) and two X signals (*small arrows*). (Kindly supplied by Dr D. K. Griffin.)

of an X or Y chromosome and gives no information on copy number. The importance of this is illustrated by a series of five couples who underwent preimplantation diagnosis by dual FISH for the avoidance of X-linked disease (Delhanty *et al.* 1993). In two cases, two female embryos were transferred and one singleton pregnancy was established, resulting in the birth of a normal girl. In the third case, all eight embryos were of very poor quality and no female diagnosis was possible. In the remaining two cases, no embryos were transferred, owing to the detection of an abnormal number of X chromosome signals in the female embryos. Investigation of the biopsied embryos that were not transferred revealed evidence of X chromosome mitotic non-disjunction in one and complete X monosomy in a second. While the first of these could have led to the production of a fetus mosaic for X trisomy or monosomy, the second was potentially more damaging as, had it survived, the X monosomic fetus would not only have developed Turner's syndrome but would have been at high risk of inheriting Duchenne muscular dystrophy from the carrier mother (Delhanty *et al.* 1993). In this situation, the use of FISH rather than PCR allowed the detection of abnormal copy numbers of X chromosomes and prevented the transfer of potentially abnormal zygotes. An additional advantage is that the risk of contamination with extraneous cellular matter is minimal with FISH, as the nucleus is observed at all stages of the process, from initial spreading to final analysis.

3 Specific diagnosis of single-gene defects

Two main strategies for preimplantation diagnosis by DNA amplification are available. The first involves analysis of one or more closely linked polymorphic markers that are suitable for PCR amplification. In this case, pedigree analysis has to be completed beforehand to establish that a particular marker is informative for the couple concerned. The second strategy involves amplifying a DNA fragment containing the gene defect itself. This is only possible if the gene has been cloned and detailed sequence information is available. Since in the case of single-gene defects, each target sequence will only be present twice in the DNA of a single blastomere nucleus, a high efficiency of amplification is required for preimplantation diagnosis. It is also important to ensure that diagnosis is successful in most of the embryos tested in order to maximize the chance of identifying unaffected embryos morphologically suitable for transfer from the limited number available.

Cystic fibrosis (CF) is the most common, severe, autosomal recessive disorder affecting the Caucasian population. Preimplantation diagnosis of CF, using nested primer sets to improve the yield of DNA and the reliability of detection, was attempted initially in three couples in which both partners carried the common $\triangle$F508 mutation. Unaffected embryos were diagnosed in two cases allowing a two-embryo transfer in both; one woman became pregnant and gave birth to a healthy girl, free of the mutation (Handyside *et al.* 1992).

Lesch–Nyhan syndrome is a severe neurological condition caused by deficiency of the X-linked linked enzyme hypoxanthine phosphoribosyl transferase and affected males generally die in their teens or early 20s. Unlike CF, the mutations causing this disease are heterogeneous. Therefore, specific diagnosis involves sequencing the gene, identifying the mutation in women carriers and developing appropriate amplification and detection methods for single-cell analysis in each case. Nevertheless, this has been attempted in two at-risk couples and a healthy girl has been delivered (Hughes *et al.* unpublished).

4 The potential for misdiagnosis

As mentioned previously, the main sources of error in the case of PCR based single-gene analysis are contaminating extraneous DNA and amplification failure. Stringent precautions are necessary to guard against contamination with normal DNA that could give rise to a spurious normal result. In the case of recessive defects, amplification failure on its own would not result in the transfer of an undiagnosed affected fetus, but would reduce the efficiency of diagnosis. However, for dominant single-gene disorders, failure to amplify one allele is a potential hazard, as it is the heterozygote that is affected and failure of the mutant allele to amplify could lead to misdiagnosis of an affected embryos as unaffected.

Chromosomal mosaicism is another potential source of error. The evidence from the application of dual FISH to embryos both normally fertilized and showing normal development shows that the incidence is not negligible (Delhanty *et al.* 1993; Munné *et al.* 1993; Harper *et al.* 1995). As with amplification failure, the most serious effects would be for single-gene dominant disorders rather than for autosomal or X-linked recessives. For recessive conditions, inadvertent sampling of a minor cell line in the embryo that was monosomic for a particular chromosome (or haploid for all chromosomes) could lead to the classification of a heterozygous unaffected embryo (*wt/m*) as one of either homozygous type (*wt/wt* or *m/m*), but would not lead to the mistaken transfer of an affected embryo since *m/−* embryos would still be classed as mutant. However, in the case of dominant conditions, either autosomal or X-linked, absence of the second homologous chromosome in the biopsied cell could lead to failure to diagnose an affected embryo (Table 5.6). Obviously, the risk of error is much higher if mosaicism involving haploidy or multiple monosomy is more frequent than that due to mitotic non-disjunction of a particular chromosome, where the chances of the non-disjoined chromosome being the chromosome that carries the gene are a *priori* only 1 in 23. Future studies with multicolour FISH on normal embryos should help to clarify the exact situation. Mosaicism in cleavage-stage embryos is also an important consideration when attempting to diagnose chromosome copy number by FISH. As previously described, evidence for chromosomal mosaicism has been obtained by this method. The effects of mosaicism on diagnosis will vary depending on whether the embryo was initially trisomic or normal (disomic) for the chromosome under study (Table 5.7).

At present, to counter problems due to allele-specific amplification failure or mosaicism of various types it would seem prudent to use two independent biopsied cells for the preimplantation diagnosis of single-gene or chromosomal defects.

5 Future developments

Research is currently focused on methods of diagnosing chromosome imbalance and on improving the specificity of single-gene diagnosis and the range of disorders amenable to single-cell diagnosis. Owing to the frequency of occurence of chromosomal disorders, both in high-risk couples and in older women, methods of reliably diagnosing abnormal chromosome copy number are urgently needed. Centromeric alphoid repeat DNA probes are now available for the majority of human chromosomes and most of these allow the detection of chromosome copy number in interphase when used as probes in the FISH technique (Pinkel *et al.* 1986). It is

Table 5.6 *The effect of monosomic or haploid chromosomal mosaicism on the single-cell diagnosis of dominant single-gene defects in the preimplantation embryo*

Inheritance	Embryo	Genotype	Minor cell line	Genotype
Autosomal dominant	Disomic (chromosomally normal)	wt/m (affected)	Monosomic for chromosome carrying the gene	|wt| or m
X-linked dominant	Disomic	wt/m (female affected) m (male affected)	X-chromosome monosomic X or Y chromosome monosomic	|wt| or m no result or m

wt: wild type (normal) allele, m: mutant allele; |wt| : misdiagnosis of an affected embryo as unaffected.

Table 5.7 *The effect of chromosomal mosaicism on the single-cell diagnosis of chromosome copy number in the preimplantation embryo*

Embryo	FISH signals per nucleus	Minor cell line	FISH signals per nucleus
Disomic	● ●	Monosomic or haploid	●
		Trisomic or triploid	● ● ●
Trisomic	● ● ●	Disomic or haploid	●● or ●
			● ● ● ● or
		Tetrasomic or tetraploid	● ● ● ● ● ●

Inadvertent sampling of the minor cell line would always lead to misdiagnosis, but this would be serious only in the case of a trisomic embryo with a disomic (or haploid) cell line, ● ●

unfortunate that centromeric homology between chromosome pairs 13/21 and 14/22 means that alphoid probes derived from these chromosomes cross-hybridize; the 13/21 alphoid probe, for example, gives four signals in a normal cell, making detection of the corresponding trisomies more difficult. The amount of alphoid sequence DNA present at the centromere also varies between individuals, so that in some samples one or more of the chromosomes may not hybridize to the probe, reducing the overall efficiency. Since trisomy 21 (causing Down syndrome) is a major risk, research is ongoing to find alternative probes that will give a recognizable signal in interphase nuclei. The alphoid centromeric probes are applicable in other trisomies common in early development, such as those for chromosomes 16 and 18 (Fig. 5.2) (Schrurs *et al.* 1993). However, in older mothers and those who have had frequent IVF failure, the risk is of trisomy in general, rather than for a specific chromosome. Since one or, at most, two cells only will be available for diagnosis in the preimplantation embryo, a choice of chromosomes for which to screen has to be made. It has already been demonstrated that multicolour FISH can be used to detect chromosomes X, Y, 18, and 13/21 (Munné *et al.* 1993), and if a probe for chromosome 21 of improved reliability can be developed this will be of considerable benefit. In theory, the application of digital imaging software to assign pseudo colours to signals generated by probes labelled with varying ratios of red and green fluorochromes will improve the range and number of chromosomes that can be monitored (Reid *et al.* 1992). It remains to be seen how efficiently such techniques can be applied to nuclei of single human blastomeres.

Many couples request preimplantation diagnosis because one parent carries a chromosomal translocation. These couples are high-risk cases and the specific chromosomes involved are known, allowing a more efficient approach to the determination of chromosome imblance. However, in the majority of these cases, centromeric probes will not be applicable, as the major risk of chromosome imbalance is usually due to adjacent I disjunction at meiotic anaphase I (homologous centromeres separate in gamete formation). The chromosome imbalance in the resulting embryo will then affect only segments of the chromosome arms distal to the translocation breakpoints. To detect this, probes that map to the chromosome regions involved and that are suitable for interphase detection must be sought. At present, the most likely candidates seem to be yeast artificial chromosomes that contain a large human DNA insert derived from the relevant chromosome region. A proportion of these will give a reliably detectable signal in interphase nuclei when used in FISH (Fig. 5.3). One problem remaining to be solved in the case of all interphase cytogenetic studies is that of overlying signals. It is thought that failure to detect the correct number of signals in up to 5 per cent of a population of apparently normal diploid cells is due to the relevant regions of the two chromosomes overlying one another when the chromatin is extended in interphase. Ideally, to circumvent this problem, two probes for the chromosome or region should be used simultaneously in dual colours, but this may not always be practicable.

An alternative approach to the detection of chromosome imbalance in an embryo due to a parental translocation carrier is to attempt to obtain one or more chromosome spreads from the biopsied cells. Even if of poor

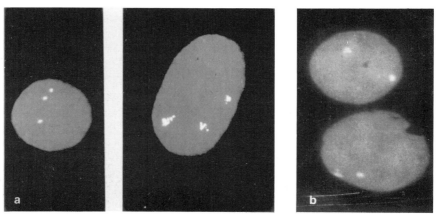

Fig. 5.2 (a) Nuclei from an uncultured chorionic villus sample from a trisomy 18 pregnancy after FISH with the DNA probe specific for chromosome 18; three signals seen in each nucleus. (b) Human blastomere nuclei after FISH with the 18 chromosome-specific probe; the normal two signals are present in each. (Kindly supplied by Dr D. K. Griffin.)

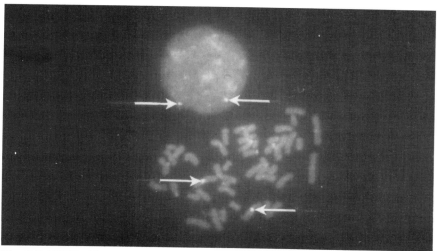

Fig. 5.3 FISH with a yeast artificial chromosome probe to interphase nuclei from human lymphocytes. Two hybridization signals are clearly visible in each nucleus. (Kindly supplied by Dr D. Wells.)

quality, it may be possible to determine the segregation of the chromosomes in the parental meiosis by the hybridization of fluorescently labelled chromosome-specific probe sets to the metaphase spread—a process known as chromosome painting, since the whole chromosome is made to fluoresce (Cremer *et al.* 1990) (Fig. 5.4).

A new technique called primer extension preamplification, in which whole genomic DNA is preamplified prior to the amplification of specific sequences in a normal PCR reaction, is applicable to single-cell diagnosis of single-gene defects (Zhang *et al.* 1992; Xu *et al.* 1993; Kristjansson *et al.* 1994). With the limited amount of DNA normally available from one or two blastomere nuclei, it is practically impossible to detect more than two affected loci using standard PCR protocols and no repeat assays can be performed to confirm results. Whole genome preamplification can generate sufficient template to perform multiple DNA analyses using PCR. Studies suggest that preamplification can be successfully applied to single human blastomeres allowing for the simultaneous detection of approximately 20 DNA sequences (Xu *et al.* 1993). In a separate investigation, the simultaneous analysis of commonly deleted exons of the dystrophin gene and the ZFX/ZFY loci in single cells derived from a variety of sources was achieved following initial whole genome amplification (Kristjansson *et al.* 1994). The 93 per cent diagnostic accuracy of the procedure in single lymphoblasts from males affected with Duchenne muscular dystrophy demonstrates its applicability to preimplantation diagnosis of this condition.

In theory, primer extension preamplification also opens the way to screening the whole karyotype for chromosome imbalance. This could be attempted by combining it with the technique of comparative genomic hybridization (Kallioniemi *et al.* 1992). Comparative genomic hybridization is used to test for genomic imbalance principally in tumour samples, where manyfold amplification of certain sequences has occurred. DNA from the test sample is fluorescently labelled and then co-hybridized with normal DNA labelled with a different fluorochrome on to normal metaphase spreads in a FISH experiment. The ratio of the hybridization of the two fluorochromes to the chromosomes is a measure of the increased or decreased copy number of the particular chromosome segment in the test sample DNA. Digital imaging software is normally required for the analysis. Results suggest that for single-copy amplification, that is one whole chromosome extra in a diploid set, the lower size limit of resolution is approximately that of the smallest human chromosome, number 21. The theory is that one could biopsy one or two blastomeres from an embryo, prepare DNA and amplify the whole genome using primer extension

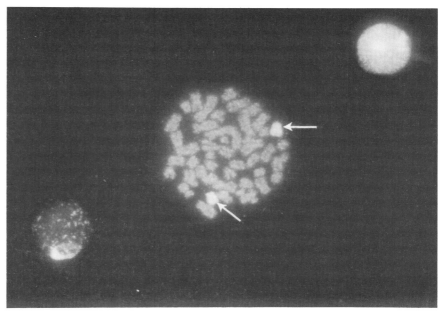

Fig. 5.4 Human lymphocyte metaphase chromosomes after FISH with a DNA probe set for chromosome 12. This procedure causes the whole chromosome to fluoresce. (Kindly supplied by Dr D. K. Griffin.)

preamplification, label this with green fluorochrome, label normal DNA with red, co-hybridize to normal metaphases by FISH and screen the whole karyotype simultaneously for any imbalance.

ACKNOWLEDGEMENTS

We gratefully acknowledge support from the Wellcome Trust for the research involving the use of FISH for diagnosis of chromosomal abnormalities in the human preimplantation embryo.

REFERENCES

Angell, R. R., Xian, J., Keith, J., Ledger, W., and Baird, D. T. (1994). First meiotic division abnormalities in human oocytes: mechanism of trisomy formation. *Cytogenetics and Cell Genetics* **65**, 194–202.

Antonarakis, S. E. and the Down Syndrome Collaborative Group. (1991). Parental origin of the extra chromosome in trisomy 21 as indicated by analysis of DNA polymorphisms. *New England Journal of Medicine*, **324**, 872–6.

Antonarakis, S. E., Petersen, M.B., McInnis, M. G., Adelsberger, P. A., Schinzel, A. A., Binkert, F., *et al.* (1992). The meiotic stage of nondisjunction in trisomy 21: determination by using DNA polymorphism. *American Journal of Human Genetics*, **50**, 544–50.

Antonarakis, S. E., Avramopoulos, D., Blouin, J-L., Talbot, C. C., Jr., and Schinzel., A. A. (1993). Mitotic errors in somatic cells cause trisomy 21 in about 4.5% of cases and are not associated with advanced maternal age. *Nature Genetics*, **3**, 146–50.

Chandley, A. C. (1991). On the parental origin of *de novo* mutation in man. *Journal of Medical Genetics*, **28**, 217–23.

Chong, S. S., Cota, J., Hardikar, S. D., Handyside, A. H., and Hughes, M. R. (1993). Preimplantation diagnosis of X-linked disease: reliable and rapid sex determination of single human cells by restriction site analysis of simultaneously amplified ZFX and ZFY sequences. *Human Molecular Genetics*, **2**, 1187–91.

Cremer, T., Popp, S., Emmerich, P., Lichter, P., and Cremer, C. (1990). Rapid metaphase and interphase detection of radiation-induced chromosome aberrations in human lymphocytes by chromosomal suppression *in-situ* hybridization. *Cytometry*, **11**, 110–8.

Delhanty, J. D. A., Griffin, D. K., Handyside, A. H., Harper, J., Atkinson, G. H. G., Pieters, M. H. E. C., *et al.* (1993). Detection of aneuploidy and chromosomal mosaicism in human embryos during preimplantation sex determination by fluorescent *in situ* hybridisation (FISH). *Human Molecular Genetics*, **2**, 1183–5.

Eiben, B., Bartels, I., Bahr-Porsch, S., Borgmann, S., Gatz, G., Gellert, G., *et al.* (1990). Cytogenetic analysis of 750 spontaneous abortions with the direct-preparation method of chorionic villi and its implications for studying genetic causes of pregnancy wastage. *American Journal of Human Genetics*, **47**, 656–63.

Fisher, J. M., Harvey, J. F., Lindenbaum, R. H., Boyd, P. A., and Jacobs, P. A. (1993). Molecular studies of trisomy 18. *American Journal of Human Genetics*, **52**, 1139–44.

Gibbs, M., Collick, A., Kelly, R. G., and Jeffreys, A. J. (1993). A tetranucleotide repeat mouse minisatellite displaying substantial instability during early preimplantation development. *Genomics*, **17**, 121–8.

Griffin, D. K., Handyside, A. H., Penketh, R. J., Winston, R. M., and Delhanty, J. D. (1991). Fluorescent *in-situ* hybridization to interphase nuclei of human preimplantation embryos with X and Y chromosome specific probes. *Human Reproduction*, **6**, 101–5.

Griffin, D. K., Wilton, L. J., Handyside, A. H., Winston, R. M., and Delhanty, J. D. (1992). Dual fluorescent *in situ* hybridisation for simultaneous detection of X and Y chromosome-specific probes for the sexing of human preimplantation embryonic nuclei. *Human Genetics*, **89**, 18–22.

Griffin, D. K., Handyside, A. H., Harper, J., Wilton, L. J., Atkinson, H. G., Soussis, I., *et al.* (1994). Clinical experience with preimplantation diagnosis of sex by dual fluorescent *in situ* hybridization. *Journal of Assisted Reproduction and Genetics*, **11**, 132–43.

Handyside, A. H. (1994). Genetic defects in the human preimplantation embryo and the diagnosis of inherited disease. In *The biological basis of early human reproductive failure* (ed. J. Van Blerkom), pp. 345–74. Oxford University Press, New York.

Handyside, A. H., Kontogianni, E. H., Hardy, K., and Winston, R. M. (1990). Pregnancies from biopsied human preimplantation embryos sexed by Y-specific DNA amplification. *Nature*, **344**, 768–70.

Handyside, A. H., Lesko, J. G., Tarin, J. J., Winston, R. M., and Hughes, M. R. (1992). Birth of a normal girl after *in vitro* fertilization and preimplantation diagnostic testing for cystic fibrosis. *New England Journal of Medicine*, **327**, 905–9.

Handyside, A. H. and Delhanty, J. D. A. (1993). Cleavage stage biopsy of human embryos and diagnosis of X-linked recessive disease. In *Preimplantation diagnosis of human genetic disease*, (ed. R. G. Edwards), pp. 239–70. Cambridge University Press.

Hardy, K., Martin, K. L., Leese, H. J., Winston, R. M., and Handyside, A. H. (1990). Human preimplantation development *in vitro* is not adversely affected by biopsy at the 8-cell stage. *Human Reproduction*, **5**, 708–14.

Hardy, K. (1993). Development of human blastocysts in vitro. In *Preimplantation embryo development*, (ed. B. Bavister), pp. 184–99. Springer-Verlag, New York.

Hardy, K., Winston, R. M. L., and Handyside, A. H. (1993). Binucleate cells in human preimplantation embryos *in vitro*: failure of cytokinesis during cleavage. *Journal of Reproduction and Fertility*, **98**, 549–58.

Harper, J. C., Coonen, E., Handyside, A. H., Winston, R. M. L., Hopman, A. H. H., and Delhanty, J. D. A. (1995). Mosaicism of autosomes and sex chromosomes in morphologically normal, monospermic preimplantation human embryos. *Prenatal Diagnosis*, **15**, 31–40.

Hartwell, L. H. and Weinert, E. A. (1989) Checkpoints: controls that ensure the order of cell cycle events. *Science*, **246**, 629–34.

Hassold, R. and Chiu, D. (1985). Maternal age-specific rates of numerical chromosome abnormalities with special reference to trisomy. *Human Genetics*, **70**, 11–7.

Hassold, T. J., (1986). Chromosome abnormalities in human reproductive wastage. *Trends in Genetics*, **2**, 105–10.

Hassold, T., Jacobs, P. A., Leppert, M., and Sheldon, M. (1987). Cytogenetic and molecular studies of trisomy 13. *Journal of Medical Genetics*, **24**, 725–32.

Hassold, T., Benham, F., and Leppert, M. (1988). Cytogenetic and molecular analysis of sex chromosome monosomy. *American Journal of Human Genetics*, **42**, 534–41.

Hassold, T. J., Sherman, S. L., Pettay, D., Page, D. C., and Jacobs, P. A. (1991). XY chromosome nondisjunction in man is associated with diminished recombination in the pseudoautosomal region. *American Journal of Human Genetics*, **49**, 253–60.

Jacobs, P. A. (1992). The chromosome complement of human gametes. *Oxford Reviews of Reproductive Biology*, **14**, 47–72.

Jacobs, P. A., Hassold, T. J., Whittington, E., Butter, G., Collyer, S., Keston, M., et al. (1988). Klinefelter's syndrome: an analysis of the origin of the additional sex chromosome using molecular probes. *Annals of Human Genetics*, **52**, 93–109.

Jacobs., P. A., Angell, R. R., Buchanan, I. M., Hassold, T. J., Matsuyama, A. M., and Manuel, B. (1978). The origin of human triploids. *Annals of Human Genetics*, **42**, 49.

Kallioniemi, A., Kallioniemi, O-P., Suder, D., Rutovitz, D., Gray, J. W., Waldman, F., et al. (1992). Comparative genomic hybridization for molecular cytogenetic analysis of solid tumours. *Science*, **258**, 818–21.

Kristjansson, K., Chong, S. S., Van den Veyver, I. B., Subramanian, S., Snabes, M. C., and Hughes, M. R. (1994). Preimplantation single cell analyses of dystrophin gene deletions using whole genome amplification. *Nature Genetics*, **6**, 19–23.

Lorda-Sanchez, I., Binkert, F., Maechler, M., Robinsn, W. P., and Schinzel, A. (1992). Reduced recombination and paternal age effect in Klinefelter syndrome. *Human Genetics*, **89**, 524–30.

Macdonald, M., Hassold, T., Harvey, J., Wang, L. H., Morton, N. E., and Jacobs, P. A. (1994). The origin of 47, XXY and 47, XXX aneuploidy: hetrogeneous mechanisms and role of aberrant recombination. *Human Molecular Genetics*, **3**, 1365–71.

Martin, R. H., Chan, K., Ko., E., and Rademaker, A. W. (1994). Detection of aneuploidy in human sperm by fluorescence *in situ* hybridization (FISH): different frequencies in fresh and stored sperm nuclei. *Cytogenetics and Cell Genetics*, **65**, 95–6.

May, K. M., Jacobs, P. A., Lee, M., Ratcliffe, S., Robinson, A., Nielsen, J., and Hassold, T. J. (1990). The parental origin of the extra X chromosome in 47, XXX females. *American Journal of Human Genetics*, **46**, 754–61.

McKusick, V. A. (1992). *Mendelian inheritance in man.* Johns Hopkins University Press, Baltimore.

Munné, S., Lee, A., Rosenwaks, Z., Grifo, J., and Cohen, J. (1993). Diagnosis of major chromosome aneuploidies in human preimplantation embryos. *Human Reproduction*, **8**, 2185–91.

Murray, A. W. (1992). Creative blocks: cell-cycle checkpoints and feedback controls. *Nature*, **359**, 599–604.

O'Farrell, P. H., Edgar, B. A., Lakich, D., and Lehner, C. F. (1989). Directing cell division during development. *Science*, **246**, 635–40.

Penketh, R. J., Delhanty, J. D., van den Berghe, J. A., Finklestone, E. M., Handyside, A. H., Malcolm, S., et al. (1989). Rapid sexing of human embryos by non-radioactive *in situ* hybridization: potential for preimplantation diagnosis of X-linked disorders. *Prenatal Diagnosis*, **9**, 489–99.

Petersen, M. A., Antonarakis, S. E., Hassold, T. J., Freeman, S. B., Sherman, S. L., Avramopoulos, D., et al. (1993). Paternal nondisjunction in trisomy 21: excess of male patients. *Human Molecular Genetics*, **2**, 1691–5.

Pinkel, D., Straume, T., and Gray, J. M. (1986). Cytogenetic analysis using quantitative high sensitivity, fluorescent hybridization. *Proceedings of the National Academy of Sciences of the USA*, **83**, 2934–8.

Reid, T., Baldini, A., Rand, T. C., and Ward, D. C. (1992). Simultaneous visualization of seven different DNA probes by *in situ* hybridization using combinatorial fluorescence and digital imaging microscopy. *Proceedings of the National Academy of Sciences of the USA*, **89**, 1388–92.

Reyniers, E., Vits, K., De Boulle, K., Van Roy, B., Van Velzen, D., de Graaff, E., (1993). The full mutation in the FMR-1 gene of male fragile X patients is absent in their sperm. *Nature Genetics*, **4**, 143–6.

Schrurs, B., Winston, R. M. L., and Handyside, A. H. (1993). Preimplantation diagnosis of aneuploidy by fluorescent *in situ* hybridization: evaluation using a chromosome 18 specific probe. *Human Reproduction*, **8**, 296–301.

Sensi, A., and Ricci, N. (1993). Mitotic errors in trisomy 21. *Nature Genetics*, **5**, 215.

Sherman, S. L., Takaesu, N., Freeman, S. B., Grantham, M., Phillips, C., Blackston, R. D., et al. (1991). Trisomy 21: association between reduced recombination and nondisjunction. *American Journal of Human Genetics*, **49**, 608–20.

Strom, C. M., Ginsberg, N, Applebaum, M., Bozorgi, N., White, M., Caffarelli, M., and Verlinsky, Y. (1992). Analyses of 95 first-trimester spontaneous abortions by chorionic villus sampling and karyotype. *Journal of Assisted Reproduction and Genetics*, **9**, 458–61.

Verkerk, A. J., Pieretti, M., Sutcliffe, J. S., Fu, Y. H., Kuhl, D. P., Pizzuti, A., et al. (1991). Identification of a gene (*FMR-1*) containing a CGG repeat coincident with a breakpoint cluster region exhibiting length variation in fragile X syndrome. *Cell*, **65**, 905–14.

Wilcox, A. J., Weinberg, C. R., O'Connor, J. F., Baird, D. D., Schlatterer, J. P., Canfield, R. E., et al. (1988). Incidence of early loss of pregnancy. *New England Journal of Medicine*, **319**, 189–94.

Williams, B. J., Ballenger, C. A., Malter, H. E., Bishop, F., Tucker, M., Zwuingaman, T. A., et al. (1993). Non-disjunction in human sperm: results of fluorescence *in situ* hybridization studies using two and three probes. *Human Molecular Genetics*, **2**, 1929–36.

Winston, N. J., Braude, P. R., Pickering, S. J., George, M. A., Cant, A., Currie, J., et al. (1991). The incidence of abnormal morphology and nucleocytoplasmic ratios in 2-, 3- and 5-day human pre-embryos. *Human Reproduction*, **6**, 17–24.

Wöhrle, D., Henning, I., Vogel, W., and Steinbach, P. (1993) Mitotic stability of fragile X mutations in differentiated cells indicates early post-conceptional trinucloetide repeat expansion. *Nature Genetics*, **4**, 140–2.

Xu, K. P., Tang, Y. X., Grifo, J. A., Rosenwaks. Z., and Cohen. J. (1993). Primer extension preamplification for detection of multiple genetic loci from single human blastomeres. *Human Reproduction*, **8**, 2206–10.

Zenzes, M. T. and Casper R. F. (1992). Cytogenetics of human oocytes, zygotes, and embryos after in vitro fertilization. *Human Genetics*, **88**, 367–75.

Zenzes, M. T., Wang, P., and Casper, R. F. (1992). Chromosome status of untransferred (spare) embryos and probability of pregnancy after *in-vitro* fertilisation. *Lancet*, **340**, 391–4.

Zhang, L., Cui, X., Schmitt, K., Hubert, R., Navidi, W., and Arnheim, N. (1992). Whole genome amplification from a single cell: implications for genetic analysis. *Proceedings of the National Academy of Sciences of the USA*, **89**, 5847–51.

6 Regulation of pituitary gonadotrophin gene expression

JULIE E. MERCER and WILLIAM W. CHIN

I **Introduction**
  1 Structures of LH and FSH
  2 Cellular localization of gonadotrophins
  3 Current questions

II **Regulation of the gonadotrophins by GnRH**
  1 Role of pulsatile GnRH
    i *In vivo* studies
    ii *In vitro* studies
    iii Genetic elements
  2 The GnRH receptor

III **Regulation of gonadotrophins by steroid hormones**
  1 Oestrogen
    i Negative-feedback effects
    ii Positive-feedback effects
  2 Progesterone
  3 Testosterone
    i *In vivo* studies
    ii *In vitro* studies

IV **Regulation of the gonadotrophins by gonadal peptides**
  1 Inhibin and activin
    i FSHβ mRNA regulation
    ii Inhibin, activin, and LH secretion
  2 Follistatin
    i FSHβ mRNA regulation
  3 Activin receptor
  4 Paracrine effects

V **New horizons**
  1 Cell Lines
  2 Transgenic animal models

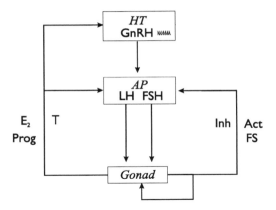

Fig. 6.1 Hypothalamo–pituitary–gonadal axis. GnRH is released from the hypothalamus (HT) in a pulsatile fashion to stimulate the secretion of LH and FSH from the anterior pituitary gland (AP). LH and FSH act, in turn, on the gonads to stimulate the production of steroid hormones, oestrogen (E_2), progesterone (Prog), and testosterone (T). The steroid hormones and the gonadal peptide hormones inhibin (Inh), activin (Act), and follistatin (FS), in turn, regulate the hypothalamo–pituitary unit.

I INTRODUCTION

The pituitary gonadotrophins luteinizing hormone (LH) and follicle-stimulating hormone (FSH) are vital to reproductive function in most vertebrates. They are secreted from the anterior pituitary gland under the control of gonadotrophin-releasing hormone (GnRH) which is secreted in a pulsatile manner from the median eminence of the hypothalamus and stimulates the pituitary gonadotroph cell to synthesize and release LH and FSH (Schally et al. 1972; Schally 1978; Clarke and Cummins 1984). LH and FSH act, in turn, on the gonads. The hypothalamo–pituitary– gonadal axis operates to fine tune the reproductive mechanisms of the organism (see Fig. 6.1).

The gonadotrophins bind to their receptors in the gonads and exert a trophic effect on the testis and ovary to direct steroidogenesis and gametogenesis. LH and FSH have distinct but complementary actions to control maturation and development of the gonads and germ cells. FSH acts on the granulosa cells of the antral follicle to stimulate oestrogen and LH receptor synthesis and the development of the follicle, and on the Sertoli cells of the testis to stimulate maturation of sperm. LH, on the other hand, acts on the granulosa cells of the pre-ovulatory follicle to activate progesterone production and to trigger ovulation, and on the

Leydig cells in the testis and the theca cells of the ovarian follicle to promote synthesis of androgens (Catt and Dufau 1978; Henderson 1979; Lincoln 1979; Means et al. 1980; Richards 1980). The gonadal steroid hormones oestrogen, progesterone, and testosterone have feedback effects on the hypothalamo–pituitary unit to regulate the release of gonadotrophins from the pituitary, either by acting on the hypothalamus to modulate production of GnRH, or by acting directly on the pituitary gland to regulate synthesis and secretion of the gonadotrophins (Knobil 1974; Karsch 1987).

In addition to the gonadal steroids, LH and FSH regulate the synthesis of the peptide hormones inhibin, activin, and follistatin in the gonads. These peptides also have selective regulatory effects on FSH secretion from the pituitary (Woodruff et al. 1988; Ying 1988). The genes for the receptors for the gonadotrophins have been isolated (Loosfelt et al. 1989; McFarland et al. 1989; Sprengel et al. 1990). It is worth noting that the gonadotrophins are also able to regulate synthesis of their own receptors in the gonads (LaPolt et al. 1991; Wang et al. 1991).

All these regulatory factors interact to exert a highly complex pattern of control over gonadotrophin secretion and thus reproductive function. The regulation of secretion of the gonadotrophins from the pituitary gland has been studied extensively over many years, but it is only in the last decade that the tools of molecular biology have made it possible to study the regulation of synthesis of these hormones at the pre-translational level.

1 Structures of LH and FSH

LH and FSH are members of the glycoprotein hormone family (Pierce and Parsons 1981). Thyroid-stimulating hormone is the third pituitary-specific member of this family, and an additional glycoprotein hormone, chorionic gonadotrophin, is found in the placentae of primates and horses (Canfield et al. 1978; Moore et al. 1980). The glycoprotein hormones each consist of two different subunits, α and β, which are glycosylated, contain multiple disulphide linkages, and are non-covalently linked to each other to form the active molecule. The α-subunit apoproteins are the same for all the glycoprotein hormones within a species and are encoded by a single gene, which in the human is located on chromosome 6 (Boothby et al. 1981). The β-subunits are different for each hormone, and determine the biological specificity of each hormone (Jutisz and Tetrin-Clary 1974; Pierce and Parsons 1981). The LHβ and FSHβ genes show similar genomic arrangements, suggesting a common ancestral origin. The human LHβ gene is on chromosome 19 and the FSHβ gene on chromosome 11 (Naylor et al. 1983). The structure of the gonadotrophin genes has been comprehensively reviewed by Gharib et al. (1990b).

The addition of the carbohydrate moieties to the apoproteins is important for correct synthesis of bioactive gonadotrophin dimers, and for achieving appropriate biological activity of the molecule (reviewed in Chin and Gharib 1986). Further studies suggest that regulatory mechanisms may operate through modifying the sugars on these hormones (Grotjan 1989; Stanton et al. 1992). The structure and biochemistry of the gonadotrophins are reviewed extensively elsewhere, and are only briefly described here. This review will focus on insights that have been achieved recently into the regulation of the genes encoding the pituitary gonadotrophins LH and FSH, using the tools available through the advent of molecular biology.

2 Cellular localization of gonadotrophins

The anterior pituitary gland in a mature animal has several different types of endocrine cells, each of which secrete particular hormones. LH and FSH are synthesized in the gonadotroph cells of the anterior pituitary, by populations of cells synthesizing LH, FSH, or both (Tixier-Vidal et al. 1975; Moriarty 1976; Lloyd and Childs 1988b). This ability of some gonadotrophs to produce more than one hormone makes them different from the other endocrine cell types in the pituitary such as the corticotrophs, thyrotrophs, lactotrophs, and somatotrophs, each of which synthesizes and secretes largely a single trophic hormone in the adult. During development, however, it is likely that the different pituitary cell types arise from a common cell type, since studies in fetal rats show the presence of multiple hormones in the same cell (Moriarty 1976; Ingraham et al.1990; Simmons et al. 1990). In addition, there are populations of gonadotrophs that exist in a quiescent state and can be recruited in response to particular stimuli, such as gonadectomy (Childs et al. 1987a, 1990), stimulation with GnRH (Lloyd and Childs 1988b), or at different stages of the oestrus cycle (Childs et al. 1987b, 1992a,b). This group has shown that there are two populations of gonadotrophs, small and large, and that it is only the small cells which respond to GnRH stimulation by increasing levels of LHβ and FSHβ mRNA. This stimulation also results in an increase in the proportion of small cells storing both hormones (Lloyd and Childs 1988a). These results suggest that the two populations of cells may respond differently to different stimuli, the smaller both synthesising and secreting gonadotrophins in response to GnRH stimulation, and the larger cells responding to stimuli by secreting gonadotrophins which are already stored in secretory granules.

3 Current questions

The regulation of the secretion of LH and FSH has been studied extensively, and the results of these studies are reviewed in Kalra and Kalra (1985) and elsewhere. The synthesis of the gonadotrophin hormones has recently been examined by molecular biological techniques to explore the differential regulation of expression of the α-subunit, LHβ, and FSHβ genes. These studies have provided additional insights into the complex regulation of the gonadotrophins by GnRH and gonadal hormones.

In order to consider the regulation of the genes for the gonadotrophin hormones, let us first consider the different levels at which regulation may occur in the gonadotroph. Firstly, a signal must be received by the cell, either via a cell membrane-bound receptor, which is the case for GnRH, or via a cytoplasmic or nuclear receptor, as for the steroid hormones. In the case of membrane-bound receptors, the signal is relayed and amplified through the cell by second messengers. Eventually the signal may reach the nucleus, where transcription of the gonadotrophin subunit genes will be modulated by nuclear *trans*-acting factors which bind to *cis*-elements on the gene. Once the gene has been transcribed into mRNA, the stability of the mRNA may also be modulated. Translation of new protein occurs, and then post-translational modification of the nascent sub-apoproteins will give rise to the mature protein products. The stability of proteins, and post-translational modification may provide further regulation of the gonadotrophins, which are stored in secretory vesicles until they are released from the cell (Fig. 6.2). In considering the specific regulation of the gonadotrophins, it is also necessary to consider that synthesis and secretion of the hormones may occur one without the other in response to certain regulatory signals, as discussed above.

In this review, we will focus on the regulation of the gonadotrophin genes at the transcriptional and post-transcriptional levels, and attempt to address some of the key questions facing this area of research. Even though we now have many sophisticated molecular biological tools, there remain large areas yet to be understood with regard to the regulation of the pituitary gonadotrophins. We believe the important questions in this field include the following:

1. How are the two gonadotrophin hormones regulated differently?
2. How do the steroid and gonadal peptide hormones regulate LH and FSH synthesis and secretion? Is it via the hypothalamus or directly at the level of the pituitary?
3. Why is there an apparent dissociation between synthesis and secretion of LH and FSH?
4. What are the important *cis*- and *trans*-acting elements in these genes that determine their tissue-specific expression and hormonal regulation?

One of the biggest problems with tackling some of these issues has been the lack of convenient *in vivo* and *in vitro* model systems. The models that have been developed so far will be discussed here, and at the end of this review we will discuss some of the promising possibilities offered by the latest approaches.

II REGULATION OF THE GONADOTROPHINS BY GnRH

1 Role of pulsatile GnRH

It is established that GnRH is secreted from the hypothalamus into the portal circulation in a pulsatile fashion, and that LH, and to a lesser extent FSH, are secreted from the pituitary in response (reviewed in Clarke 1987). Pulsatile delivery of GnRH to the pituitary gland is necessary to maintain LH and FSH secretion and, indeed, there is a close temporal relationship between GnRH and LH pulses (Clarke and Cummins 1982; Levine *et al.* 1985). Further, it has been demonstrated that both the amplitude and frequency of GnRH pulses affect qualitative and quantitative aspects of

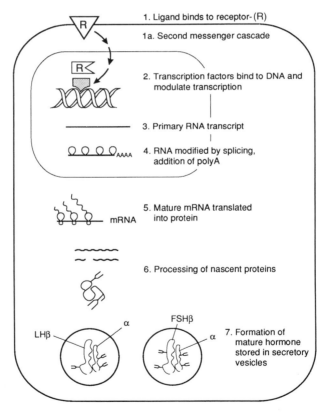

Fig. 6.2 Representation of a typical gonadotroph, showing the different levels at which synthesis of gonadotophin hormones may be regulated (see text).

secretion of the gonadotrophins (Pohl *et al.* 1983). For example, during the menstrual or oestrous cycle, there is modulation of the pattern of pulsatility of GnRH, with increased pulse frequency and amplitude at the time of the LH surge which precedes ovulation (Clarke 1987).

i In vivo *studies*

As we might expect, GnRH also regulates the synthesis of the pituitary gonadotrophins. Studies in ovariectomized, hypothalamo–pituitary disconnected (OVX/HPD) sheep, where the pituitary is surgically isolated from any hypothalamic input (Clarke *et al.* 1983), have shown that GnRH is an absolute requirement to maintain basal levels of synthesis of LH and FSH. It has been shown that levels of mRNA for LHβ, FSHβ, and α-subunit fall to very low values in these animals, and are restored to the levels in ovariectomized sheep by administration of pulsatile infusion of GnRH (Hamernik and Nett 1988*a*; Mercer *et al.* 1988, 1989). The expression of the LHβ gene in the sheep pituitary gonadotroph is sensitive to the nature of GnRH administration in that a constant infusion of GnRH does not stimulate LHβ synthesis. Pulsatile delivery of GnRH is necessary to maintain synthesis (Leung *et al.* 1987; Mercer and Clarke 1989). When OVX/HPD sheep are treated with constant GnRH, LH secretion ceases, with LHβ and FSHβ mRNA levels reduced to 50 per cent of those in GnRH-pulsed animals after seven days of infusion (Mercer and Clarke 1989). The administration of 25 ng pulses of GnRH (a tenth the size of those used to maintain physiological circulating levels of LH and FSH) to OVX/HPD sheep does not maintain secretion of LH, but is sufficient to maintain LHβ mRNA levels (Fig. 6.3).

Studies in rats and mice using GnRH agonists and antagonists and administration of GnRH antisera have further confirmed the role of GnRH in maintaining gonadotrophin subunit mRNA levels (Kim *et al.* 1988; Lalloz *et al.* 1988*a, b*; Wierman *et al.* 1989*a*) Using a castrated, testosterone-replaced rat model in which endogenous GnRH is suppressed, Marshall and co-workers have shown that both frequency and amplitude of GnRH pulses regulate the synthesis of gonadotrophin subunits at the mRNA level (Haisenleder *et al.* 1987, 1988*a,b*, 1990). Interestingly, the frequencies and amplitudes that define the maximal response of LH secretion also, in general, evoke the maximal stimulation of LHβ mRNA levels; in contrast, FSHβ and α-subunit mRNA levels respond maximally to different GnRH pulse patterns from LHβ mRNA levels. This group has also demonstrated that GnRH pulses stimulate transcription of all three gonadotrophin genes, whereas a constant infusion of GnRH does not, and that the transcription rates of the gonadotrophin subunit genes are differentially regulated by GnRH (Haisenleder *et al.* 1991).

These data suggest that there is a species difference in the way that GnRH regulates synthesis and secretion of LH. In the sheep, there is a

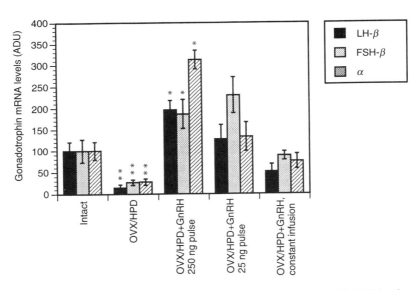

Fig. 6.3 Regulation of gonadotrophin subunit mRNA levels by GnRH in sheep. Female sheep were treated with 250 ng pulses, 25 ng pulses or a constant infusion of GnRH following ovariectomy and hypophysectomy (OVX/HPD). Gonadotrophin subunit mRNA levels were determined by northern blot analysis. Each bar represents the mean arbitrary densitometric units (ADU) +/− SEM of at least four animals, with intact levels taken as 100 per cent (*, $p < 0.05$). (Modified from Mercer et al. 1988.)

dissociation between LH synthesis and secretion, in that GnRH pulses of a particular amplitude (250 ng) are required to maintain secretion of LH, whereas synthesis, but not secretion, can be normally maintained by lower amplitude pulses of GnRH, and persists to some extent in the face of constant GnRH levels. In the rat model described above, there did not appear to be a similar dissociation between synthesis and secretion of LH at any of the pulse frequencies and amplitudes examined. The regulation of FSH does not appear to show a dissociation between synthesis and secretion in the sheep; secretion of FSH appears to follow FSHβ synthesis. In the rat, FSHβ and α-subunit mRNA levels reach a peak at lower amplitude pulses of GnRH than LHβ mRNA, so both of these animal models demonstrate a distinct difference in the way that GnRH regulates LH and FSH synthesis in the pituitary gonadotroph.

ii In vitro *studies*

In an early *in vitro* study in which rat pituitary cells were treated with GnRH in static culture, effects on release of LH were clear, but no increase in LHβ mRNA was demonstrated (Attardi et al. 1989a). Further, a study using pituitary cells in a perifusion system showed no effect of GnRH on LHβ mRNA levels (Weiss et al. 1990), and in another study, no effect was seen on LHβ mRNA transcription rate after GnRH treatment, despite an observed effect on LH release (Salton et al. 1988). Although these data again suggest a dissociation between synthesis and secretion of LH, the failure of the authors to show an effect on LHβ mRNA levels may also be due to an inappropriate pattern of GnRH administration having been used.

This proposition is supported by the studies of Jakubowiak, Haisenleder and colleagues. In the first of these experiments, pulsatile GnRH stimulated mRNA levels for all three gonadotrophin subunits in perifused pituitary cells after 20 hours of treatment. Furthermore, a constant infusion of GnRH caused increased α-subunit and reduced FSHβ mRNA levels, and had no effect on LHβ mRNA levels relative to unstimulated cells (Jakubowiak et al. 1991b). Haisenleder et al. (1993) have shown similar effects of pulsatile GnRH in perifused pituitary cells, but in this case they exposed cells to different GnRH pulse amplitudes. LHβ mRNA levels were stimulated by only one of the pulse amplitudes, while FSHβ and α-subunit mRNA levels responded to several different GnRH pulse amplitudes, demonstrating that the pattern of response of individual gonadotrophin subunits is determined by the nature of the GnRH stimulus. Shupnik (1990) has recently shown in rat pituitary quarters that pulsatile GnRH will stimulate the transcription rate of both the α-subunit and LHβ genes, while a constant infusion of GnRH stimulates α-subunit gene transcription only; in this study, GnRH stimulated LH secretion with both pulsatile and constant infusion. In perifused pituitary cells, pulsatile GnRH treatment resulted in modification of the polyadenylation of α-subunit and LHβ mRNAs, giving rise to longer mRNA transcripts, but there was no effect on the length of FSHβ mRNA. This demonstrates that, in addition to having a transcriptional effect, GnRH modifies α-subunit and LHβ mRNAs at a post-transcriptional level (Weiss et al. 1992a). Although these experiments show differing absolute effects of GnRH on the synthesis of the gonadotrophins, the common observations are that GnRH does regulate the transcription of all three gonadotrophin genes *in vitro*. Further, the GnRH signal arriving at the pituitary gonadotroph apparently regulates expression of the LHβ gene more closely than the α-subunit and FSHβ genes, including the post-transcriptional modification of LHβ and α-subunit mRNAs. In addition, these studies provide further evidence of the dissociation between the synthesis and secretion of LH with different modes of administration of GnRH.

In vitro studies on the effects of GnRH on gonadotrophin synthesis have proved more difficult, and have provided less consistent results than *in vivo* studies performed in rats and sheep. Part of the problem is the difficulty inherent in trying to duplicate accurately a very complicated physiological milieu, where many factors regulate the system. Another aspect is the use of dispersed pituitary cells which have a limited lifetime in culture conditions, and which change in character relatively quickly once they are plated out or placed into perifusion systems (Farnworth et al. 1988). However, the experiments done to date all point to the conclusion that the synthesis of LHβ is much more closely regulated by GnRH than is synthesis of FSHβ or the α-subunit. Further, it is evident that LHβ synthesis requires a particular type of GnRH signal, in terms of pulse amplitude and frequency, and that synthesis of LH may be triggered by a different signal to that which will stimulate secretion of LH. Why are LH and FSH regulated differently by GnRH? The answer may well lie in the fact that LH and FSH stimulate different cells in the gonads, or have different effects on the same cells, and thus a different reception of the signal from the common trophic factor, GnRH, at the pituitary gonadotroph is required to ensure that the appropriate LH or FSH signal reaches the end organ, the gonad, at the correct stages of development, or of the reproductive cycle. The regulation of the α-subunit is not as closely regulated as the LHβ-subunit, and this probably reflects the fact that it is synthesized in excess of either of the gonadotrophin β-subunits in the pituitary, and that it is the β-subunits which confer biological specificity on the mature hormones.

iii Genetic elements

Progress in studies of the gonadotrophin genes to determine which elements in the promoters of each gene are important in their regulation has been slow compared with the study of other hormonal genes. One of the major reasons for this has been the lack of a gonadotroph cell line that would allow expression studies of the genes. Study of the growth hormone or prolactin genes for example has been hugely facilitated by the availability of GH cells, somato-lactotroph-derived cell lines, which are relatively easy to maintain in culture, and which express the genes of interest in relatively high levels (Tashjian 1979; Inverson et al. 1990; Karin et al. 1990). Some limited studies have been done of the gonadotrophin genes in dispersed pituitary cells, a system which requires a large number of animals, and which has a limited life. The studies to date have shown that 1.7 kb and 0.8 kb of the LHβ and α-subunit genes, respectively, are sufficient to direct expression of reporter genes to pituitary cells (Burrin and Jameson 1989; Kim et al. 1990).

Choriocarcinoma cells have been used extensively to study the transcriptional regulation of the α-subunit gene by cAMP in transient transfection studies. A region in the promoter of the human α-subunit gene has been identified which is capable of conferring cAMP responsiveness. This cAMP-responsive element is an 18bp repeated palindromic sequence homologous to the elements found in other cAMP-responsive genes (Delegeane et al. 1987; Silver et al. 1987; Jameson et al. 1988). Kay and Jameson (1992) have recently used a transient transfection assay in dispersed rat pituitary cells to further investigate regulatory elements in the human α-subunit promoter involved in regulation by GnRH. They confirmed the presence of a cAMP-responsive element between −132 and −99 bp, deletion of which resulted in a marked loss of basal transcriptional activity. They also reported that a region between −346 and −244 bp conferred GnRH responsiveness on the α-subunit promoter, indicating that the cAMP-responsive sequences are distinct from those involved in GnRH-regulated expression of the α-subunit gene.

Using a different strategy to show that the α-subunit promoter is important for GnRH regulation of the expression of the α-subunit gene, Hamernik *et al.* (1992) have performed elegant experiments where they produced transgenic mice by using either 1.6 kb of the human, or 315 bp of the bovine, α-subunit gene 5′-flanking region fused to a chloramphenicol acetyltransferase (CAT) reporter gene. They then administered pulses of GnRH every two hours to these animals and showed that this stimulated expression of the CAT reporter gene in the pituitaries of the transgenic mice. The localization of a GnRH-responsive element in this region of the human and bovine genes agrees with the results of Kay and Jameson (1992). These studies open up the exciting possibility of identifying the exact region of the promoter involved in GnRH regulation, and then identifying the *trans*-acting factors which might bind to this region of the gene.

2 The GnRH receptor

The gene for the pituitary GnRH receptor has been cloned from mouse, rat, sheep, and human (Eidne *et al.* 1992; Kaiser *et al.* 1992b; Kakar *et al.* 1992; Reinhart *et al.* 1992; Tsutsumi *et al.* 1992; Chi *et al.* 1993; Illing *et al.* 1993) (see Chapter 8). The receptor gene was originally cloned from αT3 cells, a mouse pituitary cell line produced by Windle *et al.* (1990). The gene encodes a protein of 327 residues which has seven putative transmembrane domains, characteristic of members of the G-protein-coupled receptor family. However, this receptor lacks an intracellular carboxy-terminal tail which is also characteristic of these receptors (Fig. 6.4). The mRNA for the GnRH receptor has been identified in pituitary, ovary, and testis, in addition to the murine αT3 cells, with three different sizes of mRNA present in rat pituitary and ovary, and mouse αT3 cells. The predominant mRNA transcript is 4.5 kb, with minor 5.0 and 1.8 kb species. GnRH mRNA levels in the rat pituitary doubled following ovariectomy and increased fivefold following castration. When ovariectomized female rats were treated with replacement doses of oestrogen, GnRH receptor mRNA levels fell to levels similar to those in intact rats. In castrated males treated with replacement doses of testosterone, however, GnRH receptor mRNA levels only fell slightly. In perifused rat pituitary cells treated with GnRH pulses, GnRH

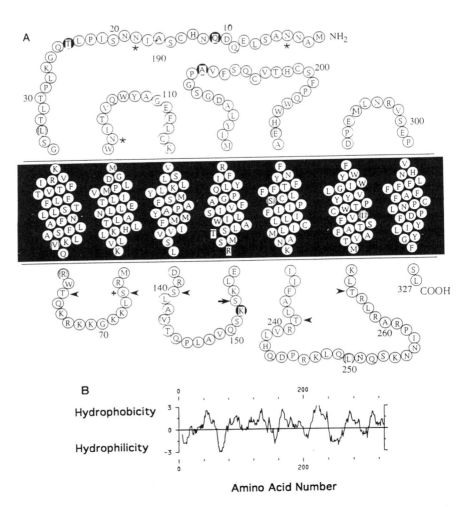

Fig. 6.4 Structure of the rat GnRH receptor. (A) Amino acid residues in black represent non-conserved amino acids between the rat and mouse receptors; shaded amino acid residues are non-identical but conserved between the two species. *Asterisks* denote potential glycosylation sites, and potential phosphorylation sites are indicated for protein kinase C (*arrowheads*), casein kinase II (*arrow*), and protein kinase A (*cross*). (B) Hydrophobicity plot for the receptor. (From Kaiser et al. 1992.)

receptor mRNA levels increased 12-fold, but did not change when the cells were treated with a constant infusion of GnRH (Fig. 6.5) (Kaiser et al. 1993). These regulatory patterns mirror the previously described changes in GnRH binding sites in the rat pituitary and in dispersed pituitary cells in culture. Further studies of how regulation of the GnRH receptor is integrated with modulation of the hypothalamic GnRH signal to effect changes in gonadotrophin synthesis will be a fascinating area of study.

The molecular mechanisms of receptor desensitization is another issue which can now be investigated using these clones. It is interesting to note the recent report of the role of the carboxy-terminal tail in the regulation of the thyrotrophin-releasing hormone (TRH) receptor (Nussenzveig et al. 1993). As the GnRH receptor lacks a carboxy-terminal tail, it is possible that desensitization of this receptor may operate via a completely different mechanism to that of the TRH receptor. Some insights into the structural relationships of the transmembrane helices 2 and 7 and their function in ligand binding have been provided by studies using a reciprocal mutation in these helices, which showed that they are probably adjacent in the cell membrane, and are involved in GnRH binding (Davison et al. 1994; Zhou et al. 1994). This group has employed molecular modelling to predict further structural interactions in the receptor molecule. Yet another area of current interest is the study of the second messenger systems involved in GnRH signal transduction. Two studies have shown that the TRH receptor is coupled to the G proteins, $G_{\alpha q}$ and $G_{\alpha 11}$ (Aragay et al. 1992; Hsieh and Martin 1992). Furthermore, Hsieh and Martin (1992) showed that the GnRH receptor is linked to these G protein regulatory subunits which modify the activity of phospholipase C.

The availability of the receptor cDNA will also open the way for cellular localization studies which should help us to understand further how GnRH has different effects on LH and FSH, and on LH synthesis and secretion. Regulation of GnRH receptor synthesis may play an important part in this differential regulation.

III REGULATION OF GONADOTROPHINS BY STEROID HORMONES

1 Oestrogen

The mechanisms by which sex steroids regulate synthesis of gonadotrophins are complex. There is an increase in oestrogen levels shortly before the pre-ovulatory surge of LH. This rise in oestrogen is necessary, along with an increase in GnRH secretion, in order to elicit an LH surge. On the other hand, during the follicular or metestrus phase of cycling animals and during anoestrus in seasonal breeding animals, oestrogen

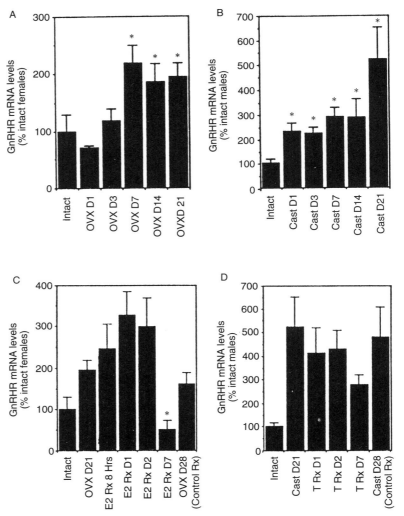

Fig. 6.5 Regulation of GnRH receptor (GnRHR) mRNA levels in the rat pituitary gland. Upper panels show the time-course of changes in rat pituitary GnRH receptor mRNA levels at one, three, seven, 14, and 21 days after ovariectomy (OVX) or castration (Cast). (A) Female; (B) male. Lower panels show the time-course of changes in rat pituitary GnRH receptor mRNA levels following sex steroid hormone replacement. (C) Female after eight hours, one, two, and seven days of oestradiol treatment (E2 Rx). (D) Male rats following one, two, and seven days of testosterone treatment. GnRH receptors mRNA levels were determined using northern blot analysis. Results are expressed as a percentage of intact values. Each bar represents the mean arbitrary densitometric units (ADU) +/− SEM for three to seven rats (*, $p < 0.01$). (From Kaiser et al. 1993.)

has an inhibitory effect on gonadotrophins (Knobil 1974; Clarke 1987; Karsch 1987).

i Negative-feedback effects

Gonadectomy leads to increases in mRNA levels for all three gonadotrophin subunits in male and female rats (Gharib *et al.* 1987), in ewes (Nilson *et al.*, 1983; Landefeld and Kepa 1984; Landefeld *et al.* 1984; Mercer *et al.* 1989; Herring *et al.* 1991), and in rhesus monkeys (Attardi *et al.* 1992). In addition, *in situ* hybridization studies have shown both that the number of LHβ and FSHβ mRNA-containing cells increases, and that the concentration of LHβ and FSHβ mRNA also increases in the rat pituitary following gonadectomy (Childs *et al.* 1987*a*, 1990). Chronic treatment of ovariectomized animals with oestrogen reduces gonadotrophin subunit mRNA to levels at least as low as those in intact animals (Counis *et al.* 1983; Nilson *et al.* 1983; Landefeld and Kepa 1984; Gharib *et al.* 1986, 1987; Mercer *et al.* 1989).

Much effort has gone into defining the site of action of steroid hormones within the hypothalamo–pituitary unit. The OVX/HPD sheep model has enabled *in vivo* studies of the surgically isolated pituitary to be undertaken. These studies have shown that oestrogen in the sheep does not have a direct negative action at the pituitary, since oestradiol has no effect on LHβ mRNA levels in OVX/HPD, GnRH-pulsed sheep (Mercer *et al.* 1988); in intact animals, oestrogen thus presumably down-regulates LHβ mRNA levels via an effect on GnRH. In contrast, FSHβ and α-subunit mRNA levels are decreased by oestrogen treatment in this model, indicating a direct effect of oestrogen on the pituitary to regulate negatively these two genes (Mercer *et al.* 1989). (Fig. 6.6).

ii Positive-feedback effects

Positive feedback of oestrogen on the hypothalamo–pituitary unit leads to the LH surge. Landefeld and co-workers have shown that LHβ mRNA levels are increased by 60 per cent during the pre-ovulatory LH surge in sheep (Leung *et al.* 1988), and the proestrus LH peak in rats is also preceded both by a rise in steady-state LHβ mRNA levels (Zmeili *et al.* 1986; Haisenleder *et al.* 1988*a*) and LHβ transcription (Shupnik *et al.* 1989*a*).

Ovariectomized ewes treated short-term with oestrogen to elicit an LH surge show either no increase (Hamernik and Nett 1988*b*; Landefeld *et al.* 1989) or a decrease (Mercer *et al.* 1993) in LHβ mRNA levels following steroid injection. These data in sheep suggest that direct positive regulation of the LHβ gene is not a major mechanism for positive oestrogen feedback in this species, but that oestrogen effects may be largely on GnRH secretion, or up-regulation of GnRH receptors to increase the sensitivity of the gonadotroph to GnRH. The recent isolation of the GnRH receptor cDNA will enable the investigation of regulation of GnRH receptor levels at the time of the LH surge, and thus will add significantly to our understanding of the manner in which oestrogen is able to give rise to the LH surge.

Interestingly, when cultured rat pituitary fragments were incubated with oestradiol for 2 or 6 hours, no effects were seen on the transcription of α-subunit or FSHβ genes; in contrast, LHβ mRNA synthesis was increased (Shupnik et al. 1989a). This effect on LHβ mRNA synthesis is consistent with the site of action for negative regulation by oestrogen being the hypothalamus, but suggests that the site of action of positive oestrogen regulation of LHβ mRNA in the rat is directly at the pituitary. An oestrogen regulatory element consensus sequence has been identified in the 5' untranslated region of the rat LHβ gene, and has been characterized by both DNA binding and functional studies (Shupnik et al. 1989b).

The issue of oestrogen effects on GnRH-producing neurons in the hypothalamus is currently controversial. There has been no identification of oestrogen receptors in GnRH neurons in the hypothalamus, although immunoreactive oestrogen receptor has been demonstrated on impinging

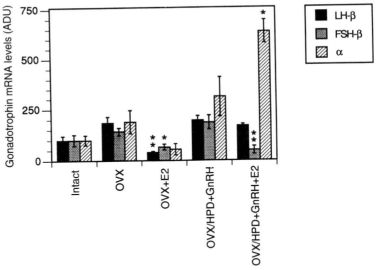

Fig. 6.6 Regulation of sheep gonadotrophin subunit mRNA levels by oestrogen. Ovariectomized (OVX) or OVX/HPD, GnRH-pulsed sheep were treated with oestrogen (E2) for 14 days. Gonadotrophin β-subunit mRNA levels were determined by northern blot analysis. Each bar represents the mean arbitrary densitometric units (ADU) +/− SEM of at least four animals. Results are expressed as a percentage of intact control levels ($*, p < 0.05, **p < 0.001$) (Modified from Mercer et al. 1988, 1989.)

interneurones in that region, and mRNA for GnRH has been also been localized to that region (Roberts *et al.* 1989 *a*; Rothfeld *et al.* 1989). The GnRH-secreting hypothalamic cell line GT-1 does not express oestrogen receptor mRNA (Mellon *et al.* 1990). However, the identification of an oestrogen response element in the GnRH gene that is able to mediate a positive response to oestrogen (Radovick *et al.* 1991), and of regions in the GnRH promoter able to confer negative regulation by oestrogen (Wierman *et al.* 1992) in heterologous cells suggests that oestrogen may be able to regulate GnRH neurones directly. Further studies are required to establish definitively whether or not GnRH neurones have receptors for oestrogen.

2 Progesterone

Progesterone has profound effects on the secretion of the gonadotrophins. There is increased progesterone secretion which accompanies the preovulatory LH surge, and this serves to synchronize and amplify the secretion of LH. During the luteal phase, high levels of progesterone exert an inhibitory effect on the gonadotrophins (Soules *et al.* 1984; Van Vugt *et al.* 1984; Marshall and Kelch 1986).

The role of progesterone in regulating the synthesis of the gonadotrophins is much less clear. Wise *et al.* (1985) demonstrated that LHβ mRNA levels fall between days 50 and 140 during gestation in the ewe, a time when oestrogen and progesterone levels are maximal. In a more direct experiment, Hamernik *et al.* (1987) showed that progesterone administration had no effect on gonadotrophin subunit mRNA levels in ovariectomized ewes. In contrast, ovine pituitary cells in culture treated with progesterone showed decreases in FSH secretion and FSHβ mRNA levels (Laws *et al.* 1990). In the rat, administration of progesterone alone had no effect on gonadotrophin subunit mRNA levels (Counis *et al.* 1983; Simard *et al.* 1988), but administration of progesterone with oestradiol resulted in a synergistic suppression of α-subunit and LHβ mRNA levels (Counis *et al.* 1983). Conversely, administration of progesterone to immature rats with oestradiol implants resulted in increased FSHβ mRNA levels (Attardi and Fitzgerald 1990).

The effects that progesterone has at the level of the pituitary remain controversial, but it is clear from recent studies that progesterone exerts major effects in modulating GnRH secretion from the hypothalamus in the rat (Kim and Ramirez 1985; Lee *et al.* 1990). Thus, it is possible that progesterone may not have a direct role in modulating synthesis of the gonadotrophins in the rat, but regulates the trophic signal from the hypothalamus. Regulation by progesterone in the sheep is less clear, since a direct effect of progesterone on the pituitary to downregulate FSHβ mRNA levels has been demonstrated (Batra and Miller 1985). It is possible that FSH is regulated at the pituitary, and LH at the hypothalamus via GnRH in the sheep.

3 Testosterone

i In vivo studies

It is well known that plasma levels of both LH and FSH are increased in males following castration. However, the response of these two hormones to replacement doses of testosterone is divergent. Secretion of both LH and FSH is lowered in castrate rats treated with testosterone, but synthesis of FSHβ is not inhibited. LHβ and α-subunit mRNA levels fall in castrates treated with testosterone, but FSHβ mRNA levels do not (Gharib et al. 1987; Wierman et al. 1989b) (Fig. 6.7). Further, when castrate rats are treated with a GnRH antagonist and then treated with testosterone, FSHβ mRNA levels increase while LHβ and α-subunit mRNA levels decrease or remain unchanged (Paul et al. 1990; Perheentupa and Huhtaniemi 1990; Wierman and Wang 1990; Dalkin et al. 1992). Castrate rats exposed to exogenous pulses of GnRH and treated with testosterone show an increase in FSHβ mRNA levels with low doses of GnRH, while higher doses of GnRH do not elicit any change in FSHβ mRNA levels (Iliff-Sizemore et al. 1990). These results indicate that this stimulatory effect of testosterone on FSHβ mRNA levels is mediated via an effect on GnRH, since it is apparently GnRH dependent. Further studies of this phenomenon have shown that the increase in steady-state levels of FSHβ mRNA occurs without any increase in transcription rate of the FSHβ gene, suggesting that this regulation may occur via a post-transcriptional mechanism (Paul et al. 1990).

ii In vitro studies

In cultured pituitary cells, testosterone treatment increases FSHβ mRNA levels while downregulating LHβ and α-subunit mRNA levels (Gharib et al. 1990a). However, when pituitary cell cultures receive pulsatile stimulation with GnRH, the effect of testosterone treatment is to reduce mRNA levels

Fig. 6.7 Regulation of gonadotrophin mRNA levels by testosterone. Time-course of changes in pituitary gonadotrophin subunit mRNA levels in male rats after castration (A) and following testosterone replacement (B). Gonadotropin mRNA levels were determined at various time points following castration by northern blot analysis (A) and after one and seven days of treatment with testosterone (B). In each panel, data points are expressed as percentages of intact control animals (0 days post-castration in panel A and group N in panel B). Each bar represents the mean arbitrary densitometric units (ADU) +/− SEM of at least three determinations. (*, $p < 0.05$; **, $p < 0.01$) (Modified from Gharib et al. 1987.)

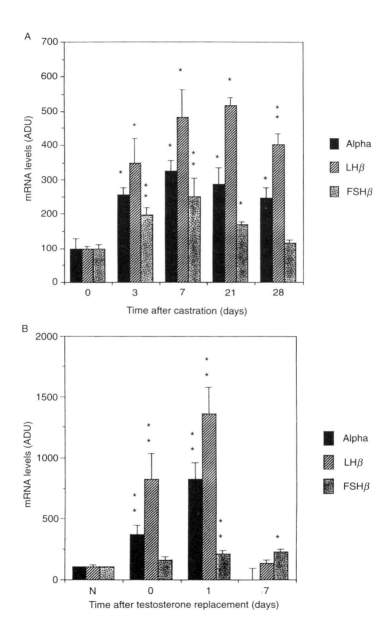

for all three subunits (Jakubowiak et al. 1991a; Winters et al. 1992). Using the α-CAT transgenic mice described earlier, Nilson and his co-workers have shown that the α-subunit promoter contains sequences which respond negatively to androgen treatment (Clay et al. 1993). They went on to identify, in vitro, a high-specificity androgen receptor binding site located at −111 to −97 in the α-subunit promoter using gel shift analysis.

These studies indicate that testosterone has a direct regulatory effect at the level of the pituitary to stimulate FSHβ and to inhibit LHβ and α-subunit, and in the case of the α-subunit that regulatory effect may be mediated, at least in part, via a direct interaction of the androgen receptor with the α-subunit promoter. In the presence of GnRH, however, it would appear that testosterone can inhibit synthesis of all three of the gonadotrophin subunits. How can this be explained? If the action of testosterone at the level of the hypothalamus is to inhibit GnRH release, we would expect to see a fall in mRNA levels for all three gonadotrophin subunits. But if, in addition to the effect on GnRH, testosterone can also stimulate FSHβ synthesis at the level of the pituitary, while exerting a negative effect on LHβ and α-subunit, then we would see a fall in LHβ and α-subunit mRNA levels with testosterone treatment, but, depending on the interaction of GnRH and testosterone at the gonadotroph, there might be stimulation of, inhibition of, or no effect on FSHβ mRNA levels. Indeed, this is exactly what is seen in in vivo studies, depending on the experimental paradigm employed.

In addition to the issue of where in the hypothalamo–pituitary axis testosterone acts, these studies provide strong evidence that there must be at least one other gonadal factor responsible for the inhibition of FSH by the gonads, since testosterone administered in vivo to castrate animals does not reduce FSHβ mRNA to levels in the intact male (Wierman et al. 1989; Dalkin et al. 1990).

IV REGULATION OF GONADOTROPHINS BY GONADAL PEPTIDES

1 Inhibin and activin

The gonadal peptides inhibin, activin, and follistatin were isolated on the basis of their effects on FSH secretion from the pituitary gland, and have now been shown to have major paracrine effects in the gonads (Findlay 1993). Inhibin and activin have opposite effects on FSH, with inhibin suppressing FSH secretion (Ling et al. 1985; Robertson et al. 1985), and activin stimulating FSH release (Ling et al. 1986). The structurally unrelated follistatin also suppresses FSH secretion (Vale et al. 1986; Ueno et al. 1987). In addition to its effects on FSH, activin also acts as a growth factor in early development, and was independently isolated on the basis of its ability to stimulate development of mesoderm in *Xenopus* embryos (Smith 1990; Van den Eijinden-Van Raaij 1990), and its effects to stimulate erythropoiesis (Eto *et al.* 1987).

Molecular biology, along with standard biochemical techniques, has helped to elucidate the structures of inhibin and activin, and shown that there are multiple forms of both molecules. Inhibin is a heterodimer consisting of an α-subunit covalently linked to one of two distinct but homologous β-subunits, βA or βB, giving rise to inhibin A or inhibin B. Activin is a homodimer of the β-subunits of inhibin; thus, there are three possible activin molecules, activin A, consisting of the homodimer βAβA, activin B (βBβB), and activin AB (βAβB) (reviewed in Ying 1988). Activin A and AB have been isolated from biological samples, but, to date, there is only one report of activin B isolated from biological tissues, although recombinant activin B has been synthesized in bacteria (Mason *et al.* 1989). On the basis of their overall structure, inhibin and activin belong to a family of peptide signalling molecules involved in growth and development that includes the TGF-β subfamily, Mullerian inhibiting substance, the decapentaplegia gene in *Drosophila*, and the *Xenopus* growth factor Vg1.

i FSHβ mRNA regulation

FSHβ mRNA levels are rapidly suppressed within 6 hours to 20 per cent of control levels in OVX/HPD, GnRH-pulsed sheep treated with an inhibin-containing follicular fluid extract. This result indicates a direct pituitary site of action for inhibin *in vivo* in the ewe. Levels of α-subunit and LHβ mRNA were not changed by inhibin in this model (Mercer *et al.* 1987). *In vitro* studies using dispersed rat pituitary cells in static cultures have also demonstrated a specific effect of inhibin to reduce FSHβ mRNA levels in both untreated (Attardi *et al.* 1989*a*; Carroll *et al.* 1989) and GnRH-stimulated cells (Attardi *et al.* 1989*b*). In pituitary cell cultures, it has been shown that activin increases, and inhibin reduces, FSHβ mRNA levels, with no change in α-subunit or LHβ mRNA levels (Attardi *et al.* 1989*a*; Carroll *et al.* 1989) (Fig. 6.8).

Additional studies in perifused cultures of pituitary cells have provided further insights. Jakubowiak *et al.* (1991*a*) have shown that, in the presence of pulsatile GnRH, inhibin suppressed FSHβ mRNA levels, but had no effect on LHβ or α-subunit mRNA levels even though secretion of LH was decreased by 50 per cent. Recently, Weiss *et al.* (1992*b*) used a perifusion system to examine the effects of activin, and found that, in cells perifused from the time of dispersion, administration of activin stimulated FSHβ mRNA levels to 20–75 times control levels, with no appreciable change in either LHβ or α-subunit mRNA levels. In addition, when they used conditioned medium from static cultures of pituitary cells, they

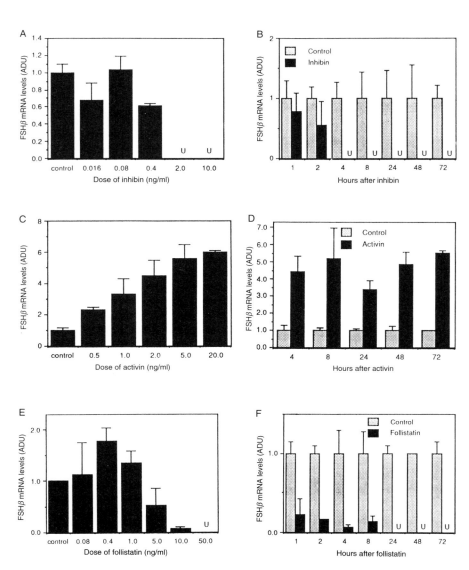

Fig. 6.8 Effects of inhibin, activin, and follistatin on gonadotrophin mRNA levels *in vitro*. Primary pituitary cultures were treated three days after dispersion with varying concentrations of purified porcine inhibin (panel A), activin (panel C), and purified porcine follistatin (panel E), and for varying times with 20ng/ml of inhibin (panel B), 20 ng/ml of activin (panel D), and 20 ng/ml of follistatin (panel F). FSHβ mRNA levels were standardized to levels observed in untreated controls. All values represent mean arbitrary densitometric units (ADU) +/− SD, $n = 2$; U, undetectable. (Modified from Carroll *et al.* 1989.)

observed only six-fold increases in FSHβ mRNA. These results suggest that cultured pituitary cells are secreting a stimulatory factor, probably activin itself, which in static cultures masks the large stimulatory effect seen in this study by maintaining gonadotrophs in a stimulated state before administration of exogenous activin. This apparent paracrine effect will be further discussed later.

It has also been shown that one mechanism by which activin increases steady-state mRNA levels for FSHβ is by altering the half-life of the mRNA species. Carroll *et al.* (1991) examined the stability of FSHβ mRNA following treatment with inhibin and activin in order to determine whether the effects of these hormones were at the transcriptional or post-transcriptional levels. In that paper, they reported that activin increases the half-life of FSHβ mRNA. Finally, it has recently been shown that inhibin reduces the transcription rate of the FSHβ gene by 50 per cent in the ewe (Clarke *et al.* 1993).

ii Inhibin, activin, and LH secretion

In addition to the effect of inhibin on FSH secretion, inhibin has been reported to decrease the basal secretion of LH from rat pituitary *in vivo* (Culler and Negro-Villar 1989) and *in vitro* (Farnworth *et al.* 1988), and to reduce the LH pulse amplitude in OVX/HPD sheep (Clarke *et al.* 1986; Mercer *et al.* 1987). In addition, sheep immunized against the α-subunit of inhibin show different patterns of responsiveness to GnRH and impairment of reproductive function (Findlay *et al.* 1989; Mann *et al.* 1989), and, when endogenous inhibin was immunoneutralized in rats, LH pulse amplitude and frequency increased, as did basal FSH secretion (Rivier and Vale 1989; Culler 1992). Clearly, the studies described above have demonstrated that these effects on LH are not exerted at the pretranslational level, since inhibin does not cause any decrease in LHβ mRNA levels both *in vivo* and *in vitro* (Mercer *et al.* 1987; Carroll *et al.* 1989). It would appear that the effects of inhibin on LH are being mediated largely at either the post-translational level, or at the level of control of secretion.

The pulsatile secretion of LH very closely follows the pulsatile release of GnRH into the portal circulation (Clarke and Cummins 1982). Recent studies of inhibin and GnRH receptor levels may provide some insight into the mechanism by which inhibin alters LH secretion. Inhibin has been reported to decrease steady-state levels of GnRH receptors on gonadotrophs in cultured rat pituitary cells (Wang *et al.* 1989) and, in particular, to block the stimulation of GnRH receptor levels by GnRH itself (Braden *et al.* 1990). The negative effect of inhibin on LH pulse amplitude is mimicked by cycloheximide (Jakubowiak *et al.* 1989), suggesting that the action of inhibin may involve the inhibition of specific protein synthesis, possibly of the GnRH receptor. Further, activin A has been shown by Braden and Conn (1992) to have a stimulatory effect on GnRH receptor levels. This effect is distinct from the mechanism by which GnRH stimulates its own receptor and is not antagonized by inhibin. In contrast to the rat studies, inhibin has been shown to increase GnRH receptor levels on ovine gonadotrophs (Laws et al. 1990), which would suggest a species difference in the role of inhibin in regulation of GnRH receptor.

The availability of a probe for the GnRH receptor will allow the further elucidation of the effect of inhibin. For instance, whether or not the effect of inhibin and activin is directly to alter the synthetic rate of the GnRH receptor itself, or whether these effects are mediated via an effect on a protein or proteins involved in the signal transduction pathway may soon be determined. Regardless of how inhibin or activin exert their effects on the secretion of LH, it is interesting to note that these results provide another example of the dissociation between effects on LH secretion and synthesis, and on the differential regulation of LH and FSH.

2 Follistatin

Follistatin was originally discovered during the purification of inhibin. It was present in a column fraction, different from that containing inhibin, of partially purified follicular fluid that was able to suppress the secretion of FSH from pituitary cells in the bioassay used for the detection of inhibin. Although similar in function to inhibin, follistatin is structurally distinct. It is a glycosylated, single polypeptide chain, with two different molecular weight forms in the human and pig, and three different forms in the cow. The cloning of the follistatin gene revealed that the different molecular weight species originally isolated from follicular fluid are the results of alternative splicing events which give rise to a truncated form of the molecule, with each species having a different carboxy-terminal end (reviewed in Ying 1988). The rat follistatin gene has now also been shown to exhibit this phenomenon of alternative splicing in the 3' end of the primary transcript (Shimasaki et al. 1989). Follistatin has several noteworthy characteristics: it is unusually cysteine rich; has a very acidic carboxy-terminal region; and is organized into three highly homologous domains, each of which is encoded by a separate exon (Shimasaki et al. 1989).

i FSHβ mRNA regulation

In addition to its effects on the secretion of FSH from cultured pituitary cells, follistatin also decreases FSH synthesis by downregulating steady-state levels of FSHβ mRNA. Carroll et al. (1989) have shown that, using dispersed rat pituitary cells in static culture, follistatin (10 ng/ml) inhibits FSHβ mRNA to 10 per cent of control levels within 4 hours. Higher doses of follistatin result in FSHβ mRNA falling below detectable levels. When activin and follistatin (20 ng/ml) are administered simultaneously, FSHβ mRNA and protein levels remain undetectable with increasing amounts of activin until activin approaches equimolar concentrations to follistatin, even though a 20-fold lower dose of activin alone will increase FSHβ mRNA levels to more than 300 per cent of control levels. These data indicate a possible interaction between activin and follistatin. Indeed, Sugino and co-workers, while trying to isolate the activin receptor in rat ovary homogenates, isolated follistatin and showed that it binds to activin (Nakamura et al. 1990). This observation might, at least in part, explain the interaction between activin and follistation in the regulation of FSH in the pituitary gland. However, another important molecule must also be considered in this system—the activin receptor.

3 Activin receptor

Several activin receptors have now been identified, and shown to be members of the TGFβ receptor superfamily. As with TGFβ receptors, activin receptors need to interact with one another to produce a cellular response: a type I and a type II receptor must associate in the cellular membrane to bind ligand and transduce an intracellular signal (Wrana et al. 1992). A cDNA encoding a type II activin receptor (ActRII) has been isolated from a mouse pituitary corticotroph cell line (AtT20) expression cDNA library (Matthews and Vale 1991). Subsequently, Attisano et al. (1992) cloned another member of the family (ActRIIB), which bears close homology to the original receptor, from the mouse fibroblast cell line BALB/c 3T3. More recently, several different TGFβ receptor superfamily members have been identified. Both the type I and the type II activin receptors consist of an extracellular domain, a short hydrophobic membrane-spanning domain, and a longer intracellular region which contains a serine–threonine kinase domain. The activin receptor type I (ActRI) (Matsuzaki et al. 1993) is widely expressed (Attisano et al. 1993); the ActRII is expressed in a wide range of tissues, including the gonads (Matthews and Vale 1991). ActRIIB, which is encoded by a different gene to ActRII, has two regions close to the membrane-spanning domain which may be alternately spliced, thus giving rise to a possible four different mRNA transcripts (ActRIIB1–4) (Attisano et al. 1992) These four transcripts were isolated by PCR from mouse testis, but it is not known if they are expressed in other tissues. The ActRII and the ActRIIB1–4 isoforms all bind inhibin in addition to activin, but with a 10-fold lower affinity. The intracellular, serine-kinase domains of the various type II receptors share a high degree (90 per cent) of amino acid identity, but there is only 40 per cent identity between type I and type II receptors in that domain. The extracellular domains of the type I and type II receptors have very little sequence similarity, apart from a cysteine cluster motif found in all superfamily members so far identified.

In addition to ActRI, a second type I receptor (TSR-I) which binds either activin or TGFβ has been identified (Attisano *et al.* 1993). This receptor is able to interact with either ActRII or the TGFβ type II receptors to bind the appropriate ligand, but when it binds activin in combination with ActRII it does not stimulate transcription of the ActRI/ActRII responsive reporter gene used in these studies, suggesting that it does not work via the same signalling pathway as the ActRI and ActRII combination of receptors (Attisano *et al.* 1993). At least three other TGFβ superfamily type I receptors have now been cloned (Franzen *et al.* 1993; Tsuchida *et al.* 1993; Bassing *et al.* 1994; ten Dijke *et al.* 1994); it is probable that there are further members of this superfamily, and that one or more of these may prove to be the inhibin receptor(s).

At present, it is unclear which of these receptors is responsible for mediating the effect of activin on FSH in the pituitary gland. The type II receptor has been shown, using oocyte injections of mRNA, to mediate mesoderm induction in *Xenopus* (Hemmati-Brivanlou and Melton 1992). Both type II receptors (ActRII and ActRIIB) have been demonstrated to bind activin and inhibin (although with a 10-fold lower affinity) (Matthews and Vale 1991; Attisano *et al.* 1992). Identification of the appropriate gonadotroph activin receptor, and the putative inhibin receptor, will help in the understanding of the interaction of these peptides in the regulation of the pituitary gonadotrophins. It is probable that the expression of activin receptors is hormonally regulated, as are many other hormone receptors. If so, this clearly provides another level at which FSHβ gene expression may be modulated. The second messenger pathways through which these receptors act have not been identified, nor have the possible transcription factors which are involved in gene regulation by these hormones. The mechanism of action of TGFβ has been much more extensively studied, with at least two different intracellular pathways proposed to mediate its effects: one pathway is believed to involve the phosphorylation of the retinoblastoma protein. However, the precise molecules involved in post-receptor events remain unidentified.

4 Paracrine effects

Inhibin, activin, and follistatin were initially purified from follicular fluid, and subsequently immunoreactive peptides were identified in both the ovary and testis. Expression of the genes for these hormones has been confirmed in the ovary and testis, and in a wide range of other tissues. Of importance for this discussion are the reports of their expression in the pituitary gland. Inhibin is an elusive molecule that was eventually discovered on the basis that it was a gonadal factor with the ability to inhibit secretion of FSH from the pituitary gland. Indeed, with the availability of radioimmunoassays to measure inhibin, it has been confirmed that circulating forms do exist and are quantifiable in serum, but it is less clear that activin and follistatin are transported in the circulation.

The identification of mRNA for the inhibin subunits and follistatin in the pituitary points to the possibility that these peptides are produced locally in the pituitary gland, where they may act as paracrine or autocrine regulators of FSH. Using S1-nuclease assays, Meunier *et al.* (1988) showed that the α- and βB-subunits of inhibin are present in the pituitary. A subsequent study from the same group localized the α- and βB-inhibin subunit polypeptides to LH and FSH immunoreactive gonadotrophs (Roberts *et al.* 1989*b*). They also reported the detection of the βA-subunit, but this staining was not confined to any particular pituitary cell type. The effect of ovariectomy on levels of the subunits was examined, and both the α- and βB-subunits were significantly lowered.

In addition, the folliculo-stellate cells of the pituitary have been reported to secrete follistatin (Gospodarowicz and Lau 1989). In this case, follistatin was isolated during the purification of vascular endothelial growth factor from these cells. This would appear to be a fortuitous occurrence based on the similar chromatographic properties of the two molecules, rather than a similarity in function. Follistatin mRNA was subsequently identified in the pituitary (Michel *et al.* 1990). Kaiser *et al.* (1992*a*) have shown that follistatin mRNA is localized to the folliculo-stellate cells and gonadotrophs, and that mRNA levels are regulated in the pituitary gland (Kaiser and Chin 1993). This regulation was shown using reverse transcription-PCR, which allows the detection of minute amounts of mRNA. By using known amounts of a very similar template in the PCR reaction it is possible to quantify the amount of mRNA of interest in the starting material. This technique showed that follistatin mRNA levels are stimulated by ovariectomy. Surprisingly, treatment of ovariectomized rats with oestrogen further increases the levels of follistatin mRNA in the pituitary gland (Kaiser and Chin 1993) (Fig. 6.9).

The identification of expression of the 'gonadal' peptides in the pituitary means that both inhibin B and activin B may be synthesized in gonadotrophs, and that follistatin is produced by gonadotrophs and folliculo-stellate cells of the pituitary gland, and thus they may play autocrine or paracrine regulatory roles in the control of FSH synthesis. This possibility was investigated by Corrigan *et al.* (1991) who used a monoclonal antibody against activin B to immunoneutralize any endogenous hormone secreted by pituitary cells in culture, and thus dissect its role in the regulation of FSH. Indeed, cells treated with the antibody secreted lower basal levels of FSH, and FSHβ mRNA levels in these cells were inhibited. Inhibin and follistatin were able to inhibit further FSH secretion from cells treated with antibody. From the results of this study, and of the study of Weiss *et al.* (1992 *b*), who showed that there is a stimulatory factor in conditioned medium from cultured pituitary cells, it seems clear that activin B is secreted and has local autocrine/paracrine actions in the pituitary gland.

The interaction of these peptides is complex, and it is still unclear whether follistatin acts simply to bind activin, and thus to serve an inhibitory action on FSH by preventing the stimulatory effect of activin, or whether it interacts directly with the gonadotroph itself. Similarly, it is not clear whether inhibin acts directly on the gonadotroph, or by interfering with the binding of activin to its receptor. Further elucidation of the precise mechanism of local action of what until now have been considered gonadal peptide hormones may lead us to rethink our concept of pituitary–gonadal feedback loops (Fig. 6.10).

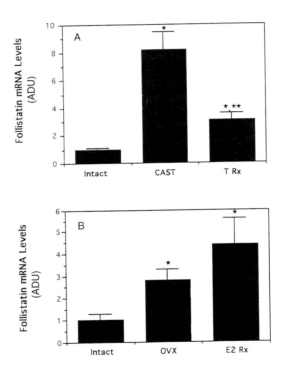

Fig. 6.9 Regulation of follistatin mRNA levels in the rat pituitary gland. (A) intact male rats, 21-day post-castration (CAST) rats, and 21-day post-castration rats treated with testosterone (T Rx) for seven days. (B) intact female rats, 21-day post-ovariectomy (OVX) rats, and 21-day post-ovariectomy rats treated with oestrogen (E2 RX) for seven days. Levels of mRNA were determined by quantitative PCR assay. Each bar represents the mean arbitrary densitometric units (ADU) +/− SEM for three to five rats. *, $p < 0.01$, different from intact rats; **, $p < 0.01$, different from gonadectomized rats. (From Kaiser and Chin 1993.)

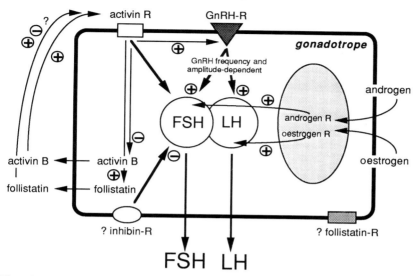

Fig. 6.10 Representation of a pituitary gonadotroph, showing the possible autocrine and paracrine interactions occuring within the pituitary to regulate gonadotrophin synthesis. GnRH-R, gonadotrophin-releasing hormone receptor.

V   NEW HORIZONS

1 Cell lines

The study of the pituitary gonadotrophin genes is currently severely hampered by the lack of an appropriate gonadotroph-derived cell line that will express these genes. Several groups have reported extensive studies of the gonadotrophin α-subunit gene in choriocarcinoma cell lines and have examined transcriptional regulation of the α-subunit gene by cyclic AMP (Deutsch *et al.* 1987; Silver *et al.* 1987; Jameson *et al.* 1988; Mellon *et al.* 1989). While these studies have provided information about how this gene is regulated in these particular cells, they do not necessarily indicate the nature of regulation of the α-subunit gene which might occur in the pituitary gonadotroph. Choriocaranoma cells do not allow the expression of the LHβ or FSHβ genes, so further definition of the role of GnRH and its possible second messengers, and of the steroid and 'gonadal' peptide hormones, will be greatly facilitated by the development of cell lines which allow the expression of the β-subunit genes.

The tumour-producing transgenic mouse model may offer some possibilities in this regard. Mellon and co-workers have used a transgenic strategy to develop the αT3 cell line (Windle et al. 1990). They used the promoter region of the human α-subunit gene fused to the T antigen of the transforming SV40 virus to make transgenic mice. Using this approach, the transforming oncogene is expressed in the cells which normally express the gene corresponding to the promoter used, in this case the α-subunit, which is normally only expressed in the thyrotroph or the gonadotroph of the pituitary gland.

These pituitary cells are apparently derived from a gonadotroph lineage, as expression of the α-subunit gene is stimulated by GnRH but not TRH (Windle et al. 1990). Two groups have used the αT3 cells to examine the α-subunit promoter and have identified cell-specific regions required for expression of that gene (Horn et al. 1992; Schoderbek et al. 1992). Further, two elements in the α-subunit promoter have been identified as mediating GnRH responsiveness using transient transfection studies in this cell line. An enhancer region at −416 to −385 which was GnRH responsive, and a region at −344 to −300 enhanced basal transcription (Schoderbek et al. 1993). The enhancer region corresponds to the GnRH responsive region identified by Kay et al.(1994) between −346 and −244 of the α-subunit promoter using primary cultures of pituitary cells. Additionally, Kay et al. (1994) have used αT3 cells which were stably transfected with a chimeric gene consisting of the α-subunit promoter fused to the CAT reporter gene to show that GnRH transiently stimulates transcription of the α-subunit gene, with desensitization to GnRH following continued exposure via downregulation of the protein kinase C pathway. This interesting approach provides a means of addressing questions regarding the regulation of the gonadotrophin subunit genes which had not been resolvable using primary pituitary cell cultures.

The preparation of transgenic cell lines opens up the possibility that there will soon be a cell line which allows the expression of the β-subunits of the gonadotrophins in a regulated fashion, and which will enable the further study of hormone-specific *cis*-regulatory elements in those genes and cell-specific *trans*-regulatory elements.

2 Transgenic animal models

The use of animals which express gonadotrophin transgenes also will expand our knowledge, as has the model of Nilson and his co-workers described earlier (Hamernik et al. 1992). Transgenic animals which overexpress the genes, or in which the particular gene has been 'knocked out', using either embryonic stem cell manipulations to generate homologous recombinants or targeted cell death strategies, will also enable us to learn much about all the genes involved in the regulation of the gonadotrophins. A transgenic mouse has been produced in which the inhibin α-subunit was knocked out

(Matzuk et al. 1992). These animals developed testicular tumours and had impaired reproductive function. The exact mechanism of this effect is unclear, but it is tantalizing to speculate on the role of inhibin as a possible repressor of gene transcription, perhaps of a gene or genes involved in testicular growth and development.

In another example, Kendall et al.(1991) have used the promoter of the bovine α-subunit fused to the diphtheria toxin gene to ablate gonadotrophs in transgenic mice. Surprisingly, these transgenic mice have normal thyrotroph function, indicating that it is only gonadotrophs which are ablated, and not thyrotrophs, which also express the α-subunit gene. These mice failed to develop gonads and exhibited reproductive failure, similar to mice homozygous for the mutation *hpg*, representing a deletion in the GnRH gene (Mason et al. 1986). These results demonstrate that the 313 base pairs of the α-subunit promoter used in this study is sufficient to direct expression of this gene to gonadotrophs, but that a separate, additional, *cis*-acting element is required for thyrotroph-specific expression of the gene.

This technology provides a tool for dissecting the interrelationships of the pituitary gonadotrophins and the 'gonadal' peptides. Other genes that might be deleted include the β-subunits of inhibin and activin, the activin receptor, as well as the gonadotrophin β-subunits. Knocking out these genes could provide us with much information about the tissue specificity of their expression, their regulation, and the inter-relationships which are occurring in the pituitary gland.

The elucidation of the extremely complex endocrine and autocrine/paracrine relationships in the anterior pituitary gland which together regulate the production of the gonadotrophin hormones will remain a vital area of study for some time. As we have outlined in this review, many sophisticated and exciting techniques are now being employed in attempts to better understand how all these molecules interact to tune finely the regulation of one of the organism's most important functions, reproduction, and thus the survival of the species.

REFERENCES

Aragay, A. M., Katz, A., and Simon, M. I. (1992). The $G_{\alpha q}$ and $G_{\alpha 11}$ proteins couple the thyrotropin-releasing hormone receptor to phospholipase C in GH3 rat pituitary cells. *Journal of Biological Chemistry*, **267**, 24983–8.

Attardi, B. and Fitzgerald, T. (1990). Effects of progesterone on the estradiol-induced follicle-stimulating hormone (FSH) surge and FSHβ messenger ribonucleic acid in the rat. *Endocrinology*, **126**, 2281–7.

Attardi, B., Keeping, H. S., Winters, S. J., Kotsuji, F., Maurer, R. A., and Troen, P. (1989*a*). Rapid and profound suppression of messenger ribonucleic acid encoding follicle-stimulating hormone β by inhibin from primate sertoli cells. *Molecular Endocrinology*, **3**, 280–7.

Attardi, B., Keeping, H. S., Winters, S. J., Kotsuji, F., and Troen, P. (1989b). Effect of inhibin from primate Sertoli cells and GnRH on gonadotropin subunit mRNA in rat pituitary cell cultures. *Molecular Endocrinology*, **3**, 1236–42.

Attardi, B., Marshall, G. R., Zorub, D. S., Winters, S. J., Miklos, J., and Plant, T. M. (1992). Effects of orchidectomy on gonadotropin and inhibin subunit messenger ribonucleic acids in the pituitary of the rhesus monkey (*Macaca mulatta*). *Endocrinology*, **130**, 1238–44.

Attisano, L., Carcamo, J., Ventura, F., Weis, F. M. B., Massagué, J., and Wrana, J. L. (1993). Identification of human activin and TGFβ type I receptors that form heteromeric kinase complexes with type II receptors. *Cell*, **75**, 671–80.

Attisano, L., Wrana, G. L., Cheifetz, S., and Massagué, J. (1992). Novel activin receptors: distinct genes and alternate splicing generate a repertoire of serine/threonine kinase receptors. *Cell*, **68**, 97–108.

Bassing, C., Yingling, J. M., Howe, D. J., Wang, T., He, W. W., Gustafson, M. L., *et al.* (1994). A transforming growth factor β type I receptor that signals to activate gene expression. *Science*, **263**, 87–9.

Batra, S. K. and Miller, W. L. (1985). Progesterone inhibits the basal production of follicle-stimulating hormone in ovine pituitary cell culture. *Endocrinology*, **117**, 2443–8.

Boothby, M., Ruddon, R. W., Anderson, C., McWilliams, D., and Boime, I. (1981). A single gonadotropin α-subunit gene in normal tissue and tumor-derived cell lines. *Journal of Biological Chemistry*, **256**, 5121–7.

Braden, T. D. and Conn, P. M. (1992). Activin-A stimulates the synthesis of gonadotropin-releasing hormone receptors. *Endocrinology*, **130**, 2101–5.

Braden, T. D., Farnworth, P. G., Burger, H. G., and Conn, P. M. (1990). Regulation of the synthetic rate of gonadotropin releasing hormone receptors in rat pituitary cell cultures by inhibin. *Endocrinology*, **127**, 2387–92.

Burrin, J. M. and Jameson, J. L. (1989). Regulation of transfected glycoprotein hormone α-gene expression in primary pituitary cell cultures. *Molecular Endocrinology*, **3**, 1643–51.

Canfield, R. E., Birken, S., Morse, J. H., and Morgan, F. J. (1978). Human chorionic gonadotrophin. In *Peptide hormones* (ed. J. A. Parson), pp. 299–315. London University Press.

Carroll, R. S., Corrigan, A. Z., Gharib, S. D., Vale, W., and Chin, W. W. (1989). Inhibin, activin and follistatin: regulation of follicle-stimulating hormone messenger ribonucleic acid levels. *Molecular Endocrinology*, **3**, 1969–76.

Carroll, R. S., Corrigan, A. Z., Vale, W., and Chin, W. W. (1991). Activin stabilizes follicle-stimulating hormone-beta messenger ribonucleic acid levels. *Endocrinology*, **129**, 1721–6.

Catt, K. J. and Dufau, M. L. (1978). Gonadotropin receptors and regulation of interstitial cell function in the testis. *Receptors and Hormone Action*, **3**, 291–339.

Chi, L., Zhou, W., Prikhozhan, A., Flanagan, C., Davidson, J. S., Golembo, M., Illing, N., Millar, R. P., and Sealfon, S. C. (1993). Cloning and

characterization of the human GnRH receptor. *Molecular and Cellular Endocrinology*, **91**, R1–6.
Childs, G. V., Lloyd, J. M., Unabia, G., Gharib, S. D., Wierman, M. E., and Chin, W. W. (1987 *a*). Detection of luteinizing hormone beta messenger ribonucleic acid (mRNA) in individual gonadotropes after castration: use of a new *in situ* hybridization method with a photobiotinylated complementary RNA probe. *Molecular Endocrinology*, **1**, 926–32.
Childs, G. V., Unabia, G., Tibolt, R., and Lloyd, J. M., (1987*b*). Cytological factors that support nonparallel secretion of luteinizing hormone and follicle-stimulating hormone during the estrous cycle. *Endocrinology*, **121**, 1801–13.
Childs, G. V., Unabia, G., Wierman, M. E., Gharib, S. D., and Chin, W. W. (1990). Castration induces time-dependent changes in the follicle-stimulating hormone β-subunit messenger ribonucleic acid-containing gonadotrope cell population. *Endocrinology*, **126**, 2205–13.
Childs, G. V., Unabia, G., Lee, B. L., and Rougeau, D. (1992*a*). Heightened secretion by small and medium-sized luteinizing hormone (LH) gonadotropes late in the cycle suggests contributions to the LH surge or possible paracrine interactions. *Endocrinology*, **130**, 345–52.
Childs, G. V., Unabia, G., and Lloyd, J. (1992*b*). Recruitment and maturation of small subsets of luteinizing hormone gonadotropes during the estrous cycle. *Endocrinology*, **130**, 335–44.
Chin, W. W. and Gharib, S. D. (1986). Organization and expression of gonadotropin genes. In *Molecular and cellular aspects of reproduction* (ed. D. S. Dhindsa and O. P. Bahl), pp. 245–65. Plenum Press, New York.
Clarke, I. J. (1987). GnRH and ovarian feedback. *Oxford Reviews of Reproductive Biology*, **9**, 54–95.
Clarke, I. J. and Cummins, J. T. (1982). The temporal relationship between gonadotropin releasing hormone (GnRH) and luteinizing hormone (LH) in ovariectomised ewes. *Endocrinology*, **111**, 1737–9.
Clarke, I. J. and Cummins, J. T. (1984). Direct pituitary effects of estrogen and progesterone on gonadotropin secretion in the ovariectomized ewe. *Neuroendocrinology*, **39**, 267–4.
Clarke, I. J., Cummins, J. T., and de Kretser, D. M. (1983). Pituitary gland function after disconnection from direct hypothalamic influences in the sheep. *Neuroendocrinology*, **36**, 376–84.
Clarke, I. J., Findlay, J. K., Cummins, J. T., and Ewens, W. J. (1986). Effects of ovine follicular fluid on plasma LH and FSH secretion in ovariectomized ewes to indicate the site of action of inhibin. *Journal of Reproduction and Fertility*, **77**, 575–85.
Clarke, I. J., Rao, A., Fallest, P. C., and Shupnik, M. A. (1993). Transcription rate of the FSHβ gene is reduced by inhibin in sheep but this does not fully explain the decrease in mRNA. *Molecular and Cellular Endocrinology*, **91**, 211–6.
Clay, C. M., Keri, R. A., Finicle, A. B., Heckert, L. L., Hamernik, D. L., Marschke, K. M., *et al.* (1993). Transcriptional regulation of the glycoprotein hormone α subunit gene by androgen may involve direct binding of androgen receptor to the proximal promoter. *Journal of Biological Chemistry*, **268**, 13556–64.
Corrigan, A. Z., Bilezikjian, L. M., Carroll, R. S., Bald, L. N., Schmelzer, C.

H., Fendly, B. M., et al. (1991). Evidence for an autocrine role for activin B within rat anterior pituitary cultures. *Endocrinology*, **128**, 1682–4.

Counis, R., Corbani, M., and Jutisz, M. (1983). Estradiol regulates mRNAs encoding precursors to rat lutropin (LH) and follitropin (FSH) subunits. *Biochemical and Biophysical Research Communications*, **114**, 65–73.

Culler, M. D. (1992). In vivo evidence that inhibin is a gonadotropin surge-inhibiting/attenuating factor. *Endocrinology*, **131**, 1556–8.

Culler, M. D. and Negro-Vilar, A. (1989). Endogenous inhibin suppresses only basal follicle-stimulating hormone secretion but suppresses all parameters of pulsatile luteinizing hormone secretion in the diestrous rat. *Endocrinology*, **124**, 2944–53.

Dalkin, A. C., Haisenleder, D. J., Ortolano, G. A., Suhr, A., and Marshall, J. C. (1990). Gonadal regulation of gonadotropin subunit gene expression: evidence for regulation of follicle-stimulating hormone-β messenger ribonucleic acid by nonsteroidal hormones in female rats. *Endocrinology*, **127**, 798–806.

Dalkin, A. C., Paul, S. J., Haisenleder, D. J., Ortolano, G. A., Yasin, M., and Marshall, J. C. (1992). Gonadal steroids effect similar regulation of gonadotrophin subunit mRNA expression in both male and female rats. *Journal of Endocrinology*, **132**, 39–45.

Davison, J. S., Flanagan, C. A., Becker, I. I., Illing, N., Sealfon, S. C., and Millar, R. P. (1994). Molecular function of the gonadotrophin-releasing hormone receptor: insights from site-directed mutagenesis. *Molecular and Cellular Endocrinology*, **100**, 9–14.

Delegeane, A. M., Ferland, L. H., and Mellon, P. L. (1987). Tissue-specific enhancer of the human glycoprotein hormone α-subunit gene: dependence on cyclic AMP-inducible elements. *Molecular and Cellular Biology*, **7**, 3994–4002.

Deutsch, P. J., Jameson, J. L., and Habener, J. F. (1987). Cyclic AMP responsiveness of human gonadotropin-α gene transcription is directed by a repeated 18-base pair enhancer. *Journal of Biological Chemistry*, **262**, 12169–74.

Eidne, K. A., Sellar, R. E., Couper, G., Anderson, L., and Anderson, P. L. (1992). Molecular cloning and characterisation of the rat pituitary gonadotrophin-releasing hormone (GnRH) receptor. *Molecular and Cellular Endocrinology*, **90**, R5–9.

Eto, Y., Tsuji, T., Takezawa, M., Takano, S., Yokogawa, Y., and Shibai, H.(1987). Purification and characterization of erythroid differentiation factor (EDF) isolated from human leukemia cell line THP-1. *Biochemical and Biophysical Research Communications*, **142**, 1095–103.

Farnworth, P. G., Robertson, D. M., deKretser, D. M., and Burger, H. G. (1988). Effect of 31 kilodalton bovine inhibin on follicle-stimulating hormone and luteinizing hormone in rat pituitary cells in vitro: actions under basal conditions. *Endocrinology*, **122**, 207.

Findlay, J. K. (1993). An update of the roles of inhibin, activin and follistatin as local regulators of folliculogenesis. *Biology of Reproduction*, **48**, 15–23.

Findlay, J. K., Tsonis, C. G., Doughton, B., Brown, R. W., Bertram, K. C., Braid, G. H., et al. (1989). Immunisation against the amino-terminal peptide (α_N) of the alpha$_{43}$ subunit of inhibin impairs fertility in sheep. *Endocrinology*, **124**, 3122–4.

Franzen, P., ten Dijke, P., Ichijo, H., Yamashita, H., Schulz, P., Heldin, C., *et al.* (1993). Cloning of a TGFβ type I receptor that forms a heteromeric complex with the TGFβ type II receptor. *Cell*, **75**, 681–92.

Gharib, S. D., Bowers, S. M., Need, L. R., and Chin, W. W. (1986). Regulation of rat LH subunit mRNAs by gonadal steriod hormones. *The Journal of Clinical Investigation*, **77**, 582–9.

Gharib, S. D., Wierman, M. E., Badger, T. M., and Chin, W. W. (1987). Sex steroid hormone regulation of follicle-stimulating hormone subunit messenger ribonucleic acid (mRNA) levels in the rat. *Journal of Clinical Investigation*, **80**, 294–9.

Gharib, S. D., Leung, P. C. K., Carroll, R. S., and Chin, W. W. (1990*a*). Androgens positively regulate follicle-stimulating hormone β subunit mRNA levels in cultured rat pituitary cells. *Molecular Endocrinology*, **4**, 1620–6.

Gharib, S. D., Wierman, M. E., Shupnik, M. A., and Chin, W. W. (1990*b*). Molecular biology of the pituitary gonadotropins. *Endocrine Reviews*, **11**, 177–99.

Gospodarowicz, D. and Lau, K. (1989). Pituitary follicular cells secrete both vascular endothelial growth factor and follistatin. *Biochemical and Biophysical Research Communications*, **165**, 292–8.

Grotjan, H. E. (1989). Oligosaccharide structures of the anterior pituitary and placental glycoprotein hormones. In *Microheterogeneity of glycoprotein hormones* (ed. B. A. Keel and H. E. Grotjan), pp. 23–52. CRC Press, Boca Raton.

Haisenleder, D. J., Khoury, S., Zmeili, S. M., Papavasiliou, S., Ortolano, G. A., Dee, C., *et al.* (1987). The frequency of gonadotropin-releasing hormone secretion regulates expression of α and luteinizing hormone β-subunit messenger ribonucleic acids in male rats. *Molecular Endocrinology*, **1**, 834–8.

Haisenleder, D. J., Barkan, A. L., Papavasiliou, S., Zmeili, S. M., Dee, C. Jameel, M. L., *et al.* (1988*a*). LH subunit mRNA concentrations during LH surge in ovariectomized estradiol-replaced rats. *American Journal of Physiology*, **254**, E99–106.

Haisenleder, D. J., Katt, J. A., Ortolano, G. A., El-Gewely, M. R., Duncan, J. A., Dee, C., and Marshall, J. C. (1988*b*). Influence of gonadotropin-releasing hormone pulse amplitude, frequency, and treatment duration on the regulation of luteinizing hormone (LH) subunit messenger ribonucleic acids and LH secretion. *Molecular Endocrinology*, **2**, 338–43.

Haisenleder, D. J., Ortolano, G. A., Dalkin, A. C., Ellis, T. R., Paul, S. J., and Marshall, J. C. (1990). Differential regulation of gonadotropin subunit gene expression by gonadotropin-releasing hormone pulse amplitude in female rats. *Endocrinology*, **127**, 2869–75.

Haisenleder, D. J., Dalkin, A. C., Ortolano, G. A., Marshall, J. C. and Shupnik, M. A. (1991). A pulsatile gonadotropin-releasing hormone stimulus is required to increase transcription of the gonadotropin subunit genes: evidence for differential regulation of transcription by pulse frequency *in vivo*. *Endocrinology*, **128**, 509–17.

Haisenleder, D. J., Ortolano, G. A., Yasin, M., Dalkin, A. C., and Marshall, J. C. (1993). Regulation of gonadotropin subunit messenger ribonucleic acid expression by gonadotropin-releasing hormone pulse amplitude *in vitro*. *Endocrinology*, **132**, 1292–6.

Hamernik, D. L. and Nett, T. M. (1988a). Gonadotropin-releasing hormone increases the amount of messenger ribonucleic acid for gonadotropins in ovariectomized ewes after hypothalamic-pituitary disconnection. *Endocrinology*, **122**, 959–66.

Hamernik, D. L. and Nett, T. M. (1988b). Measurement of the amount of mRNA for gonadotropins during an estradiol-induced preovulatory-like surge of LH and FSH in ovariectomized ewes. *Domestic Animal Reproduction*, **5**, 129–9.

Hamernik, D. L., Kim, K. E., Maurer, R. A., and Nett, T. M. (1987). Progesterone does not affect the amount of mRNA for gonadotropins in the anterior pituitary gland of ovariectomized ewes. *Biology of Reproduction*, **37**, 1225–32.

Hamernik, D. L., Keri, R. A., Clay, C. M., Clay, J. N., Sherman, G. B., Sawyer, H. R., et al. (1992). Gonadotrope- and thyrotrope- specific expression of the human and bovine glycoprotein hormone α-subunit genes is regulated by distinct *cis*-acting elements. *Molecular Endocrinology*, **6**, 1745–55.

Hemmati-Brivanlou, A. and Melton, D. A. (1992). A truncated activin receptor inhibits mesoderm induction and formation of axial structures in *Xenopus* embryos. *Nature*, **359**, 609–14.

Henderson, K. M. (1979). Gonadotropic regulation of ovarian activity. *British Medical Bulletin*, **35**, 161–6.

Herring, R. D., Hamernik, D. L., Kile, J. P., Sousa, M. E., and Nett, T. M. (1991). Chronic administration of estradiol produces a triphasic effect on serum concentrations of gonadotropins and messenger ribonucleic acid for gonadotropin subunits, but not on pituitary content of gonadotropins, in ovariectomized ewes. *Biology of Reproduction*, **45**, 151–6.

Horn, F., Windle, J. J., Barnhart, K. M., and Mellon, P. L. (1992). Tissue-specific gene expression in the pituitary: the glycoprotein hormone α-subunit is regulated by a gonadotrope-specific protein. *Molecular and Cellular Biology*, **12**, 2143–53.

Hsieh, K-P. and Martin, T. F. J. (1992). Thyrotropin-releasing hormone and gonadotropin-releasing hormone receptors activate phosphlipase C by coupling to the guanosine triphosphate-binding proteins G_q and G_{11}. *Molecular Endocrinology*, **6**, 1673–81.

Iliff-Sizemore, S. A., Ortolano, G. A., Haisenleder, D. J., Dalkin, A. C., Krueger, K. A., and Marshall, J. C. (1990). Testosterone differentially modulates gonadotropin subunit messenger ribonucleic acid responses to gonadotropin-releasing hormone pulse amplitude. *Endocrinology*, **127**, 2876–83.

Illing, N., Jacobs, G. F., Becker, I. I., Flanagan, C. A., Davidson, J. S., Eales, A., et al. (1993). Comparative sequence analysis and functional characterization of the cloned sheep gonadotrophin-releasing hormone receptor reveal differences in primary structure and ligand specificity among mammalian receptors. *Biochemical and Biophysical Research Communications*, **196**, 745–51.

Ingraham, H., Albert, V. R., Chen, R., Crenshaw, E. B., Elsholtz, H. P., He, X., et al. (1990). A family of Pou-domain and Pit-1 tissue specific transcription factors in pituitary and endocrine development. *Annual Review of Physiology*, **52**, 773–91.

Inverson, R. A., Day, K. H., d'Emden, M., Day, R. M., and Maurer, R. A. (1990). Clustered point mutation analysis of the rat prolactin promoter. *Molecular Endocrinology*, **4**, 1564–71.

Jakubowiak, A., Janecki, A., and Steinberger, A. (1989). Similar effects of inhibin and cyclohexamide on gonadotropin release in superfused pituitary cell cultures. *Biology of Reproduction*, **41**, 454–63.

Jakubowiak, A., Janecki, A., Tong, D., Sanborn, B. M., and Steinberger, A. (1991a). Effects of recombinant human inhibin and testosterone on gonadotrophin secretion and subunit mRNA in superfused male rat pituitary cell cultures stimulated with pulsatile gonadotropin-releasing hormone. *Molecular and Cellular Endocrinology*, **82**, 265–73.

Jakubowiak, A., Tong, D., Janecki, A., Sanborn, B. M., and Steinberger, A. (1991b). Pulsatile GnRH stimulation increases steady-state mRNA levels for FSHβ, LHβ, and α-subunits in superfused pituitary cell cultures. *Molecular and Cellular Neurosciences*, **2**, 277–83.

Jameson, J. L., Jaffe, R. C., Deutsch, P. J., Albanese, C., and Habener, J. F. (1988). The gonadotropin α-gene contains multiple protein binding domains that interact to modulate basal and cAMP-responsive transcription. *Journal of Biological Chemistry*, **263**, 9879–86.

Jutisz, M. and Tetrin-Clary, C. (1974). Luteinizing hormone and chorionic gonadotrophin: structure and activity. *Current Topics in Experimental Endocrinology*, **2**, 195–246.

Kaiser, U. B. and Chin, W. W. (1993). Regulation of follistatin messenger ribonucleic acid levels in the rat pituitary. *Journal of Clinical Investigation*, **19**, 2523–31.

Kaiser, U. B., Jakubowiak, A., Steinberger A., and Chin, W. W. (1993). Regulation of rat pituitary gonadotropin-releasing hormone receptor mRNA levels *in vivo* and *in vitro*. *Endocrinology*, **133**, 931–34.

Kaiser, U. B., Lee, B. L., Carroll, R. S., Unabia, G., Chin, W. W., and Childs, G. V. (1992a). Follistatin gene expression in the pituitary: localization in gonadotroph and folliculostellate cells in diestrous rats. *Endocrinology*, **130**, 3048–56.

Kaiser, U. B., Zhao, D., Cardona, G. R., and Chin, W. W. (1992b). Isolation and characterization of cDNAs encoding the rat pituitary gonadotropin-releasing hormone receptor. *Biochemical and Biophysical Research Communications*, **189**, 1645–52.

Kakar, S. S., Musgrove, L. C., Devor, D. C., Sellers, J. C., and Neill, J. D. (1992). Cloning, sequencing and expression of human gonadotropin releasing hormone (GnRH) receptor. *Biochemical and Biophysical Research Communications*, **189**, 289–95.

Kalra, P. S. and Kalra, S. P. (1985). Control of gonadotropin secretion. In *The pituitary gland* (ed. H. Imura), pp. 189-220. Raven Press, New York.

Karin, M., Castrillo, J. L., and Theil, L. E. (1990). Growth hormone gene regulation: a paradigm for cell-type specific gene activation. *Trends in Genetics*, **6**, 92–6.

Karsch, F. J. (1987). Central actions of ovarian steroids in the feedback regulation of pulsatile secretion of luteinizing hormone. *Annual Review of Physiology*, **49**, 365–82.

Kay, T. W. H., Chedrese, P. J., and Jameson, J. L. (1994). Gonadotropin-releasing hormone causes transcriptional stimulation followed by desensitization of the glycoprotein hormone α promoter in transfected αT3 gonadotrope cells. *Endocrinology*, **134**, 568–73.

Kay, T. W. H. and Jameson, J. L. (1992). Identification of a gonadotropin-releasing hormone-responsive region in the glycoprotein hormone α-subunit promoter. *Molecular Endocrinology*, **6**, 1767–73.

Kendall, S. K., Saunders, T. L., Jin, L., Lloyd, R. V., Glode, L. M., Nett, T. M., et al. (1991). Targeted ablation of pituitary gonadotropes in transgenic mice. *Molecular Endocrinology*, **5**, 2025–36.

Kim, K. and Ramirez, D. (1985). *In vitro* luteinizing hormone release from superfused rat hypothalami: site of action of progesterone and effect of estrogen priming. *Endocrinology*, **116**, 252–7.

Kim, K. E., Day, K. H., Howard, P., Salton, S. R. J., Roberts, J. L., and Maurer, R. A. (1990). DNA sequences required for expression of the LHβ promoter in primary cultures of rat pituitary cells. *Molecular and Cellular Endocrinology*, **74**, 101–7.

Kim, W. H., Yuan, Q. X., Swerdloff, R. S., and Bhasin, S. (1988). Regulation of alpha and luteinizing hormone beta subunit messenger ribonucleic acids during stimulatory and downregulatory phases of gonadotropin-releasing hormone action. *Biology of Reproduction*, **39**, 847–53.

Knobil, E. (1974). On the control of gonadotropin secretion in the rhesus monkey. *Recent Progress in Hormone Research*, **30**, 1–35.

Lalloz, M. R. A., Detta, A., and Clayton, R. N. (1988a). Gonadotropin-releasing hormone desensitization preferentially inhibits expression of the luteinizing hormone β-subunit gene *in vivo*. *Endocrinology*, **122**, 1689–94.

Lalloz, M. R. A., Detta, A., and Clayton, R. N. (1988b). Gonadotropin-releasing hormone is required for enhanced luteinizing hormone subunit gene expression *in vivo*. Endocrinology, **122**, 1681–8.

Landefeld, T. D. and Kepa, J. (1984). Regulation of LH beta subunit mRNA in the sheep pituitary gland during different feedback states of estradiol. *Biochemical and Biophysical Research Communications*, **122**, 1307–13.

Landefeld, T. D., Kepa, J., and Karsch, F. J. (1984). Estradiol feedback effects on the alpha subunit mRNA in the sheep pituitary gland: correlation with serum and pituitary luteinizing hormone concentrations. *Proceedings of the National Academy of Science of the USA*, **81**, 1322–6.

Landefeld, T. D., Bagnell, T., and Levitan, I. (1989). Effects of estradiol on gonadotropin subunit messenger ribonucleic acid amounts during an induced gonadotropin surge in anestrous ewes. *Molecular Endocrinology*, **3**, 10–14.

LaPolt, P. S., Jia, X-C., Sincich, C., and Hsueh, A. J. W. (1991). Ligand-induced down regulation of testicular and ovarian luteinizing hormone (LH) receptors is preceded by tissue-specific inhibition of alternately processed LH receptor transcripts. *Molecular Endocrinology*, **5**, 397–403.

Laws, S. C., Beggs, M. J., Webster, J. C., and Millie, W. L. (1990). Inhibin increases and progesterone decreases receptors for gonadotropin-releasing hormone in ovine pituitary culture. *Endocrinology*, **127**, 373–80.

Lee, W. S., Smith, M. S., and Hoffman, G. E. (1990). Progesterone enhances

the surge of luteinizing hormone by increasing the activation of luteinizing-hormone releasing hormone neurons. *Endocrinology*, **127**, 2604–6.
Leung, K., Kaynard, A. H., Negrini, B. P., Kim, K. E., Maurer, R. A., and Landefeld, T. D. (1987). Differential regulation of gonadotropin subunit messenger ribonucleic acids by gonadotropin-releasing hormone pulse frequency in ewes. *Molecular Endocrinology*, **1**, 724–8.
Leung, K., Kim, K. E., Maurer, R. A., and Landefeld, T. D. (1988). Divergent changes in the concentrations of gonadotropin β-subunit messenger ribonucleic acid during the estrous cycle of sheep. *Molecular Endocrinology*, **2**, 272–6.
Levine, L. E., Norman, R. L., Gliessman, P. M., Oyama, T. T., Bangsberg, D. R., and Spies, H. G. (1985). *In vivo* gonadotropin-releasing hormone release and serum luteinizing hormone measurements in ovariectomized, estrogen-treated rhesus monkeys. *Endocrinology*, **102**, 1008–14.
Lincoln, G. A. (1979). Pituitary control of testicular activity. *British Medical Bulletin*, **35**, 167–72.
Ling, N., Ying, S. Y., Ueno, N., Esch, F., Denoroy, L., and Guillemin, R. (1985). Isolation and characterization of a MW 32,000 protein with inhibin activity from porcine follicular fluid. *Proceedings of the National Academy of Sciences of the USA*, **82**, 7217–21.
Ling, N., Ying, S. Y., Ueno, N., Shimasaki, S., Esch, F., Hotta, M., *et al.* (1986). A homodimer of the subunits of inhibin A stimulates the secretion of pituitary follicle-stimulating hormone. *Biochemical and Biophysical Research Communication*, **138**, 1129–37.
Lloyd, J. M. and Childs, G. V. (1988a). Changes in the number of GnRH-receptive cells during the rat estrous cycle: biphasic effects of estradiol. *Neuroendocrinology*, **47**, 210–22.
Lloyd, J. M. and Childs, G. V. (1988b). Differential storage and release of luteinizing hormone and follicle-releasing hormone from individual gonadotropes separated by centrifugal elutriation. *Endocrinology*, **122**, 1282–90.
Loosfelt, H., Mishrahi, M., Atger, M., Salesse, R., Thi, M., Jolivet, A., *et al.* (1989). Cloning and sequencing of porcine LH-hCG receptor cDNA: variants lacking transmembrane domain. *Science*, **245**, 525–7.
Mann, G. E., Campbell, B. K., McNeilly, A. S., and Baird, D. T. (1989). Passively immunizing ewes against inhibin during the luteal phase of the oestrous cycle raises the plasma concentration of FSH. *Journal of Endocrinology*, **123**, 383–91.
Marshall, J. C. and Kelch, R. P. (1986). Gonadotropin-releasing hormone: role of pulsatile secretion in the regulation of reproduction. *New England Journal of Medicine*, **315**, 1459–63.
Mason, A. J., Hayflick, J. S., Zoeller, R. T., Young, W. S., Phillips, H. S., Nikolics, K., *et al.* (1986). A deletion truncating the gonadotropin-releasing hormone gene is responsible for hypogonadism in the hpg mouse. *Science*, **234**, 1366–71.
Mason, A. J., Berkemeier, L. M., Schmelzer, C. H., and Schwall, R. H. (1989). Activin B: precursor sequences, genomic structure and *in vitro* activities. *Molecular Endocrinology*, **3**, 1352–8.
Matsuzaki, K., Xu, J., Wang, F., McKeehan, W. L., Krummen, L., and Kan, M.

(1993). A widely expressed transmembrane serine/threonine kinase that does not bind activin, inhibin, transforming growth factor β, or bone morphogenic factor. *Journal of Biological Chemistry*, **268**, 12719–23.

Matthews, L. S. and Vale, W. W. (1991). Expression cloning of an activin receptor, a predicted transmembrane serine kinase. *Cell*, **65**, 973–82.

Matzuk, M. M., Finegold, M. J., Su, J. G., Hsueh, A. J., and Bradley, A. (1992). α-inhibin is a tumor-supressor gene with gonadal specificity in mice. *Nature*, **360**, 313–9.

McFarland, K. C., Sprengel, R., Phillips, H. S., Kohler, M., Rosemblit, N., Nikolics, K., *et al.* (1989). Lutropin-choriogonadotropin receptor: an unusual member of the G protein-coupled receptor family. *Science*, **245**, 494–9.

Means, A. R., Dedman, J. R., Tash, J. S., Tindall, D. J., van Sickle, M., and Welsh, M. J. (1980). Regulation of the testis sertoli cell by follicle stimulating hormone. *Annual Review of Physiology*, **42**, 59–70.

Mellon, P. L., Clegg, C. H., Correll, L. A., and McKnight, G. S. (1989). Regulation of transcription by cyclic AMP-dependent protein kinase. *Proceedings of the National Academy of Sciences of the USA*, **86**, 4887–91.

Mellon, P. L., Windle, J. J., Goldsmith, P. C., Padula, C. A., Roberts, J. L., and Weiner, R. I. (1990). Immortalization of hypothalamic GnRH neurons by genetically targeted tumorigenesis. *Neuron*, **5**, 1–10.

Mercer, J. E. and Clarke, I. J. (1989). Regulation of anterior pituitary gonadotrophin subunit mRNA levels by gonadotrophin-releasing hormone in the ewe. *Journal of Neuroendocrinology*, **1**, 327–31.

Mercer, J. E., Clements, J. A., Funder, J. W., and Clarke, I. J. (1987). Rapid and specific lowering of pituitary FSHβ mRNA levels by inhibin. *Molecular and Cellular Endocrinology*, **53**, 251–4.

Mercer, J. E., Clements, J. E., Funder, J. W., and Clarke, I. J. (1988). Luteinizing hormone-β mRNA levels are regulated primarily by gonadotropin-releasing hormone and not by negative estrogen feedback on the pituitary. *Neuroendocrinology*, **47**, 563–6.

Mercer, J. E., Clements, J. A., Funder, J. W., and Clarke, I. J. (1989). Regulation of follicle-stimulating hormone β and common α-subunit messenger ribonucleic acid by gonadotropin releasing hormone and estrogen in the sheep pituitary. *Neuroendocrinology*, **50**, 280–285.

Mercer, J. E., Phillips, D. J., and Clarke, I. J. (1993). Short-term regulation of gonadotrophin subunit mRNA levels by estrogen: studies in the hypothalamo-pituitary intact and hypothalamo-pituitary disconnected ewe. *Journal of Neuroendocrinology*, **5**, 591–6.

Meunier, H., Rivier, C., Evans, R. M., and Vale, W. (1988). Gonadal and extragonadal expression of inhibin α, $β_A$, and $β_B$ subunits in various tissues predicts diverse functions. *Proceedings of the National Academy of Sciences of the USA*, **85**, 247–51.

Michel, U., Albiston, A., and Findlay, J. K. (1990). Rat follistatin: gonadal and extragonadal expression and evidence for alternate splicing. *Biochemical and Biophysical Research Communications*, **173**, 401–7.

Moore, W. T., Burleigh, B. D., and Ward, D. N. (1980). Chorionic gonadotropins. comparative studies and comments on relationships to other glycoprotein hormones. In *Chorionic gonadotropin* (ed. S. J. Segal), pp. 89–126. Plenum Press, New York.

Moriarty, G. C. (1976). Immunocytochemistry of the pituitary glycoprotein hormones. *Journal of Histochemistry and Cytochemistry*, **24**, 846–63.

Nakamura, T., Takio, K., Eto, Y., Shibai, H., Titani, K., and Sugino, H. (1990). Activin-binding protein from rat ovary is follistatin. *Science*, **247**, 836–8.

Naylor, S. L., Chin, W. W., Goodman, H. M., Lalley, P. A., Grzeschik, K. H., and Sakaguchi, A. Y. (1983). Chromosome assignment of genes encoding the α and β subunits of glycoprotein hormone in man and mouse. *Somatic Cell Genetics*, **9**, 757–70.

Nilson, J. H., Nejedlik, M. T., Virgin, B. J., Crowder, M. E., and Nett, T. M. (1983). Expression of α subunit and luteinizing hormone β genes in the ovine anterior pituitary: estradiol suppresses accumulation of mRNAs for both α subunit and luteinizing hormone β. *Journal of Biological Chemistry*, **258**, 12087–90.

Nussenzveig, D. R., Heinflink, M., and Gershengorn, M. C. (1993). Agonist-stimulated internalization of the thyrotropin-releasing hormone receptor is dependent on two domains in the receptor carboxyl terminus. *Journal of Biological Chemistry*, **268**, 2389–92.

Paul, S. J., Ortolano, G. A., Haisenleder, D. J., Stewart, J. M., Shupnik, M. A., and Marshall, J. C. (1990). Gonadotropin subunit messenger RNA concentrations after blockade of gonadotropin-releasing hormone action: testosterone selectively increases follicle-stimulating hormone β-subunit messenger RNA by post-transcriptional mechanisms. *Molecular Endocrinology*, **4**, 1943–55.

Perheentupa, A. and Huhtaniemi, I. (1990). Gonadotropin gene expression and secretion in gonadotropin-releasing hormone agonist-treated male rats: effect of sex steroid replacement. *Endocrinology*, **126**, 3204–9.

Pierce, J. G. and Parsons, T. F. (1981). Glycoprotein hormones: structure and function. *Annual Reviews of Biochemistry*, **50**, 465–95.

Pohl, C. R., Richardson, D. W., Hutchison, J. S., Germak, J. A., and Knobil, E. (1983). Hypophysiotropic signal frequency and the functioning of the pituitary-ovarian system in the Rhesus monkey. *Endocrinology*, **39**, 214–9.

Radovick, S., Ticknor, C. M., Nakayama, Y., Notides, A. C., Rahman, A., Weintraub, B. D., *et al.* (1991). Evidence for direct estrogen regulation of the human gonadotropin-releasing hormone gene. *Journal of Clinical Investigation*, **88**, 1649–55.

Reinhart, J., Mertz, L. M., and Catt, K. J. (1992). Molecular cloning and expression of cDNA encoding the murine gonadotropin-releasing hormone receptor. *Journal of Biological Chemistry*, **267**, 21281–4.

Richards, J. S. (1980). Maturation of ovarian follicles: actions and interactions of pituitary and ovarian hormones on follicular cell differentiation. *Physiological Reviews*, **60**, 51–89.

Rivier, C. and Vale, W. (1989). Immunoneutralization of endogenous inhibin modifies hormone secretion and ovulation rate in the rat. *Endocrinology*, **125**, 152–7.

Roberts, J. L., Dultow, C. M., Jakubowski, M., Blum, M., and Millar, R. P.

(1989a). Estradiol stimulates preoptic area-anterior hypothalamic pro-GnRH-GAP gene expression in ovariectomized rats. *Molecular Brain Research*, **6**, 127–34.

Roberts, V., Meunier, H., Vaughan, J., Rivier, J., Rivier, C., Vale, W., et al. (1989b). Production and regulation of inhibin subunits in pituitary gonadotropes. *Endocrinology*, **124**, 552–4.

Robertson, D. M., Foulds, L. M., Leversha, L., Morgan, F. J., Hearn, M. T. W., Burger, H. G., et al. (1985). Isolation of inhibin from bovine follicular fluid. *Biochemical and Biophysical Research Communications*, **126**, 220–6.

Rothfeld, J., Hejtmancik, J. F., Conn, P. M., and Pfaff, D. W. (1989). In situ hybridization for LHRH mRNA following estrogen treatment. *Molecular Brain Research*, **6**, 121–5.

Salton, S. R. J., Blum, M., Jonassen, J. A., Clayton, R. N., and Roberts, J. L. (1988). Stimulation of pituitary luteinizing hormone secretion by gonadotropin-releasing hormone is not coupled to β-luteinizing hormone gene transcription. *Molecular Endocrinology*, **2**, 1033–42.

Schally, A. V. (1978). Aspects of hypothalamic regulation of the pituitary gland. *Science*, **202**, 18–28.

Schally, A. V., Kastin, A. J., and Arimura, A. (1972). FSH-releasing hormone and LH-releasing hormone. *Vitamins and Hormones*, **30**, 83.

Schoderbek, W. E., Kim, K. E., Ridgway, E. C., Mellon, P. L., and Maurer, R. A. (1992). Analysis of DNA sequences required for pituitary-specific expression of the glycoprotein hormone α-subunit gene. *Molecular Endocrinology*, **6**, 893–903.

Schoderbek, W. E., Robertson, M. S., and Maurer, R. A. (1993). Two different DNA elements mediate gonadotropin releasing hormone effects on expression of the glycoprotein hormone α-subunit gene. *Journal of Biological Chemistry*, **268**, 3903–10.

Shimasaki, S., Koga, M., Buscaglia, M. L., Simmons, D. M., Bicsak, T. A., and Ling, N. (1989). Follistatin gene expression in the ovary and extragonadal tissues. *Molecular Endocrinology*, **3**, 651–9.

Shupnik, M. A. (1990). Effects of gonadotropin-releasing hormone on rat gonadotropin gene transcription in vitro: requirement for pulsatile administration for luteinizing hormone-β gene stimulation. *Molecular Endocrinology*, **4**, 1444–50.

Shupnik, M. A., Gharib, S. D., and Chin, W. W. (1989a). Divergent effects of estradiol on gonadotropin gene transcription in pituitary fragments. *Molecular Endocrinology*, **3**, 474–80.

Shupnik, M. A., Weinmann, C. M., Notides, A. C., and Chin, W. W. (1989b). An upstream region of the rat luteinizing hormone beta gene binds estrogen receptor and confers estrogen responsiveness. *Journal of Biological Chemistry*, **264**, 80–6.

Silver, B. J., Bokar, J. A., Virgin, J. B., Vallen, E. A., Milsted, A., and Nilson, J. H. (1987). Cyclic AMP regulation of the human glycoprotein hormone α-subunit gene is mediated by an 18-base-pair element. *Proceedings of the National Academy of Science of the USA*, **84**, 2198–202.

Simard, J., Labrie, C., Hubert, J. F., and Labrie, F. (1988). Modulation by sex steroids and [D-TRP^6, des-Gly-NH$_2$^{10}] luteinizing hormone (LH)-releasing hormone ethylamide of α subunit and LHβ messenger ribonucleic acid levels in the rat anterior pituitary gland. *Molecular Endocrinology*, **2**, 775–84.

Simmons, D. M., Voss, J. W., Ingraham, H. A., Hollaway, J. M., Broide, R. S., Rosenfeld, M. G., and Swanson, L. W. (1990). Pituitary cell phenotypes involve cell-specific Pit-1 mRNA translation and synergistic interactions with other classes of transcription factors. *Genes and Development*, **4**, 695–711.

Smith, J. C. (1990). Identification of a potent *Xenopus* mesoderm-inducing factor as a homologue of activin A. *Nature*, **345**, 729–31.

Soules, M. R., Steiner, R. A., Clifton, D. K., Cohen, N. L., Aksel, S., and Bremner, W. J. (1984). Progesterone modulation of pulsatile luteinizing hormone secretion in normal women. *Journal of Clinical Endocrinology and Metabolism*, **58**, 378–83.

Sprengel, R., Braun, T., Nikolics, K., Segaloff, D. L., and Seeburg, P. H. (1990). The testicular receptor for follicle stimulating hormone: structure and functional expression of cloned cDNA. *Molecular Endocrinology*, **4**, 525–30.

Stanton, P. G., Robertson, D. M., Burgon, P. G., Schmauk-White, B., and Hearn, M. T. (1992). Isolation and physicochemical characterization of human follicle-stimulating hormone isoforms. *Endocrinology*, **130**, 2820–32.

Tashjian, A. H. (1979). Clonal strains of hormone-producing pituitary cells. *Methods in Enzymology*, **58**, 527–35.

ten Dijke, P. T., Yamashita, H., Ichijo, H., Franzen, P., Laiho, M., Miyazono, K., *et al.* (1994). Characterization of type I receptors for transforming growth factor-β and activin. *Science*, **264**, 101–4.

Tixier-Vidal, A., Tougard, C., Kerdelhue, B., and Jutisz, M. (1975). Light and electron microscopic studies on immunocytochemical localization of gonadotropic hormones in the rat pituitary gland with antisera against ovine FSH, LH, LHα, and LHβ. *Annals of the New York Academy of Science*, **254**, 433–60.

Tsuchida, K., Matthews, L. S., and Vale, W. W. (1993). Cloning and characterization of a transmembrane serine kinase that acts as an activin type I receptor. *Proceedings of the National Academy of Science of the USA*, **90**, 11242–6.

Tsutsumi, M., Zhou, W., Millar, R. P., Mellon, P. L., Roberts, J. L., Flanagan, C. A., *et al.* (1992). Cloning and functional expression of a mouse gonadotropin-releasing hormone receptor. *Molecular Endocrinology*, **6**, 1163–9.

Ueno, N., Ling, N., Ying, S. Y., Esh, F., Shimasaki, S., and Guillemin, R. (1987). Isolation and partial purification of follistatin, a novel Mr 35,000 monomeric protein which inhibits the release of follicle-stimulating hormone. *Proceedings of the National Academy of Sciences of the USA*, **84**, 8282–6.

Vale, W., Rivier, J., Vaughan, J., McClintock, R., Corrigan, A., Woo, W., *et al.* (1986). Purification and characterization of an FSH releasing protein from porcine ovarian follicular fluid. *Nature*, **321**, 776–8.

Van den Eijinden-Van Raaij, A. J., van Zoelent, E. J., van Nimmen, K., Koster, C. H., Snoek, G. T., Durston, A. J., *et al.* (1990). Activin-like factor from a *Xenopus leavis* cell line responsible for mesoderm induction. *Nature*, **345**, 732–4.

Van Vugt, D. A., Lam, N. Y., and Ferin, M. (1984). Reduced frequency of pulsatile luteinizing hormone secretion in the luteal phase of the Rhesus monkey: involvement of endogenous opiates. *Endocrinology*, **115**, 1095–101.

Wang, H., Segaloff, D. L., and Ascoli, M. (1991). Lutropin/choriogonadotropin down regulates its receptor by both receptor-mediated endocytosis and a cAMP-dependent reduction in receptor mRNA. *Journal of Biological Chemistry*, **266**, 780–5.

Wang, Q. F., Farnworth, P. G., Findlay, J. K., and Burger, H. G. (1989). Inhibitory effect of pure 31-kilodalton bovine inhibin on gonadotropin-releasing hormone (GnRH)-induced up-regulation of GnRH binding sites in cultured rat anterior pituitary cells. *Endocrinology*, **124**, 363–8.

Weiss, J., Jameson, J. L., Burrin, J. M., and Crowley, W. F. J. (1990). Divergent responses of gonadotropin subunit messenger RNAs to continuous versus pulsatile gonadotropin-releasing hormone *in vitro*. *Molecular Endocrinology*, **4**, 557–64.

Weiss, J., Crowley, W. F., and Jameson, J. L. (1992*a*). Pulsatile gonadotropin-releasing hormone modifies polyadenylation of gonadotropin subunit messenger ribonucleic acids. *Endocrinology*, **130**, 415–20.

Weiss, J., Harris, P. E., Halvorson, L. M., Crowley, W. F., and Jameson, J. L. (1992*b*). Dynamic regulation of follicle-stimulating hormone-β messenger ribonucleic acid levels by activin and gonadotropin-releasing hormone in perifused rat pituitary cells. *Endocrinology*, **131**, 1403–8.

Wierman, M. E., Rivier, J., and Wang, C. (1989*a*). Gonadotropin releasing hormone (GnRH)-dependent regulation of gonadotropin subunit mRNA levels in the rat. *Endocrinology*, **124**, 272–8.

Wierman, M. E., Gharib, S. D., Rovere, J. M. L., Badger, T. M., and Chin, W. W. (1989*b*). Selective failure of androgens to regulate follicle-stimulating hormone β mRNA levels in the male rat. *Molecular Endocrinology*, **2**, 492–8.

Wierman, M. E. and Wang, C. (1990). Androgen selectively stimulates follicle-stimulating hormone-β mRNA levels after gonadotropin-releasing hormone antagonist administration. *Biology of Reproduction*, **42**, 563–71.

Wierman, M. E., Kepa, J. K., Sun, W., Gordon, D. F., and Wood, W. M. (1992). Estrogen negatively regulates rat gonadotropin releasing hormone (rGnRH) promoter activity in transfected placental cells. *Molecular and Cellular Endocrinology*, **62**, 1–10.

Windle, J. J., Weiner, R. I., and Mellon, P. L. (1990). Cell lines of the pituitary gonadotrope lineage derived by targeted oncogenesis in transgenic mice. *Molecular Endocrinology*, **4**, 597–603.

Winters, S. J., Ishizaka, K., Kithahara, S., Troen, P., and Attardi, B. (1992). Effects of testosterone on gonadotropin subunit messenger ribonucleic acids in the presence or absence of gonadotropin-releasing hormone. *Endocrinology*, **130**, 726–34.

Wise, M. E., Nilson, J. H., Nejedlik, M. T., and Nett, T. M. (1985). Measurement of messenger RNA for luteinizing hormone β-subunit and α-subunit during gestation and the postpartum period in ewes. *Biology of Reproduction*, **33**, 1009–13.

Woodruff, T. K., Meunier, H., Jones, P. B. C., Hsueh, A. J. W., and Mayo, K. E. (1988). Dynamic changes in inhibin messenger RNAs in rat ovarian follicles during the reproductive cycle. *Science*, **239**, 1296–9.

Wrana, J. L., Attisano, L., Carcamo, J., Zentella, A., Doody, J., Laiho, M., *et al.*

(1992). TGFβ signals through a heteromeric protein kinase receptor complex. *Cell*, **71**, 1003–14.
Ying, S. (1988). Inhibins, activins and follistatins: gonadal proteins modulating the secretion of follicle stimulating hormone. *Endocrine Reviews*, **9**, 267–93.
Zhou, W., Flanagan, C., Ballesteros, J. A., Konvicka, K., Davidson, J. S., Weinstein, H., et al. (1994). A reciprocal mutation supports helix 2 and helix 7 proximity in the gonadotrophin-releasing hormone receptor. *Molecular Pharmacology*, **45**, 165–70.
Zmeili, S. M., Papvasiliou, S. S., Thorner, M. O., Evans, W. S., Marshall, J. C., and Landefeld, T. D. (1986). Alpha and luteinizing hormone beta subunit mRNAs during the rat estrous cycle. *Endocrinology*, **119**, 1867–9.

7 Neural and endocrine mechanisms underlying the synchrony of sexual behaviour and ovulation in the sheep
DOMINIQUE BLACHE and GRAEME B. MARTIN

I   Introduction

II The oestrous cycle of the ewe: the experimental model
    1 General observations
    2 The sexual behaviour of the ewe
    3 Gonadotrophin secretion in the ewe

III Preovulatory events in the cycle
    1 The luteal phase as a preovulatory event
        i   Secretion of sex steroids
        ii Secretion of gonadotrophins: a dual action of oestrogen and progesterone
        iii Sexual behaviour: a dual action of oestrogen and progesterone
    2 The early follicular phase
        i   Secretion of sex steroids and gonadotrophins
        ii Sexual behaviour
    3 The late follicular phase
        i   Secretion of sex steroids and gonadotrophins
        ii Sexual behaviour
    4 Summary: endocrinology of the synchrony between ovulation and oestrus

IV Neural mechanisms
    1 Localization of the protagonists
        i   Gross anatomical studies
        ii GnRH secretion during the preovulatory surge
        iii The GnRH neurones
            a General distribution
            b Which GnRH neurones are associated with ovulation?
        iv Sites of steroid action
            a Oestrogen binding sites
            b Sites of oestrogen action

             c Progesterone binding sites
             d Sites of progestagen action
        2 The role of other neuromediators
             i   Monoaminergic cells
             ii   γ-aminobutyric acid cells
             iii GnRH and peptidergic cells
             iv   Other peptidergic cells

   V **General considerations and conclusions**
        1 Questions remaining
        2 Conclusion

I   INTRODUCTION

Internal fertilization in mammals, particularly those that are seasonally polyoestrous, such as the sheep, is only possible if the sexual partners are of compatible reproductive status and if they can adequately express this compatibility through sexual behaviour. In most male sheep, gamete production varies only slightly with the seasons and overt sexual behaviour can be displayed at any time. By contrast, gamete production in the female varies temporally, not only with season, but also with the time of the cycle. The ewe, for example, is only fertile for a brief period around the time of ovulation, once each oestrous cycle—perhaps on only five days, at 17-day intervals, each year. Clearly, tight synchrony of behavioural and gametogenic events—the display of the right behaviour at the right time—is an absolutely essential component of reproductive strategy. This synchrony involves several neural and endocrine components of the reproductive system.

The time of ovulation is the result of the exchange of a complex series of reciprocal hormonal messages between the central nervous system (CNS), the pituitary gland, and the gonads. Gonadotrophin-releasing hormone (GnRH), a neurohormone produced in the brain, controls the synthesis and release of the gonadotrophins, luteinizing hormone (LH), and follicle-stimulating hormone (FSH), by the anterior pituitary gland. The gonadotrophins act on the ovary and control the maturation and release of the oocyte by the follicle. In turn, ovarian hormones (progesterone and oestradiol) control gonadotrophin secretion by stimulating or inhibiting the activity of the hypothalamo–pituitary axis. A further consequence of this exchange of endocrine signals between ovary and brain is that the sex steroids prepare the genital tract for the meeting of the gametes and the ensuing pregnancy, and also elicit behavioural responses, namely the motivation, postural organization, and communication that characterize oestrus (Feder 1981).

Thus, females periodically exhibit a specifically sexual behaviour which

Beach (1976) has dissected into three phases: physical changes to attract the attention of the male (attractivity, not present in all species); active search for and attraction towards the male (proceptivity); and acceptance of mating attempts (receptivity). These phases are present, with variable degrees of complexity, throughout the animal kingdom.

Mechanisms leading the meeting of the sexual partners must be rigorously synchronized with those leading to ovulation. The females of some mammalian species (for example, rabbits, camelids, cats) achieve this through the phenomenon of 'reflex ovulation', in which coitus acts as an external synchronizing agent (reviewed by Morali and Beyer 1979). Other mammals are classified as 'spontaneous ovulators', because the external synchronizer is replaced by an internal synchronizer—a peak in the production of sex steroids by the ovarian structures producing the ova that are to be fertilized (Morali and Beyer 1979).

This review focuses on the neural and endocrine mechanisms involved in controlling the synchrony of sexual behaviour with ovulation during the oestrous cycle in female sheep. In temperate and Mediterranean regions, the sheep is a seasonally polyoestrous species that displays alternate periods of sexual activity (the breeding season) and sexual quiescence (the non-breeding season or anoestrus). During the breeding season, female sheep ovulate spontaneously, though the ewes of some breeds, can be induced to ovulate during the non-breeding season simply by the introduction of males (Martin and Scaramuzzi 1983). The neuroendocrine mechanisms underlying the seasonality of sheep, including their responses to photoperiodic, social or nutritional cues, have been reviewed elsewhere (Martin 1984; Karsch *et al.* 1984; Goodman 1988; Thiéry and Martin 1991; Fabre-Nys *et al.* 1993*b*). Similarly, the functions of ovarian structures (the follicles and corpora lutea) that secrete sex steroids and are closely involved in the synchrony of ovulatory and behavioural events have been reviewed recently (Scaramuzzi *et al.* 1993). None of these topics will be dealt with here, as we will restrict the discussion to the processes that occur within the breeding season, when the ewe ovulates spontaneously. Our emphasis will be on brain mechanisms, though we will begin by briefly providing background information that outlines the critical systemic events leading up to ovulation and oestrus in the sheep, and emphasizes the unique advantages of this species as an experimental model.

II THE OESTROUS CYCLE OF THE EWE: THE EXPERIMENTAL MODEL

The ewe presents several advantages as a model for studying the synchrony of endocrine and behavioural events that lead to successful fertilization. The sheep was one of the first species in which the physiology and endocrinology of reproduction was studied, and a vast body of data and knowledge has been collected over the past 90 years. This was partly due to the availability of large quantities of tissue from abattoirs, which allowed researchers to purify a wide variety of hormones and develop specific assays for them. For this review, the most relevant example is mammalian GnRH, which was first isolated from sheep hypothalamic fragments (Burgus et al. 1971; Schally et al. 1971). The placid temperament of the animal and its large blood volume are critical aids to endocrine work, allowing, for example, the simultaneous sampling of blood every 10 minutes from the jugular vein and the hypothalamic–pituitary portal system for up to 60 hours. This allowed the first demonstration in a conscious animal of the temporal relationship between the secretion of GnRH and the gonadotrophins (Clarke and Cummins 1982; Moenter et al. 1991). The wide variation between animals in the shape of the skull has limited neurophysiological studies based on stereotactic surgery, but this is partly compensated for by the large brain volume, which allows the use of sophisticated techniques for long periods, including microdialysis (sampling every 30 minutes for up to 54 hours), lesion, deafferentation, infusion or implantation into anatomically discrete regions, and the regular collection of cerebrospinal fluid (1 ml every 4 hours). The fact that these techniques can be used on conscious, freely moving animals is critical, because it means that endocrine and behavioural events can be observed at the same time.

Finally, not only is the sheep an economically important farm animal, but the cycle in the ewe is not too far removed from that in the females of many non-rodent species, including humans, if only because the luteal phase plays such an important role in the processes controlling fertility. Findings in sheep are relevant to human reproductive problems.

1 General observations

The normal oestrous cycle of the ewe is remarkably constant in length (17 ± 1 days) and the slight variations that have been described can be attributed to differences in breed and age (McKenzie and Terrill 1937). The cycle is divided into two major phases:

1 The progestational or luteal phase is the longest phase of the cycle (about 14 days) and follows ovulation and persists until luteolysis; the endocrine and behavioural events during this part of the cycle are dominated by the secretions of the corpus luteum.
2 The follicular phase can be conveniently subdivided into early and late phases; in the early follicular phase, ovarian follicles enter their final stages of maturation (Scaramuzzi et al. 1993); the late follicular phase begins with the onset of oestrous behaviour and ends with ovulation and the termination of oestrus, and is thus characterized by the activity

of Graafian follicles in the ovaries. These follicles have been selected to ovulate and produce the large amounts of oestradiol that play a critical role in the induction of ovulation and oestrus (Baird and Scaramuzzi 1976*b*; Scaramuzzi *et al.* 1993).

2 The sexual behaviour of the ewe

In each cycle, the ewe shows a pattern of sexual behaviour that can be broken down into two major components: proceptivity, the active search for and an attraction towards the ram; and receptivity, or acceptance of mating attempts by the ram (Banks 1964). Oestrous behaviour in sheep is part of a continuum of interactions between the sexes which includes mutual sniffing and licking, as well as motor patterns that are conducive to the mating act (for review see Fabre-Nys *et al.* 1993*b*). Attractivity, the third component of sexual behaviour defined by Beach (1976), is difficult to identify in the ewe.

Every 16–18 days, ewes become oestrous, search for males and accept their mounting attempts (Fig. 7.1). Oestrus lasts about 1.8 days, although the duration depends on the breed: 47 hours in Prealpes du Sud; 38 hours in Ile de France; 29.2 hours in Merino; and 35.5 hours in Awassi (Joubert 1962; Banks 1964; Schindler and Amir 1972; Fabre-Nys and Venier 1989). However, the accuracy with which the duration of sexual activity can be measured is critically dependent on the type of behavioural test that is used. Nearly all the values in the literature were derived using qualitative tests based on the presence or absence of receptivity, detected by a standing response to mating attempts by a male. This field was revolutionized when Fabre-Nys and Venier (1987) developed quantitative behavioural tests which allowed them to measure both proceptivity and receptivity in ewes with greatly increased precision and accuracy. These techniques allowed them to show that oestrous behaviour begins abruptly, with the ewe becoming fully proceptive and receptive within eight hours, remains intense for a variable period, and then terminates slowly (Fabre-Nys and Venier 1989). This has led to a detailed re-assessment of the endocrine control of sexual behaviour in this species, especially with regard to the roles of the sex steroids.

3 Gonadotrophin secretion in the ewe

The secretion of both gonadotrophins, LH and FSH, is controlled primarily by GnRH, a decapeptide synthesized by neurones in the preoptico-hypothalamic continuum. Some of these neurones terminate in the median eminence where they release the GnRH into the hypophyseal portal blood stream which carries it down to the pituitary gonadotrophs. Clarke and Cummins (1982) developed a system for sampling blood directly from the hypophyseal portal system of the conscious ovariectomized ewe and showed that GnRH is released in pulses. This finding has been extended to include castrated rams and intact ewes (Clarke et al. 1987; Caraty and Locatelli 1988). A very elegant study, involving sampling of pituitary portal blood every 30 sec, has described the GnRH profile in great detail, showing that each secretory pulse approximates to a square-wave, with an abrupt

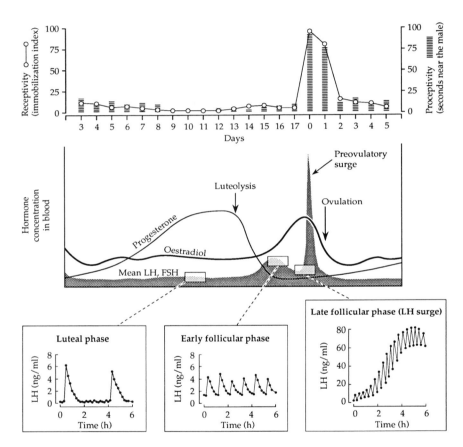

Fig. 7.1 Schematic representation of the temporal relationships between the major endocrine and behavioural events of the oestrous cycle of the ewe (modified after Thiéry and Martin 1991). Upper panel: the two principal components of sexual behaviour, receptivity as measured by the receptivity index (percentage of immobilization responses to male courtships), and proceptivity, as measured by the time spent near a group of rams in a choice test (Fabre-Nys and Venier 1987). Middle panel: the general patterns of the circulating concentrations of gonadotrophins and ovarian steroids. Lower panel: the pulsatile secretion of LH during the three phases of the oestrous cycle.

increase followed by a plateau (with some fluctuations) lasting about 5.5 minutes, and then a rapid decrease in secretion (Moenter *et al.* 1992*a*).

An important development from this work was the finding of a virtually perfect relationship (Clarke and Cummins 1982; Caraty *et al.* 1987; Moenter *et al.* 1992*a*) between GnRH pulses entering the pituitary gland and the LH pulses that had been observed in jugular blood since the early 1970s (reviewed by Martin 1984). By contrast, there is little evidence for pulsatile release of FSH, so FSH secretion reflects the secretion of GnRH relatively poorly in the short term (minutes), although long-term (days) changes in the frequency of GnRH pulses will lead to changes in circulating FSH concentrations (Lincoln 1979; Martin *et al.* 1986).

Classically, gonadotrophin secretion in females has been considered to be characterized by two modes of secretion, a tonic mode, which is evident during the luteal phase and early follicular phase, and a surge mode, which is evident in the late follicular phase (Fig. 7.1). As we shall see later, the distinction between these is not as clear as was thought, particularly for LH, primarily because both modes are driven by pulsatile secretion of GnRH. In this review of neuroendocrine control of oestrus and ovulation, we will be primarily interested in LH. There are two reasons for this: first, the one-to-one relationship between GnRH and LH pulses allows us to use LH pulse patterns in the peripheral blood stream to bioassay the activity of the hypothalamic neurons controlling GnRH; second, the surge in the secretion of LH during the late luteal phase is responsible for ovulation of the Graafian follicle(s). There is also a surge in the secretion of FSH at this time, but this phenomenon does not appear to be necessary for ovulation. Indeed, its function is not clear.

III PREOVULATORY EVENTS IN THE CYCLE

A characteristic of the endocrine profile of the cycle is the inverse relationship between the concentrations of progesterone and oestradiol (Fig. 7.1). This relationship and the way in which it drives gonadotrophin secretion is critical to our understanding of the processes leading to ovulation and oestrus. The endogenous steroid patterns can be easily reproduced in ovariectomized ewes. For several decades, this was accomplished simply by using a sequence of daily injections of progestagen (for five to 12 days) followed by a single injection of oestrogen one day after the last injection of progestagen. Oestrous behaviour was exhibited about 16 hours later. This system, in association with the acceptance of mounting, led to much of our understanding of the endocrine control of oestrus and ovulation.

However, the limitations of this model became apparent in the 1970s, especially for studies of control of tonic gonadotrophin secretion. A more refined approach was developed, primarily by the Michigan group (Karsch

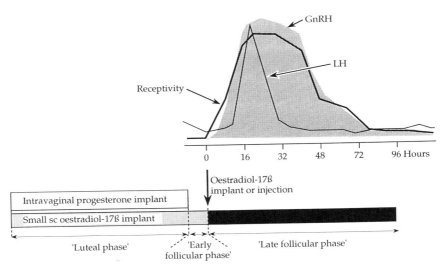

Fig. 7.2 The steroid treatments used to induce an artificial oestrous cycle in ovariectomized ewes, replicating changes in the balance of endogenous oestradiol and progesterone during the three phases of the cycle and inducing the endocrine and behavioural phenomena similar to those observed in intact ewes.

et al. 1977, 1980, 1983), who used subcutaneous implants to provide consistent and controlled doses of exogenous hormones that more closely mimicked the natural steroid sequence than did injections. This approach was also adopted by workers interested in periovulatory events, particularly the endocrine control of oestrus and the preovulatory surge (Fabre-Nys and Martin 1991; Moenter *et al.* 1993). The system developed by Fabre-Nys and Martin (1991) has proved very satisfactory: an intravaginal device containing progesterone and a 1 cm long subcutaneous silastic oestradiol implant are inserted for one week (the 'luteal phase'); 16 hours after the progesterone is removed (early follicular phase), an additional 1 cm oestradiol implant is inserted to produce a rise of oestradiol similar to that observed during the late-follicular phase (Fig. 7.2). The use of small implants allows accurate tailoring of the doses for different breeds and different body sizes, so the above sequence is almost always successful for induction of a preovulatory-type of LH surge, and the intensity and duration of the oestrous behaviour is very similar to that expressed by spontaneously ovulating, intact ewes. These behavioural responses in particular were greatly improved over those induced in the injection-based model (Fabre-Nys and Martin 1991). Importantly, this implant model provided an alternative to the rat, where the steroid sequence is basically reversed with oestrogen priming followed by a progesterone trigger, rather than progesterone priming followed by an oestradoil trigger.

1 The luteal phase as a preovulatory event

It is usual in reviews of this type to discuss the processes leading to ovulation before describing the endocrine systems of the luteal phase, simply because the corpus luteum is the result of ovulation. However, the ewe usually experiences a luteal phase before a fertile ovulation and oestrus. This is because the first ovulation of the breeding season is often silent, in that it is not accompanied by overt sexual behaviour, and it often produces a non-functional corpus luteum (reviewed by Martin 1984). Thus, the luteal phase is a critical pre-determinant of the expression of sexual behaviour and the fertility of the ovulation that accompanies that behaviour. We have therefore chosen to treat the luteal phase as the first of the preovulatory events.

i Secretion of sex steroids

During the luteal phase, the dominant steroid is progesterone produced by the corpus luteum (Short *et al.* 1963; Bjersing *et al.* 1972). From day 3 after ovulation (typically defined as day 0; Fig. 7.1), progesterone secretion increases until peak concentrations of 3–5 ng/ml are reached on days 8–10. After this there is a plateau until luteolysis on day 14, when there is a rapid decrease to values similar to those on the day of ovulation (Stabenfeldt *et al.* 1969; Thorburn *et al.* 1969; Hauger *et al.* 1977). Ovarian follicles grow and regress throughout the luteal phase and, although they do not enter the final phases of development, they still produce small but significant amounts of oestrogen (Baird 1978). Thus, the circulating concentrations of oestradiol are low (1–5pg/ml) but, as we shall see below, there is still sufficient to play a critical role in negative feedback, one of the processes controlling gonadotrophin secretion.

ii Secretion of gonadotrophins: a dual action of oestrogen and progesterone

During the luteal phase, only the tonic mode of gonadotrophin secretion is observed and GnRH and LH pulses are relatively infrequent (every 3–4 hours), especially when progesterone concentrations are highest. Thus, concentrations of LH are inversely correlated with the plasma concentrations of progesterone (Baird and Scaramuzzi 1976*a*; Hauger *et al.* 1977). The concentrations of FSH are also lower at this time than they are during the early follicular phase, though they show wave-like patterns over period lasting a few days, due to waves of follicular development in the ovary (Scaramuzzi *et al.* 1993). This low tonic secretion of the gonadotrophins is sufficient to maintain progesterone secretion by the corpus luteum, and to sustain the development of follicles, at least to the gonadotrophin-dependent stage (Scaramuzzi *et al.* 1993), thus maintaining a low, but physiologically significant, basal rate of oestradiol secretion.

Tonic gonadotrophin secretion is maintained in equilibrium by the process of negative feedback. This process has two components:

1. In the hypothalamus, progesterone and oestradiol act synergistically to inhibit the activity of the systems controlling the tonic mode of the GnRH cells, so that a low pulse frequency is maintained (Hauger *et al.* 1977; Karsch *et al.* 1977; Karsch *et al.* 1980; Martin *et al.* 1983);
2. At the pituitary level, oestradiol and inhibin, also acting synergistically, inhibit the secretion of FSH (Martin *et al.* 1988). Oestradiol and progesterone may also reduce the responsiveness of the gonadotrophs to GnRH, helping to control LH pulse amplitude, though the evidence for this is not convincing (Martin 1984).

Both of these negative-feedback systems involve oestrogen, and the small amounts present during the luteal phase are sufficient for that purpose. However, the synergism is critical, for without luteal progesterone to help control GnRH pulse frequency, or follicular inhibin to help control FSH secretion, gonadotrophin secretion would be out of control. These effects constrain follicle development during the luteal phase, but allow sufficient gonadotrophin secretion to sustain luteal and follicular activity and thus ensure the maintenance of circulating concentrations of oestrogen and progesterone. It should be remembered that the luteal phase may in fact be the early stages of pregnancy, and that this steroidal environment is essential for uterine and mammary gland development.

The constraint of follicle development during the luteal phase is probably enough to prevent ovulation, but there is also an insurance policy—inhibition of the preovulatory gonadotrophin surge. In the hypothalamus, progesterone completely blocks the activity of the systems controlling the surge mode of the GnRH cells, which would normally be induced by the positive-feedback effects of oestradiol (Martin 1984; Kasa-Vubu *et al.* 1992).

iii Sexual behaviour: a dual action of oestrogen and progesterone

In sheep, progesterone alone does not induce female sexual behaviour. As with the surge mode of gonadotrophin secretion, sexual behaviour is induced by oestrogen and this effect is completely blocked by progesterone (Moore and Robinson 1957; Signoret 1969). Consequently, very little proceptivity or receptivity is displayed during the luteal phase (Fig. 7.1).

However, this inhibitory effect does not mean that luteal progesterone has no role to play in the control of oestrous behaviour. On the contrary, progesterone pretreatment is essential if complete responses to physiological doses of oestrogen are to be elicited in ovariectomized ewes. A single injection of oestrogen can induce oestrus, but a high dose is required and the efficacy of the treatment decreases when the same dose is injected every few days (Robinson 1954b; Scaramuzzi et al. 1972). Progesterone pretreatment overcomes this refractoriness, reduces the dose of oestradiol required and decreases the latency between oestrogen injection and the onset of oestrus (Robinson 1954a). These priming effects of progesterone are not dose-dependent, but the duration of the priming treatment is important (Robinson et al. 1956; Scaramuzzi et al. 1971).

Similarly, the small amounts of oestradiol that are secreted by the ovaries during the luteal phase play an important facilitatory role in the induction of oestrous behaviour in the subsequent follicular phase. This was evident from studies by Fabre-Nys and Martin (1991). When a low dose of oestradiol was added to the progesterone used to prime ovariectomized ewes during the artificial luteal phase, the dose of oestradiol needed to trigger oestrous behaviour could be reduced while still ensuring that more than 90 per cent of the animals responded, and that the duration and the intensity of the behaviour was similar to that observed in intact ewes.

It is clear that the temporal relationship between the fall in circulating progesterone concentrations and the rise in circulating oestradiol concentrations, and not just the absolute concentrations, is critical for the induction of sexual behaviour. The artificial cycle proposed by Fabre-Nys and Martin (1991) respects this chronology, and has been used for many of the studies of the neurophysiological processes underlying the expression of sexual behaviour in female sheep (Fig. 7.2). During the priming period of at least five days, progesterone is delivered by an intravaginal implant that provides physiologically normal concentrations in the circulation (about 5 ng/ml), while oestradiol is delivered by a silastic subcutaneous implant (1 cm long). Between 14 and 16 hours after progesterone removal, a single injection of about 16 μg of oestradiol-17β will induce a complete oestrous behaviour and a preovulatory type surge of LH (Fig. 7.2). This regime is effective in Ile-de-France ewes during the breeding season, but can be extended to other breeds and seasons after subtle modification of the doses of oestradiol.

2 The early follicular phase

i Secretion of sex steroids and gonadotrophins

The follicular phase begins with luteolysis, when progesterone concentration falls rapidly from the luteal phase peak (3–5 ng/ml) to reach values of < 0.5 ng/ml that are characteristic of the six-day period between the demise of one corpus luteum and the initiation of detectable progesterone secretion by the next (Fig. 7.1). As progesterone concentrations fall, the amount of negative feedback exerted at the hypothalamic level also falls, so GnRH pulse frequency increases. This is, of course, reflected in an increase in LH pulse frequency and FSH concentrations, usually evident by about eight hours after the induction of luteolysis with exogenous prostaglandin (Wallace et al. 1988). This strong gonadotrophic signal drives one to three ovarian follicles into the final stages of maturation. Ultimately, the ovulatory follicle develops LH receptors on its granulosa cells and these receptors become linked to aromatase, the main source of oestradiol (Scaramuzzi et al. 1993). The addition of this resource to oestradiol production leads to a rapid increase in the concentrations of this steroid in the blood, which reach a peak of 5–10 pg/ml about 48 hours after luteolysis, when the late follicular phase begins (Scaramuzzi and Land 1978; Herriman et al. 1979). Despite rising concentrations of oestradiol, a high frequency of GnRH/LH pulses (about one pulse every 45 minutes) persists throughout the early follicular phase because, in the absence of progesterone, oestradiol exerts a very weak inhibitory effect on the activity of the hypothalamic centres controlling GnRH release during the breeding season. In fact, there is evidence that GnRH pulse frequency is slightly stimulated under these conditions (Karsch et al. 1983).

The gonadotrophs in the pituitary gland are still inhibited by oestradiol, however, and this is evident in the fall in pituitary responsiveness to GnRH, leading to a fall in LH pulse amplitude, as the follicular phase progresses and oestradiol concentrations rise (Caraty et al. 1990). The secretion of FSH is initially stimulated by the high GnRH pulse frequency immediately after luteolysis, but then concentrations of FSH fall steadily under the influence of the oestradiol and inhibin secreted by the maturing follicles (Wallace et al. 1988). Ovarian follicles in the later stages of maturity no longer require FSH and need only small amounts of LH, but immature follicles do need FSH and under these conditions they undergo atresia (Scaramuzzi et al. 1993).

During this period, the absence of progesterone theoretically removes the block on positive feedback by oestradiol, so the preovulatory surge can be initiated. However, the surge mechanism includes several built-in delays: the time required for the development of mature follicles that can satisfy the need for concentrations of oestradiol that exceed the threshold for positive feedback (Fabre-Nys et al. 1993a); a lag in the stimulatory response to oestradiol once the threshold is passed (Goding et al. 1969; Scaramuzzi et al. 1971); and an additional delaying effect of progesterone on positive feedback that is carried over from the luteal phase (Karsch et al. 1980).

ii Sexual behaviour

In the early follicular phase, neither intact ewes nor steroid-supplemented ovariectomized ewes show detectable proceptivity or receptivity. In the absence of progesterone, the block to induction of oestrus by oestradiol has been removed, but there is again a delay mechanism, similar to the one controlling the preovulatory surge because it includes the high threshold for the concentrations of oestradiol and the lag in the stimulatory response to oestradiol once the threshold is passed (Robinson 1954 a, b; Fabre-Nys and Martin 1991). The additional carryover effect by luteal progesterone is in complete contrast, however, with the delaying role of progesterone on oestradiol-induced positive feedback, because progesterone priming decreases the latency between oestrogen injection and the onset of oestrus (Robinson 1954a). These opposing effects are absolutely critical in controlling the temporal relationship between sexual activity and ovulation–in the absence of the priming effects of a luteal phase, these two events are very poorly synchronized (Karsch et al. 1983).

3 The late follicular phase

The late follicular phase starts with the abrupt initiation of two tightly synchronized events—oestrus and the preovulatory surge of gonadotrophin.

i Secretion of sex steroids and gonadotrophins

At this stage of the oestrous cycle, progesterone concentrations are still very low and the dominant ovarian structures are the very large Graafian follicles that are producing large amounts of oestradiol (Baird and Scaramuzzi 1976 b; Scaramuzzi et al. 1993). The peak oestradiol concentration of 5–10 pg/ml is reached at the onset of the LH surge, which immediately initiates massive and irreversible biochemical changes in the cells of the mature follicles—the process of luteinization. Consequently, oestradiol secretion falls rapidly over the next 24 hours until it reaches the lowest levels observed in the cycle (Scaramuzzi and Land 1978; Herriman et al. 1979; Fig. 7.1).

The blood concentration of LH increases rapidly and within 4–8 hours after the onset of oestrus it reaches a peak, with values that are 50–100 times greater than those secreted tonically, before declining equally rapidly to baseline values within a further 8 hours, so that the average duration of the surge is about 12 hours (Geschwind and Dewey 1968; Niswender et al. 1968; Pelletier et al. 1968; Goding et al. 1969). The secretion of LH during the preovulatory surge has been shown to be pulsatile in the ewe, though the pulses are so frequent that they are difficult to distinguish from each other, even with sampling every 4 minutes (Martin et al. 1987). Despite this technical problem, these studies did reveal the rapidity of the changes involved, as the onset of the surge was easily detected as an abrupt fourfold change in the amplitude of successive pulses that were separated by an interval of only 10–15 minutes. Mechanistically, this can be explained by a complex series of events, including:

(1) accumulation of gonadotrophin by the pituitary gland over the final 24 hours of the early follicular phase, permitted by the inhibitory

effects of oestradiol on pituitary responsiveness to GnRH (see Martin *et al.* 1987);

(2) the sudden reversal of this effect of oestradiol on pituitary responsiveness, through an unknown mechanism that appears to increase GnRH receptor number (Clarke *et al.* 1988), such that the gonadotrophs become hypersensitive to GnRH (Caraty *et al.* 1984; Clarke and Cummins 1984);

(3) an increase in the amount of GnRH secreted by the hypothalamus (see below).

ii Sexual behaviour

Within a few hours of the onset of the late follicular phase, the behaviour of the ewe changes radically. Intact ewes express a great interest in rams and start actively to search for them, and they exhibit behavioural responses to the courtship of the male (Banks 1964; Fabre-Nys and Venier 1989). This continues for 30–48 hours, depending on the breed (Fabre-Nys and Venier 1989). The appearance of these behaviours is triggered by oestrogen; this is demonstrated by the facts that oestrogen injection elicits sexual behaviour in castrated progesterone-primed ewes (Robinson 1954*a, b*; Fabre-Nys and Martin 1991) and that oestrus can be blocked in intact ewes by immunization against oestradiol (Fairclough *et al.* 1976; Rawlings *et al.* 1978). Increasing the dose of oestrogen decreases the latency to the onset of oestrus (Fabre-Nys *et al.* 1993*a*; Goodman *et al.* 1981), but the duration of receptivity is controlled by the time during which high concentrations of oestradiol are maintained (Fabre-Nys *et al.* 1993*a*). Intially, it was thought that the duration of receptivity was also controlled by dose, but this seems to have been a misinterpretation that can be attributed to the extra time required for clearance of high doses.

4 Summary: endocrinology of the synchrony between ovulation and oestrus

As we emphasized in the introduction, the coincidence of sexual behaviour and ovulation are required for mating and fertilization. Ovulation follows the LH surge, after a delay that appears very constant (22–26 hours; Cumming *et al.* 1973) under most circumstances, so the simultaneous onset of oestrus and the LH surge at the start of the late follicular phase effectively ensures the correct temporal relationship between ovulation and mating behaviour.

Generally, it has been considered that these two events were synchronized within the limits of the error of measurement (Wheatley and Radford 1969; Mauer *et al.* 1972), which was determined by the detection of acceptance of mounting, a qualitative test prone to error. With the advent of the quantitative behavioural tests, the accuracy and precision of

LH measurements were finally matched and we now know that, in fact, proceptivity and receptivity begin to be expressed 8 hours before the LH surge is triggered (Fabre-Nys and Martin 1991, Fabre-Nys *et al.* 1993*a*). The coincidence of these events depend mainly on the action of the ovarian steroids. During the first cycle of the breeding season after the anoestrous period, ovulation is not accompanied by oestrous behaviour (Wheeler and Land 1977). This so-called silent ovulation is preceded by a peak in oestradiol secretion that itself is not preceded by a period of elevated blood concentrations of progesterone (Robinson 1954*b*). Scaramuzzi *et al.* (1971), using peripheral injection of steroids in ovariectomized ewes, showed that oestradiol triggers both the LH surge and oestrous behaviour, but only if the oestradiol injection is not too close to the final priming injection of progesterone. This alone suggests that progesterone is important in the synchronization of oestrous behaviour and ovulation. Karsch *et al.* (1980), on the basis of experiments using subcutaneous implants which allow accurate control of dose and timing of steroid exposure, subsequently suggested that progesterone also controls the latency to the onset of the surge. Most recently, by using the implant technique and a quantitative measurement of receptivity, Fabre-Nys and Martin (1991) underlined the dominant nature of the role played by progesterone. They showed that the timing of periovulatory behaviour and endocrine events is primarily determined by the time of progesterone withdrawal, with both beginning 24–30 hours later, independently of the quantity and time of oestradiol treatment. Thus, although oestrogen acts as a trigger, in terms of timing, it serves only to modulate the effects of progesterone or its withdrawal.

IV   NEURAL MECHANISMS

1 Localization of the protagonists

i Gross anatomical studies

The first question about the anatomical structures involved in ovulation and oestrus concerned the relative importance of the hypothalamus and the pituitary gland. Clegg *et al.* (1958) used electrolytic lesions to show that the hypothalamus was involved in the control of sexual behaviour in the ewe. Subsequently, Radford (1967) and Domanski *et al.* (1972*b*) observed that lesions in the mediobasal hypothalamus disrupted the behaviour but not the ovarian cyclicity, although Przekop (1978) did a similar study and found that both phenomena were blocked. This disagreement is probably explained by differences in the size and position of the lesions relative to the location of the GnRH cells and the sites of steroid action, due to the lack of a good stereotactic atlas for the sheep brain and the wide variation between animals.

Despite this, the primary importance of the hypothalamic and preoptic areas in both ovulation and oestrus was clearly evident from these early studies, and has since been confirmed by work demonstrating the anatomical relationship between the key protagonists: the neurones synthesizing GnRH and the neurones sensitive to ovarian steroids (discussed below). For the control of sexual behaviour, the dominant role of neural systems was never in doubt. Similarly, it was clear that some sort of neural input was needed for ovulation, but here there was considerable uncertainty: neural systems could play an active and dominant role, eliciting the preovulatory surge of LH simply by secreting a similar surge of GnRH, or they could play a passive role, simply supplying a tonic signal to maintain the activity of the pituitary gonadotrophs which, independently of any changes in the hypothalamic signal, responded directly to oestrogen by secreting the LH surge.

In the ewe, the evidence supporting the hypothesis of a passive hypothalamus and active hypophysis is centred on the increase in pituitary responsiveness to GnRH that is induced by oestrogen and coincides with the onset of the LH surge (Jackson 1975; Coppings and Malven 1976; Clarke and Cummins 1982). This effect is mediated through an increase in the number of GnRH receptors (Clarke *et al.* 1988; Gregg *et al.* 1990; Gregg and Nett 1989; Laws *et al.* 1990). Certainly, Knobil's group were convinced that this system explained the preovulatory surge in primates (Krey *et al.* 1975; Hess *et al.* 1977; Knobil 1980; Wildt *et al.* 1980). Despite an impressive list of compelling observations to support this view, other groups had long disagreed and advocated the hypothesis that both the hypothalamus and the hypophysis played active roles, not only in the primates (see Norman *et al.* 1982) but also in other species (reviewed by Martin 1984). Clearly, the only way to resolve the conflict was to test whether a preovulatory surge of GnRH was secreted into the hypophyseal portal system.

ii GnRH secretion during the preovulatory surge

An increase in GnRH secretion during the preovulatory surge that is induced in the ovariectomized ewe by oestrogen was demonstrated firstly using push–pull perfusion of the median eminence (Schillo *et al.* 1985) and later confirmed by sampling hypophyseal portal blood (Clarke *et al.* 1987). The hypophyseal portal samples revealed a biphasic response of GnRH secretion after a single injection of oestradiol, with changes that paralleled the secretion of LH into the jugular blood stream (negative feedback followed by positive feedback). At the start of the surge, GnRH secretion increased until it reached values about 40-fold higher than those that control basal or tonic secretion. In the ovariectomized ewe, Caraty *et al.* 1989) showed that the onsets of both GnRH and LH surges were concomitant and started 12 hours after an injection of oestradiol (25 μg intravenously and 25 μg intramuscularly). In the intact ewe showing an endogenously driven surge, the dual onset was observed (Moenter *et al.* 1990, 1991) except that the GnRH surge lasted about twice as long as the LH surge (Moenter *et al.* 1991), in much the same way that the electrophysiological signal from hypothalamic cells that are activated at this time persists long after the LH surge has ended (Thiéry and Pelletier 1981). The reason for the persistence of GnRH secretion after the end of the LH surge is not known.

Recent work with rhesus monkeys has shown that there is an increase in the release of GnRH pulses into CSF at the onset of the surge, so it is clear that primates use a very similar mechanism (Xia *et al.* 1992). Thus, it seems that we can generalize across species. The complete expression of the preovulatory gonadotrophin surge involves increases in the activity of both hypothalamic and pituitary sites.

There is little doubt, then that GnRH secretion increases during the preovulatory surge, but detailed conclusions about the mode of GnRH secretion do differ in one important aspect. This is best illustrated by two studies which are comparable in that they both used the same variation on the technique of Clarke and Cummins (1982) for sampling pituitary portal blood in sheep. Caraty *et al.* (1989) analysed the GnRH pattern derived from samples taken every 5 minutes from ovariectomized ewes and suggested that the onset of the surge was at least partly due to an increase in GnRH pulse frequency. On the other hand, Moenter *et al.* (1992*b*) took samples at very short intervals, as short as 30 seconds, from intact ewes and concluded that the surge was due to a switch from the pulsatile mode to a continuous mode of secretion of GnRH. These data are perhaps more conclusive, in which case we are led to the suggestion that the tonic and surge modes of GnRH secretion are not only mechanistically different but perhaps even anatomically different, involving two distinct populations of GnRH cells. We still have more work to do to unravel the physiological processes underlying the surge and the way that neural centres respond to oestrogen, particularly the switch from negative to positive feedback.

iii The GnRH neurones

Following the development of specific antibodies against the GnRH decapeptide, or part of it, Dubois and Barry (1974) and Kozlowski and Zimmerman (1974) began tracing the neurones containing GnRH in the sheep brain. Because of the uncertainty surrounding the specificity of the antibodies that are used in immunohistochemistry, cells traced with this method are often termed GnRH-immunoreactive cells. The same applies to any cell type that is detected immunohistochemically. It is our view that the risks associated with the method are well understood, and that reader should be reminded of them, but persistent inclusion of 'immunoreactive' is both pedantic and clumsy (for example, in tyrosine hydroxylase-immunoreactive). This opinion is supported by recent studies showing that the distribution of GnRH cells that is revealed by *in situ* hybridization histochemistry is virtually identical to distributions originally described by immunohistochemistry (Shivers *et al.* 1986; Roberts *et al.* 1989; Selmanoff *et al.* 1991; Bergen *et al.* 1991; Ronchi *et al.* 1992). For this reason, in the remainder of this review, we will mention the methodology used in anatomical studies, but then use simple terminology to describe the cells.

a General distribution The first immunohistochemical studies in sheep located GnRH fibres and terminals in the median eminence, around the septal area, in the organum vasculosum of the lamina terminalis (OVLT), and in the posterior portion of the mamillary bodies (Dubois and Barry 1974; Hoffman *et al.* 1978; Polkowska *et al.* 1980). In contrast with laboratory rodents, it was difficult to stain the GnRH cell bodies in the brain of mature sheep with early antibodies (Polkowska 1981) and more extensive studies, giving a picture of the distribution in the whole brain, were not feasible until new antibodies and histochemical techniques were developed (Dees *et al.* 1981; Advis *et al.* 1985; Glass *et al.* 1986; Lehman *et al.* 1986; Polkowska *et al.* 1987; Caldani *et al.* 1988). We now have an accurate picture of the cellular distribution, the major pathways taken by the fibres, and the locations of the terminal fields, from early embryonic life through to adulthood. Arguably, the major area that still requires investigation is the possibility of differences between the sexes.

In mature female sheep, 95 per cent of the GnRH cell bodies (Caldani *et al.* 1988) are located in the preoptic–hypothalamic complex, including the diagonal band of Broca, the lateral and medial preoptic area, medial septum, ventrolateral anterior hypothalamus, lateral hypothalamus, and mediobasal hypothalamus (Fig. 7.3b). About half of the cells in the brain are concentrated in the medial preoptic area, adjacent to the OVLT, about 15 per cent in ventromedial hypothalamus, and only a few (1–2 per cent) are seen in the arcuate nucleus (Dees *et al.* 1981; Lehman *et al.* 1986; Caldani *et al.* 1988). Some GnRH perikarya (5 per cent) are located in extrahypothalamic structures, including the accessory olfactory bulbs, nervus terminalis, pars tuberalis, and medial amygdalae (Lehman *et al.* 1986; Caldani *et al.* 1988; Skinner *et al.* 1992).

GnRH terminals have been described in all the areas containing GnRH perikarya, plus the supraoptic nucleus, median eminence, dorsal hypothalamus, mamillary bodies, and mesencephalic structures including fasciculus retroflexus, habenula, perichiasmatic cistern, and mesencephalic central grey matter (Caldani *et al.* 1988). Fibres originating from the preoptic–hypothalamic continuum project mainly to the OVLT, where they fulfil an unknown function, and to the median eminence where they are involved in the control of pituitary function. In the female sheep, two different routes have been traced from the preoptic cell bodies to the median eminence: a medial tract containing some fibres that traverse the periventricular region next to the third ventricle; and a lateral–ventral tract that is more prominent than the medial tract and reaches the external layer of the median eminence (Lehman et al. 1986). These anatomical descriptions support the data from deafferentation studies showing that most of the GnRH in the median eminence comes from the preoptic area and anterior hypothalamus (Polkowska 1981).

This anatomical picture can be aligned with the data presented by Wheaton (1979), who followed the dynamic variations in regional brain content of GnRH during the oestrous cycle of the ewe. In stalk-median eminence tissue, the highest levels of GnRH were detected during early and late follicular phase and the lowest concentrations during the early luteal phase, immediately after the preovulatory surge. The GnRH content of preoptic–suprachiasmatic tissue followed a similar pattern during the luteal phase, being low after ovulation (days 3–8) and doubling as the luteal phase progressed (to days 10–14) but, in contrast with the median eminence, it then fell during the early and late follicular phases.

The secretory pathways for GnRH and their activity during the oestrous cycle can thus be summarized:

1. During the luteal phase, when the pulse frequency is low, the GnRH involved in pituitary regulation is synthesized in the preoptic area and accumulates in those cells and their terminals in the median eminence.
2. During the early follicular phase, much of it is transported to the median eminence.
3. During the late follicular phase (at oestrus), it is released into the hypophyseal portal blood to initiate the preovulatory surge of gonadotrophin.

These changes in GnRH concentration clearly follow the pattern of changes in sex steroids, and this is no surprise considering the control that the feedback systems exert over GnRH secretion. However, whether the synthesis, accumulation, and transport are also driven by oestradiol and progesterone, or whether the changes in tissue content are merely passive responses to changes in secretory activity of the GnRH cells, remains to be determined. The available information is consistent with progesterone stimulating synthesis and blocking transport during the luteal phase, and with oestradiol stimulating transport during the follicular phase, but Occam's razor dictates that we accept the simplest hypothesis, namely passive responses to changes in secretion, until evidence to the contrary is collected.

b Which GnRH neurones are associated with ovulation? Apart from the preoptic–median eminence tracts, there are many other sites of synthesis, storage, and liberation of GnRH, such as the olfactory bulbs or mesencephalic terminals, where the function of the neurohormone has not been identified, and may be more or less concerned with reproduction.

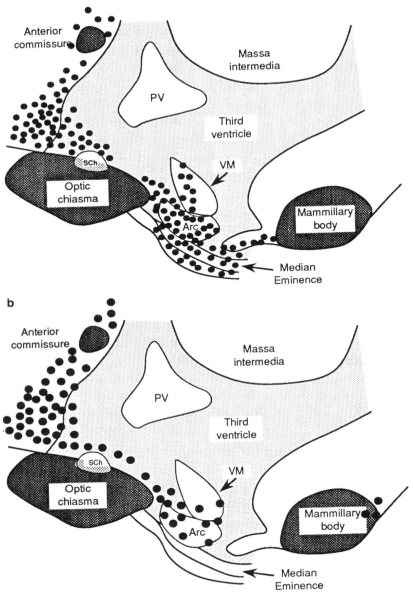

Fig. 7.3 Schematic drawings of a sagittal section of the preoptico–hypothalamus continuum of the sheep. (a) Oestrogen-sensitive cells stained immunohistochemically, using a monoclonal antibody (Abbott H222) against oestrogen receptors (adapted from Skinner *et al.* 1992, Lehman *et al.* 1993, Blache *et al.* 1994. (b) GnRH-containing cells revealed by immunohistochemistry (adapted from Lehman *et al.* 1986 and Caldani *et al.* 1988). (c) Localization of the sites where bilateral oestradiol implants will trigger proceptivity, receptivity,

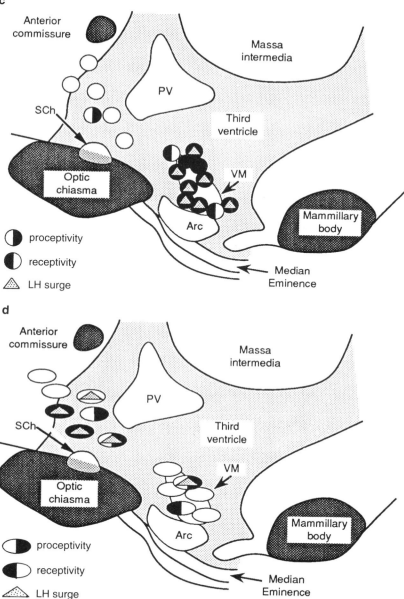

and the preovulatory LH surge in ovariectomized ewes primed with an artificial luteal phase, as shown in Fig. 7.2 (adapted from Blache *et al.* 1991*b*). (d) Locations where bilateral progesterone implants inhibit proceptivity, receptivity, and the preovulatory LH surge induced by an intramuscular injection of 8 μg oestradiol-17β, in ovariectomized ewes undergoing an artificial cycle as described in Fig. 7.2 (adapted from Blache 1991*a*). Arc, arcuate nucleus; PV, paraventricular nucleus; SCh, suprachiamatic nucleus; VM, ventromedial nucleus.

We have been able to narrow the field down to ovulatory sites through several techniques, including deafferentation and electrolytic lesions. These experiments confirmed the importance of the anterior hypothalamus by showing that interruption of pathways posterior to the suprachiasmatic nucleus blocked both ovulation in intact ewes and the oestrogen-induced LH surge in ovariectomized ewes (Jackson et al. 1978; Thiéry et al. 1978; Radford 1979; Przekop and Domanski 1980). Deafferentation at the level of suprachiasmatic nucleus decreased the amplitude of the LH surge but did not block ovulation (Jackson et al. 1978), but lesion of the suprachiasmatic nucleus itself had no effect on the LH surge (Przekop and Domanski 1980). These observations are consistent with disruption of the GnRH fibres passing from the preoptic area to the median eminence. An elegant alternative approach to this problem, resulting from developments in molecular biology, was recently used by Moenter et al. (1993) to define the GnRH cells that are involved in positive feedback. This technique involves immunohistochemical co-localization of GnRH and *fos* protein, an indicator of increased cell activity. In ovariectomized ewes treated with exogenous steroids, 41 per cent of the GnRH cells, evenly distributed throughout the preoptice–hypothalamic continuum, expressed *c-fos* during the oestrogen-induced LH surge. This contrasted with ovariectomized ewes that were not treated with steroids or that were treated with an artificial luteal phase, in which less than 0.01 per cent of the GnRH cells contained *fos* protein (Moenter et al. 1993). This work contained two other interesting observations: numerous non-GnRH neurones also expressed *c-fos* during the LH surge and these cells were located in the same region as cells containing oestradiol receptors, suggesting that they may be involved in the exertion of positive feedback by oestradiol, perhaps as afferents to the GnRH cells; and 26–70 per cent of GnRH cells expressed *c-fos* in the ewes perfused in the second half of the LH surge, whereas in the one ewe perfused during the onset on the surge only 4 per cent of GnRH cells contained *fos*.

The onset of the surge does not seem to be dependent on a massive activation of the GnRH cell bodies but, as suggested above, to a release of stored hormone by the terminals in the median eminence. On the other hand, the maintenance of high levels of GnRH secretion during the surge, which lasts up to 48 hours (Fig. 7.2), requires genomic activation in the GnRH cell bodies.

The next question we need to ask is where steroids act in order to control GnRH cellular activity.

iv Sites of steroid action
For steroids to act on their target cells, they must first bind to specific intracellular receptors (Gorski et al. 1986), and the identification of these binding sites can provide the anatomical substrate needed to unravel the

feedback effects on GnRH secretion. Autoradiography, binding site assays, immunohistochemistry of receptors, and *in situ* hybridization of mRNA encoding the steroid receptors give information about the localization. Information from *in situ* hybridization for progesterone or oestrogen receptors on sheep tissues is lacking, whereas information from other techniques is abundant. Once the steroid-sensitive cells have been located, the next step is to determine which ones are involved in the control of oestrous events.

a Oestrogen binding sites The large body volume of the sheep makes it expensive to inject labelled hormone, so there are few studies of steroid binding based on autoradiography. The early attempt by Robinson (1965) failed to reveal any binding of tritiated oestradiol to neural tissue, but Tang and Adams (1978) found it in homogenates of anterior and posterior hypothalamic tissue after injection into the jugular vein. At the same time, Perry and Lopez (1978) found that tritiated oestradiol receptor binds to chromatin isolated from hypothalamus of mature female sheep.

The original view of steroid–receptor dynamics was that unoccupied receptors are present in the cytosol and are translocated into the cell nucleus when occupied by their ligand. Early immunological localization of steroid receptors showed that all the receptors are concentrated in the nuclei (Welshons *et al.* 1984), although more recent investigations have demonstrated specific cytoplasmic staining in several species, including sheep (Skinner *et al.* 1992; Lehman *et al.* 1993; Blache *et al.* 1994). We will not argue the biological relevance of the receptors present in each cellular compartment, but only recommend to the reader a book dedicated to this topic (Clark 1987). Because most of the few studies in sheep have involved separation and assay of the nuclear and cytoplasmic compartments, we will need to retain the term cytosolic receptors for those receptors in a high-speed supernatant of the tissue homogenate, and nuclear receptors for those in a *KCl* extract, according to the method described by Roy and McEwen (1977) and used in most of the studies cited.

Few studies of oestradiol receptor concentration in the CNS sheep are available, but work with cyclic and pregnant ewes (Wise *et al.* 1986; Clarke *et al.* 1981) and with rams (Pelletier and Caraty 1981) has shown that oestradiol receptors are detectable in the preoptic area, the hypothalamus, and the pituitary gland. Variations in sensitivity to oestradiol during the cycle, during recovery of cyclicity after parturition, and with the season are correlated with variations in the number of oestradiol receptors in pituitary and hypothalamic tissue (Wise *et al.* 1986; Clarke *et al.* 1981). Moreover, one study measuring the total number of oestrogen receptors (nuclear plus cytosolic fractions) has shown that peripheral administration of oestradiol increases nuclear oestrogen receptor number in the preoptic area and mediobasal hypothalamus of ovariectomized ewes (Bittman and

Blaustein 1990). Thus, oestradiol action on its own receptors does not seem to be limited to a specific region of the sheep brain.

Autoradiographic and receptor assay studies are limited by the precision of dissection—often, the smallest volume of brain tissue that can be assayed contains a wide range of structures. Most of the receptor concentration studies are also limited by their dependence on measurement of cytosolic oestrogen receptors, the existence of which, let alone their relevance, is debatable. The problem of anatomical precision was overcome with the development of immunohistochemical techniques following the arrival of monoclonal antibodies raised against oestradiol receptors, such as H222 from Abbott Laboratories (Greene *et al.* 1980).

In the female sheep (Fig. 7.3a), cells containing oestradiol receptors were found in high density in the medial preoptic area and the mediobasal hypothalamus, and lower density in several of the limbic structures, including amygdala, bed nucleus of the stria terminalis, and lateral septum (Herbison *et al.* 1993; Lehman *et al.* 1993; Blache *et al.* 1994). This distribution is very similar to that in other mammals (Cottingham and Pfaff 1986). Oestrogen receptors are present in specific areas of the preoptic–hypothalamic continuum and especially in those involved in control of reproductive events (see below).

b Sites of oestrogen action The lesion and deafferentation studies described above have been useful, but they have drawn a confused picture of the neural network involved in the control of ovulation by steroids, because the techniques do not differentiate between pathways and sites of action. The central implantation of steroids provides a more powerful tool, described by Barfield *et al.* (1983) as a refinement of the classical castration–supplementation in castrated animal. In the ewe, the first studies based on the use of implantation of oestradiol suggested that the anterior hypothalamus and mediobasal hypothalamus were the primary sites of action (Signoret 1970; Domanski *et al.* 1972a; Malven and Coppings 1977). Blache (1991a) used a more accurate implantation technique to show that oestradiol induces proceptivity, receptivity, and the LH surge only when it is placed in the ventromedial part of the hypothalamus (Fig. 7.3). A connection running from the ventromedial hypothalamus, either forward to the GnRH cell bodies in the preoptic area, or downward to the GnRH terminals (median eminence) is implicated. The disruption of the LH surge and ovulation by lesions and cuts, as described above, could be explained by damage to these tracts as well as the GnRH pathway descending from the preoptic area to the median eminence.

Taken together, these data suggest that:

(1) Oestradiol acts primarily on the mediobasal hypothalamus to trigger both oestrous behaviour and the preovulatory surge;

(2) the next site activated in the process leading to the elaboration of the preovulatory surge is the preoptic area, where most of the GnRH cell bodies that are stimulated by oestrogen are located;
(3) for oestrous behaviour, oestradiol first acts on the ventromedial hypothalamus to trigger the expression of complex behaviours such as proceptivity, and also inhibits motor centres, leading to the immobilization response to male courtship, as it has been proposed in rodents (Pfaff and Schwartz-Giblin 1988).

The use of a single site of action for oestradiol permits, in the simplest way, the synchronization of both the behavioural and ovulatory components of the oestrous cycle.

c Progesterone binding sites Neural sites that bind labelled progesterone had never been reported for sheep. However, in one study, progestin receptors were measured in the cytosolic fraction of tissue from the preoptic area and the mediobasal hypothalamus in ovariectomized ewes. It was found that peripheral oestradiol injections increased the cytosolic concentration of progestin receptors in both regions (Bittman and Blaustein 1990). These regions also contain progestin receptors in rodents (Cottingham and Pfaff 1986), and in sheep they contain many GnRH cells. The effect of oestrogen on progestin receptors could be responsible for the synergistic interaction between the two steroids in the control of GnRH pulse frequency by negative feedback during the luteal phase.

Histochemical studies are required further to localize cells containing progestin receptors, and to test for the existence of neuroanatomical connections between them and GnRH cells. Blache (1991*a*) attempted to develop an immunohistochemical approach for the sheep brain and was unsuccessful, despite trying a large variety of monoclonal and polyclonal antibodies against progesterone receptors of various species and a variety of different techniques for tissue conservation, and the use of tissue from ewes given a high dose of oestradiol, a treatment which induces high concentrations of progestin receptors in uterine tissue (Stone *et al.* 1978). At this stage, it seems that *in situ* hybridization might be the only option.

d Sites of progestagen action The central site of action of progesterone on oestrous events has not been widely investigated in sheep. Indeed, we have been unable to find any literature on the central effect of progesterone on tonic gonadotrophin secretion, but the inhibitory effect of progesterone on both sexual behaviour and the preovulatory surge has been shown to involve central mechanisms. In their early study of ovariectomized ewes, Domanski *et al.* (1972*a*) found that oestrous behaviour could be induced by placement of oestradiol implants in the hypothalamus. These implants were never effective during peripheral progesterone treatment, but elicited the strongest behavioural response two days after withdrawal of the progesterone. These effects reflect almost perfectly the responses to the subcutaneous and intravaginal treatments with oestrogen and progesterone that led to the development of the artificial cycle (Fig. 7.2) and they suggest that progesterone acts centrally to facilitate or inhibit the action of oestradiol on sexual behaviour.

The blockage of the oestradiol-induced preovulatory surge by progesterone involves mechanisms and pathways involved in the control of GnRH secretion (Currie et al. 1993). Progesterone fails to block the oestradiol-induced LH surge in ovariectomized ewes with hypothalamo–pituitary disconnection (Clarke and Cummins 1984), showing that the pituitary component is not blocked by progesterone. Moreover, in ovariectomized Suffolk ewes receiving progesterone to suppress tonic GnRH secretion, oestradiol alone fails to induce an LH surge unless high-frequency exogenous pulses of GnRH (two pulses per hour) are infused to replace the hypothalamic surge signal (Kaynard 1984). More recently, the direct blockade of GnRH secretion into the pituitary portal blood system by progesterone has been demonstrated in ovariectomized ewes (Kasa-Vubu et al. 1992).

In ovariectomized ewes receiving a single subcutaneous injection of oestradiol (8 μg), we have shown that implants of progesterone in both the ventromedial hypothalamus and the preoptic area, will block the expression of sexual behaviour (Fig. 7.3d). In contrast, the LH surge was suppressed only by progesterone implantation into the ventromedial hypothalamus (Blache 1991a). On a mass basis, progesterone is less efficacious than oestradiol, because we had to use a double implant system to deliver more steroid into a larger area, to inhibit sexual behaviour and the preovulatory surge. This might be related to differences in the polarity of the two hormones, and thus their solubility in extra cellular fluid, but it is also compatible with the 1000-fold difference in their circulating concentrations during the luteal phase. Also, the implant studies suggest that progesterone seems to act more broadly within the preoptico–hypothalamic continuum than oestradiol but, surprisingly, the ventromedial hypothalamus is the only site where progesterone increase the number of cells containing oestradiol receptors (Blache et al. 1994).

The role of progesterone in synchronizing oestrous behaviour and ovulation seems to be mediated by its inhibitory effect as well as by its induction of oestradiol receptors in the ventromedial hypothalamus. High concentrations of progesterone, typical of the luteal phase, which reach the CNS appear to ready the whole system to respond to oestradiol. This theoretical consideration is supported by:

(1) the absolute necessity for progesterone priming for oestradiol-induced sexual behaviour, as shown by the silent ovulation at the start of the breeding season, and by the ability of progesterone priming to reduce the dose of oestradiol needed to induce oestrus;
(2) the reversal of the desensitization to oestradiol that is induced by repeated injection (Scaramuzzi et al. 1972).

2 The role of other neuromediators

The simplest explanation of the neuroendocrine control of the oestrous cycle might involve only cells that produce both GnRH and steroid receptors. However, several observations lead inevitably to the conclusion that other cell types must also play key roles (Fig. 7.4):

1. The sites involved are anatomically separate, with oestradiol acting in the ventromedial hypothalamus to induce the prevulatory surge, whereas the GnRH cells eliciting the surge are located in the preoptic area. Moreover, less than 0.1 per cent of the preoptic GnRH cells appear to contain oestradiol receptors (Herbison et al. 1993; Lehman and Karsch 1993), so there is at least one interneurone between the oestradiol-receptive cells and the GnRH cells controlling ovulation.
2. Oestradiol acts mainly in the ventromedial hypothalamus to induce sexual behaviour, but progesterone acts in both the preoptic area and the ventromedial hypothalamus to block this effect (Blache et al. 1991b).
3. Oestrus in the sheep is an elaborate behaviour, not a simple reflex, so it must involve a polysynaptic system with a range of connections to centres controlling posture.
4. Gonadotrophin secretion and sexual behaviour are affected by a variety of exteroceptive factors such as photoperiod, stress, socio-sexual cues, and nutrition (reviewed by Martin 1984). These inputs must also be mediated by a wide range of neural connections (Tillet et al. 1993).

Thus, the neural pathway controlling oestrus and ovulation is not a closed system, but rather an open one which includes interneurones that connect steroid-receptive cells with GnRH cells, and also allows the modulation of steroid effects on GnRH cellular activity (Fig. 7.4). We have already described the key roles played by the steroids and GnRH, so here we will turn to the inputs by cells using other neuromediators.

Again, anatomical and physiological data have allowed us to list the candidates involved and to develop hypotheses about the organization of these neural systems. Synaptic contact between two types of neurones is the best evidence of functional connections between them, and immunohistochemical and autoradiographic studies have shown that GnRH cells are in contact with monoaminergic, amino-acidergic, and

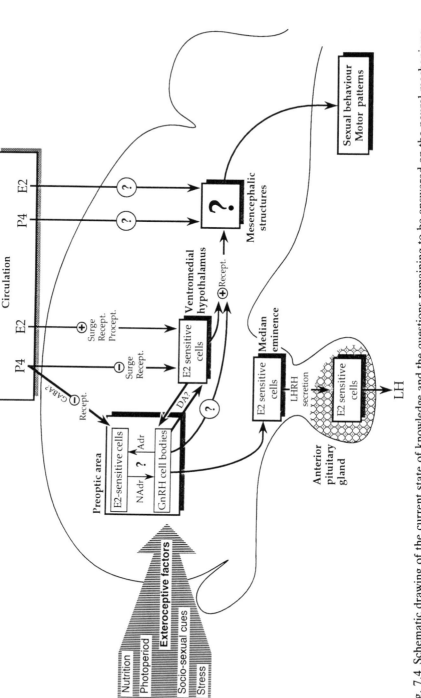

Fig. 7.4 Schematic drawing of the current state of knowledge and the questions remaining to be answered on the neural mechanisms involved in the synchronization of oestrus and ovulation during the reproductive cycle in sheep. Adr, adrenaline; NAdr, noradrenaline; DA, dopamine; E2, oestradiol; P4, progesterone; Surge, pathway affecting the preovulatory surge of GnRH and gonadotrophins;

peptidergic neurons. Such morphological studies provide us with the physical basis of the neural connections that might be involved, but that alone is insufficient for the construction of hypothetical pathways. We also need to know if these systems actually affect oestrus and ovulation, and whether they are inhibitory or stimulatory. In sheep interactions between steroids and neurotransmitters have been studied little in relation to the control of sexual behaviour and gonadotrophin secretion during the oestrous cycle, but some hypothetical considerations arise from the data available.

i Monoaminergic cells

In sheep, dopaminergic, noradrenergic, and serotonergic terminals have been observed on the surface of GnRH perikarya in the preoptic area, the anterior and lateral hypothalamus, and on GnRH terminals in the median eminence (Lehman *et al.* 1988; Kuljis and Advis 1989; Tillet *et al.* 1989, 1993). Consequently, monoaminergic neurores are strong candidates for links between oestrogen-sensitive cells and GnRH cells. However, few of the neurones that contain oestrogen receptors (less than 10 per cent in the arcuate nucleus) also contain tyrosine hydroxylase, a key enzyme in the synthesis of catecholamines (Batailler *et al.* 1992; Lehman and Karsch 1993). This suggests that monoaminergic systems are not the only links between oestrogen-sensitive cells and GnRH cells, and that many of the monoaminergic terminals on GnRH cells are mediating effects that are not oestrogen-dependent. The dopaminergic fibres arise from the diencephalon and constitute the incerto-hypothalamic bundle and the tubero-infundibulary system, whereas the serotonergic fibres arise from the median raphe, and the noradrenergic fibres from the myelencephalon and the locus coeruleus (Tillet 1987; Tillet and Thibault 1989; Thiéry and Martin 1991). This variety of origins and the differences between the dopaminergic and serotonergic systems, provides a range of possibilities for monoaminergic inputs into various aspects of the system controlling reproductive events. In fact, even from early pharmacological studies, it was evident that all of these monoamines are either inhibitors or activators of oestrous behaviour. For instance, ovulation can be blocked by the dopamine inhibitor chlorpromazine, or reserpine, which depletes catecholaminergic systems (Robertson and Rakha 1965; Jackson 1975), whereas adrenergic blocking drugs (phenoxybenzamine, pimozide) have little or no effect on the oestrogen-induced LH surge (Jackson 1977). Intracisternal infusion of serotonin inhibits cyclicity in the intact ewe (Domanski *et al.* 1975). These studies are broadly based, using peripheral or intracisternal injection of large doses of drugs and, in any case, give little indication of neuroanatomical site of action. However, Wheaton *et al.* (1972) reported that tissue concentrations of monoamines in the stalk-eminence median of sheep decrease just before the start of the preovulatory LH surge, and Wheaton *et al.* (1975) subsequently found that injection of oestrogen altered monoamine levels in the same area. This is good evidence that variations of monoaminergic activity are either directly steroid-dependent or affected by other actions of steroids.

Recently, these findings have been supported by studies coupling microdialysis with HPLC to allow measurement of extracellular concentrations of monoamines (or neuroactive amino acids) in tightly delimited regions of the brain (reviewed by Kendrick 1991). This approach has been used in the model based on ovariectomized ewes supplemented with steroids to test for changes in monoaminergic activity associated with oestrous behaviour and the preovulatory surge.

In the dorso-ventromedial hypothalamus, there is an increase in extracellular concentrations of dopamine following the withdrawal of the progesterone priming (Fabre-Nys et al. 1994). Importantly, the high levels dopamine are maintained indefinitely (at least 40 hours) unless the ewes are given a single injection of oestradiol following progesterone withdrawal, after which dopamine concentrations fall abruptly (Fig. 7.5). No change in adrenaline was observed in the same animals and the dopamine surge was absent in ovariectomized ewes not given the artificial luteal phase treatment. With regard to the induction of oestrus and ovulation, the

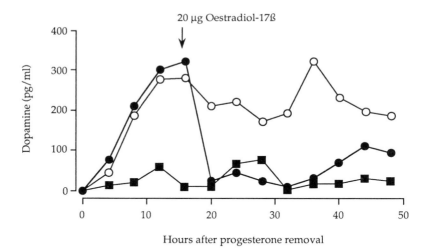

Fig. 7.5 Changes in the extracellular concentration of dopamine in the ventromedial hypothalamus of six ewes during three different artificial cycles. *Open circles*: artificial luteal phase only (intravaginal implant of progesterone plus subcutaneous implant of oestradiol), no injection of oestradiol. *Squares*: injection of oestrogen trigger (20 μg oestradiol-17β) without priming by an artificial luteal phase. *Closed circles*: full artificial cycle, with an artificial luteal phase followed by an oestrogen trigger (20 μg oestradiol-17β) 12 hours after the removal of the progesterone implant.

critical observation here is that the dopamine surge takes place in the 14–16 hour period after progesterone withdrawal, when oestradiol is unable to exert positive feedback or induce receptivity (Moore and Robinson 1957; Fabre-Nys et al. 1993a). This means that it could be involved in either the facilitatory effects of progesterone on induction of sexual behaviour by oestradiol, or the delaying effects of progesterone on the induction of the preovulatory GnRH surge by oestradiol. In either case, it would be playing a role in synchronizing the ovulatory signal with the behaviour, a phenomenon that we have attributed to progesterone priming.

In the same study, Fabre-Nys et al. (1994) recorded a transient increase in dopamine and noradrenaline in the dorso-ventromedial hypothalamus, beginning within 30 minutes of contact with the rams used to test sexual behaviour, but only when the females were receptive. These transient variations in monoaminergic activity might reflect the neural mechanism through which rams can modify GnRH secretion (Martin and Scaramuzzi 1983) and the duration of oestrus (Fletcher and Lindsay 1971).

In the preoptic-septal area, an increase in the extracellular concentrations of noradrenaline and adrenaline, but no change in dopamine concentrations, was observed during induction of the preovulatory LH surge by oestradiol (Robinson et al. 1991). Inhibition of the oestradiol-induced LH surge by progesterone is associated with a decrease in extracellular concentrations of noradrenaline but no changes in dopamine, serotonin, or adrenaline (Robinson and Kendrick 1992). Together, these microdialysis studies show that, during an artificial cycle in ovariectomized ewes, the increase in dopaminergic activity in the dorso-ventral part of the hypothalamus is apparently related to the behavioural events, whereas the increases in adrenergic and noradrenergic activity in the preoptic area are related to the induction of the preovulatory surge. These postulated roles of monoaminergic neurons during the follicular phase need further investigation.

ii γ-aminobutyric acid cells

γ-Aminobutyric acid (GABA) a neuroactive amino acid, has generally been considered to inhibit the secretion of GnRH pulses, especially in the context of changes in the steroid-dependent and -independent control of tonic GnRH secretion that are associated with photoperiod. However, GABA-ergic cells in the vicinity of the preoptic GnRH cells also appear to play an inhibitory role in periovulatory events. The evidence for this is the decrease in extracellular concentrations of GABA associated with induction of the preovulatory surge by oestradiol, and the increase in GABA concentrations associated with inhibition of this response to oestradiol by progesterone (Robinson et al. 1991; Robinson and Kendrick 1992). These observations are supported by anatomical data, based on immunohistochemical detection of glutamate decarboxylase, the enzyme that synthesizes GABA from glutamate, showing that a large proportion (40 per cent) of the GABA neurons in the preoptic area contain oestrogen receptors (Herbison et al. 1993). The steroid-dependent action of GABA cells on preoptic sites controlling GnRH secretion appears to be specific to the preovulatory surge because Scott and Clarke (1993) have suggested that it is not involved in negative feedback on tonic GnRH secretion by oestradiol. This contrasts with steroid-independent control of tonic GnRH secretion during the breeding season, for which the role of GABA- ergic neurons is supported by data from studies based on the application of GABA agonists or antagonists to the preoptic area.

iii GnRH and peptidergic cells

The observation that intraventricular injections of GnRH will inhibit the endogenous release of GnRH suggests that this decapeptide can regulate its own secretion through an ultra-short inhibitory feedback loop (Naylor et al. 1989a). This pathway is independent of opioidergic mechanisms (Naylor et al. 1989b), but may involve GABA-ergic neurone. Administration of GnRH via a microdialysis probe into the preoptic region of the ewe, where there is a major concentration of GnRH cells, raises the extracellular concentration of GABA in that region within 20 minutes (Robinson et al. 1993). Moreover, GnRH fibres contact GnRH cell bodies in the preoptic area (Caldani et al. 1988), so there appears to be a local GABA–GnRH circuitry that plays an important role in the control of GnRH pulse frequency. It may even be integral to the pulse-generating mechanisms that control the tonic and preovulatory release of GnRH.

Other actions of GnRH neurones, apart from secretion of the neurohormone into the pituitary portal system, have been known since the early 1970s. The first reports came from work with ovariectomized, oestrogen-primed rats, in which lordosis was facilitated by GnRH injected peripherally or directly into the ventromedial hypothalamus or mesencephalic structures (Moss and McCann 1973; Pfaff 1973; Moss 1977). Similar effects were subsequently reported for the courtship behaviour of female ring doves (Cheng 1977), sexual behaviour in male rats and horses (Dorsa et al. 1981; McDonnell et al. 1989), and proceptivity in female primates (Kendrick and Dixson 1985). In many species, the ventromedial hypothalamus is a major site of action of oestrogen on sexual behaviour (Barfield et al. 1983) and in the rat it is the site where GnRH affects single unit activity, in a fashion that appears to be oestrogen-dependent (Moss 1977; Chan et al. 1983). Importantly, in horses and birds, the effect of GnRH on sexual behaviour is synergistically enhanced by sex steroids (Cheng 1977; McDonnell et al. 1989). So, in all of these species at least, it is clear that GnRH is a major facilitator of the action of oestrogen on sexual behaviour.

This appears to contrast with the progesterone-primed, ovariectomized ewe, in which doses of up to 180 µg GnRH (intravenous) failed to induce receptivity or to potentiate the response to oestrogen (separate unpublished observations by G. B. Martin, T. J. Robinson, and I. A. Cumming). Intravenous treatment might be blocked by the blood–brain barrier in the ewe, so GnRH released within the brain might yet be found to play a role in the control of sexual behaviour in this species. Indeed, a potentiative effect is supported by two observations by the Michigan group: first, the duration of the GnRH surge in the pituitary portal blood of the ewe (16–24 hours) is about double that of the LH surge in peripheral blood, but similar to the duration of receptivity (Moenter *et al.* 1991; see Fig. 7.2); and second 46–70 per cent of GnRH cells are still expressing *fos* as the LH surge terminates and 57 per cent continue to do so even after the surge has ended (Moenter *et al.* 1993). The prolonged peak of GnRH secretion and the persistent activity of the GnRH cells may thus be responsible for the maintenance of sexual behaviour, which terminates at about the same time as the GnRH surge, many hours after the circulating concentrations of oestradiol have fallen to basal levels (Fig. 7.2).

Other exciting work with the female rat has demonstrated the existence of a 70 kilodalton hormone-induced protein (HIP-70) induced by oestrogen in the ventromedial hypothalamus and by GnRH in the pituitary gland (Mobbs *et al.* 1989, 1990*a*). The HIP-70 in the ventromedial hypothalamus is transported to the mesencephalic central grey (Mobbs *et al.* 1988), a site where GnRH cells are found and where GnRH affects lordosis in the rat (Shivers *et al.* 1983). HIP-70 is homologous with the phosphoinositol-specific phospholipase-C-α (Mobbs *et al.* 1990*b*) and may be part of a second messenger system activated by oestrogen (Kaplitt *et al.* 1993). This protein has not been described in sheep, but it could be associated with the extended behavioural action of oestradiol.

iv Other peptidergic cells

There is a large body of anatomical and pharmacological evidence showing that opioids, particularly β-endorphin, play a role in the control of GnRH secretion and sexual behaviour in many species, including the sheep (reviewed by Thiéry and Martin 1991). Opioid effects on the tonic secretion of GnRH pulses are generally dependent on the presence of sex steroids. Thus in ovariectomized ewes it has been suggested that opioidergic neurons mediate negative feedback by progesterone (Brooks *et al.* 1986, Yang *et al.* 1988) during the luteal phase, and negative feedback by oestradiol during the follicular phase (Curlewis *et al.* 1991). During the breeding season, injection of an opioid antagonist into the lateral cerebral ventricle of ovariectomized ewes increase LH pulse frequency only during an artificial luteal phase, when the ewes are treated with progesterone and oestradiol (Horton *et al.* 1989). This is supported by the fact that, in the arcuate nucleus of the female sheep, 15–20 per cent of the neurons that have oestrogen receptors also contain β-endorphin (Lehman and Karsch

1993). Fuentes (1989) has reported that naloxone facilitates the expression of oestrus in ewes that have been induced to ovulate during lactation or anoestrous season by treatment with exogenous progesterone and gonadotrophin. This suggests that opioid systems are involved in the inhibition of sexual behaviour, but this needs further study with ewes in other physiological states.

Oxytocin also seems to be involved in regulation of sexual behaviour. Using microdialysis, Kendrick et al. (1993) observed an increase in extracellular concentrations of oxtocin in the mediobasal hypothalamus during natural or artificial vaginocervical stimulation, and found that infusion of oxytocin into the same site inhibits receptivity. The release of oxytocin may therefore be responsible for the shortening of sexual receptivity that is observed when ewes and rams remain together, compared with when the two sexes are in intermittent contact (Parsons and Hunter 1967; Van der Westhuysen et al. 1970; Fletcher and Lindsay 1971).

In the rabbit and rat, neuropeptide Y inhibits sexual behaviour and regulates centrally the steroid-induced LH surge (Clark et al. 1985; Khorram et al. 1987; Wehrenberg et al. 1989). In sheep, neuropeptide Y neurons are found in close proximity to GnRH neurons in the preoptic area and OVLT, and are distributed similarly to the dopaminergic neurons in these regions, suggesting colocalization. Neuropeptide Y cells are also present in the mediobasal hypothalamus (Antonopoulos et al. 1989) where oestrous behaviour is controlled. Other neuropeptides, such as somatostatin and neurotensin have been localized in the infundibular nucleus, a region that contains GnRH terminals (Antonopoulos et al. 1989), and cholecystokinin, vasoactive intestinal polypeptide, and somatostatin are present in the hypothalamus (Ebling et al. 1987) as well as in mesencephalic structures (Antonopoulos et al. 1987). No physiological studies have been reported for these peptides in sheep, but they all appear to be involved in the control of sexual behaviour or LH secretion in rats (Babcock et al. 1988; Alexander et al. 1989; Weick and Stobie 1992). They are all candidates as neuromodulators of oestrous events and, like the opioids, their actions appear to be steroid-dependent and often modulated by the classical neurotransmitters, at least in rodents (reviewed by Kow and Pfaff 1988).

V GENERAL CONSIDERATIONS AND CONCLUSIONS

During the luteal phase, the neural and neuroendocrine systems controlling oestrus and ovulation are primed by progesterone and, to a lesser extent, oestrogen. Without progesterone priming, sexual behaviour is not expressed, and a small amount of oestradiol during the luteal phase promotes the priming effect of progesterone. Progesterone acts on the preoptic and hypothalamic sites where oestradiol receptors are located and where oestradiol acts to trigger the behavioural and ovulatory responses. Oestradiol also acts in other areas, such as mesencephalic structures, to control behaviour. Most of the classical neurotransmitters, such as dopamine, adrenaline, noradrenaline, and GABA, and also some of the neuropeptides are involved in the anatomical connections between these sites, and are thus involved in the regulation of the endocrine and behavioural events around ovulation. Most of their actions appear to be steroid-dependent.

The central site in the ewe where oestradiol triggers both the endocrine and the behavioural events—the mediobasal hypothalamus—contrasts with the site in rodents where steroidal effects are partitioned between the ventromedial hypothalamus (behaviour) and preoptic area (preovulatory surge). This difference is not a critical consideration to the overall theme of this review. More important is the existence of a single site of action for the trigger, whether it exists as an anatomically discrete locus or as concise network of neurons. A single site is the simplest way to ensure synchrony between ovulation and mating. Whether this simple view is the correct one for the sheep, or the only one in all species, remains to be established.

1 Questions remaining

Regarding the roles of the steroids during the luteal phase:

1. We do not understand the role of the small amount of oestrogen and its interaction with progesterone in the synchrony of behavioural and endocrine events.
2. We have not studied the neural pathway underlying the facilitatory effect of progesterone on sexual behaviour, nor have we confirmed whether progesterone induces oestradiol receptors in the ventromedial hypothalamus. There is a basic conflict in the concepts here: high levels of oestrogen receptors will facilitate the actions of oestradiol, but progesterone inhibits this action. Does progesterone operate through two separate mechanisms? This might be solved by tracing progesterone receptors in the sheep brain using molecular biology, but a cDNA probe specific to the sheep is not yet available.
3. We do not know whether the synchrony between the ovulatory and behavioural events is the direct result of inhibition by progesterone.
4. We need much more information about the GnRH pulse generator and the way it is modulated by steroids.

Regarding the roles of the steroids during the early follicular phase:

1. We need to determine how the inhibitory effect of progesterone is terminated. It cannot be due merely to its disappearance (the half-life

is much shorter than the duration of the inhibition); it may be due to actions exerted through a neuromediator, as the microdialysis studies suggest.
2. We do not know whether the synchrony of the ovulatory and behavioural events is really controlled during this period by delaying the LH surge, the onset of behaviour, or both.

Regarding the roles of the steroids during the late follicular phase there are many unanswered questions:

1. How is the biphasic action of oestradiol (negative and positive feedback) on GnRH secretion mediated? How does the hypothalamic–hypophyseal axis switch from pulsatile secretion to the surge? Why is gonadotrophin secretion low towards the end of the early follicular phase? Does this transient decrease in LH production permit accumulation of gonadotrophin for the surge?
2. Sexual behaviour depends on the steroid balance, but it is always expressed in response to male cues. At which level of control is this input integrated? Is it integrated in CNS centres, as has been suggested in rodents (Pfaff 1980), where CNS centres seem to facilitate a basic reflex at the spinal cord level that is induced by male courtship?
3. How is the steroid signal linked to the changes in motor pattern and interest that comprise the behavioural events? In sheep, do the ventromedial hypothalamus neurones produce a specific protein that is transported to the midbrain, as in the rat?
4. The duration of receptivity is similar to the duration of GnRH surge, and both are largely superior to the duration of the oestradiol peak and the preovulatory LH surge. Does GnRH play a role in the facilitation of sexual behaviour in sheep?
5. The termination of oestrous behaviour and the GnRH surge do not seem to be due to increase in circulating progesterone concentrations as the new luteal phase begins, because LH secretion and oestrus have the same duration in intact and ovariectomized, steroid-treated ewes. How are oestrous behaviour and the GnRH surge terminated? Is it simply exhaustion or is there a special termination mechanism?

2 Conclusion

With the female sheep model, we have a detailed knowledge of the secretory patterns of GnRH and LH, and the physiological relationship between these patterns. Our appreciation of role of the pulsatile GnRH signal, and the value of peripheral measurements of LH as a bioassay of GnRH cellular activity, coupled with precise control over the exogenous steroid protocols needed to induce sexual behaviour and the preovulatory surge of gonadotrophins, have allowed us to make considerable progress towards understanding the physiological processes that ensure synchrony between ovulation and sexual behaviour. We have been able to demonstrate the importance of the hypothalamus and the preoptic area in the elaboration of the primary signal, and we have begun to unravel the links mediating the dialogue between these neural sites, and with the rest of the brain—links that are the key to reproductive success. There are many questions remaining, but we now have the framework on which to construct research programmes that will answer them.

REFERENCES

Advis, J. P., Kuljis, R. O., and Dey, G. S. (1985). Distribution of LHRH content and total LHRH-degrading activity (LHRH-DA) in the hypothalamus of the ewe. *Endocrinology*, **116**, 2410–8.

Alexander, M. J., Mahomet, P. D., Ferris, C. F., Carraway, R. E., and Leeman, S. E. (1989). Evidence that neurotensin participates in the central regulation of the preovulatory surge of luteinizing hormone in the rat. *Endocrinology*, **124**, 783–8.

Antonopoulos, J., Papadopoulos, G. C., Karamanlidis, A. N., and Michaloudi, H. (1989). Distribution of neuropeptides in the infundibular nucleus of the sheep. *Neuropeptides*, **14**, 121–8.

Antonopoulos, J., Papadopoulos, G. C., Karamanlidis, A. N., Parnavelas, J., Dinopoulos, A., and Michaloudi, H. (1987). VIP- and CCK-like-immunoreactive neurons in the hedgehog (*Erinaceus europaeus*) and sheep (*Ovis aries*) brain. *Journal of Comparative Neurology*, **263**, 290–307.

Babcock, A. M., Bloch, G. J., and Micevych, P. E. (1988). Injections of cholecystokinin into the ventromedial hypothalamic nucleus inhibit lordosis behavior in the rat. *Physiology and Behavior*, **43**, 195–9.

Baird, D. T. (1978). Pulsatile release of LH and ovarian estradiol during the follicular phase of the sheep estrous cycle. *Biology of Reproduction*, **18**, 359–64.

Baird, D. T. and Scaramuzzi, R. J. (1976*a*). Changes in secretion of ovarian steroids and pituitary luteinizing hormone in the peri-ovulatory period in the ewe: the effect of progesterone. *Journal of Endocrinology*, **70**, 237–45.

Baird, D. T. and Scaramuzzi, R. J. (1976*b*). The source of ovarian oestradiol and androstenedione in sheep during the luteal phase. *Acta Endocrinologica (Copenhagen)*, **83**, 402–9.

Banks, E. M. (1964). Some aspects of sexual behaviour in domestic sheep, *Ovis aries*. *Behavior*, **23**, 249–79.

Barfield, J., Rubin, B. S., Glaser, J. H., and Davis, P. G. (1983). Sites of action of ovarian hormones in the regulation of oestrous responsiveness in rats. In *Hormones and behaviours in higher vertebrates*, (ed. J. Balthazart, E. Prove, and R. Gilles), pp. 2–17. Springer Verlag, Berlin.

Batailler, M., Blache, D., Thibault, J., and Tillet, Y. (1992). Immunohistochemical colocalization of tyrosine hydroxylase and estradiol receptors in the sheep arcuate nucleus. *Neurosciences Letters*, **146**, 125–30.

Beach, F. (1976). Sexual attractivity, proceptivity, and receptivity in female mammals. *Hormones and Behavior*, **7**, 105–38.

Bergen, H. T., Hejtmancik, J. F., and Pfaff, D. W. (1991). Effects of gamma-aminobutyric acid receptor agonists and antagonist on LHRH-synthesizing neurons as detected by immunocytochemistry and *in situ* hybridization. *Experimental Brain Research*, **87**, 46–56.

Bittman, E. L. and Blaustein, J. D. (1990). Effects of day length on sheep neuroendocrine estrogen and progestin receptors. *American Journal of Physiology*, **258**, R135–42.

Bjersing, L., Hay, M. F., Kann, G., Moor, R. M., Naftolin, F., Scaramuzzi, R. J., *et al.* (1972). Changes in gonadotrophins, ovarian steroids and follicular morphology in sheep during oestrus. *Journal of Endocrinology*, **52**, 465–79.

Blache, D. (1991a). *Etude neurobiologique du comportement sexuel et de la sécretion de l'hormone lutèinisante chez la brebis castrée; mécanismes centraux d'action des stéroïdes ovariens*. PhD Thesis Université de Bordeaux II.

Blache, D., Fabre-Nys, C., and Venier, G. (1991b). Ventromedial hypothalamus as a target for oestradiol action on proceptivity, receptivity and LH surge of the ewe. *Brain Research*, **546**, 241–9.

Blache, D., Batailler, M., and Fabre-Nys, C. (1994). Oestrogen receptors in the preoptico-hypothalamic continuum: immunohistochemical study of the distribution and cell density during oestrous cycle in ovariectomized ewe. *Journal of Neuroendocrinology*, **6**, 329–39.

Brooks, A. N., Haynes, N. B., Yang, K., and Lamming, G. E. (1986). Ovarian steroid involvement in endogenous opioid modulation of LH secretion in seasonally anoestrous mature ewes. *Journal of Reproduction and Fertility*, **76**, 709–15.

Burgus, R., Butcher, M., Ling, N., Monahan, M., Rivier, J., Fellows, R., *et al.* (1971). Structure moléculaire du facteur hypophysaire (LRF) d'origine ovine contrôlant la sécrétion de l'hormone gonadotrope hypophysaire de lutéinisation. *Comptes Rendus Hebdomadaires des Séances, Academie des Sciences (Paris)*, **D273**, 1611–3.

Caldani, M., Batailler, M., Thiéry, J. C., and Dubois, M. P. (1988). LHRH-immunoreactive structures in the sheep brain. *Histochemistry*, **89**, 129–39.

Caraty, A. and Locatelli, A. (1988). Effect of time after castration on secretion of LHRH and LH in the ram. *Journal of Reproduction and Fertility*, **82**, 263–9.

Caraty, A., Martin, G. B., and Montgomery, G. W. (1984). A new method for studying pituitary responsiveness *in vivo* using pulses of LH-RH analogue in ewes immunized against native LH-RH. *Reproduction, Nutrition, Développement*, **24**, 439–48.

Caraty, A., Locatelli, A., and Schanbacher, B. D. (1987). Augmentation par la naloxone de la fréquence et de l'amplitude des pulses de LH-RH dans le sang porte hypothalamo-hypophysaire chez le bélier castré. *Comptes Rendus Hebdomadaires des Séances, Academie des Sciences, (Paris)*, **305** (III), 369–74.

Caraty, A., Locatelli, A., and Martin, G. B. (1989). Biphasic response in the secretion of gonadotrophin-releasing hormone in ovariectomized ewes injected with oestradiol. *Journal of Endocrinology*, **123**, 375–82.

Caraty, A., Moenter, S. M., Locatelli, A., Martin, G. B., and Karsch, F. J. (1990). GnRH secretion during the preovulatory LH surge in the ewe. In *Recent progress on GnRH and gonadal peptides: proceedings of an international Symposium on GnRH and Gonadal Peptides* (ed. P. Bouchard, F. Haour, P. Franchimont, and B. Schatz), pp. 59–70. Elsevier, Paris.

Chan, A., Dudley, C. A. and Moss, R. L. (1983). Action of prolactin, dopamine and LHRH on ventromedial hypothalamic neurons as a function of ovarian hormones. *Neuroendocrinology*, **6**, 397–403.

Cheng, M-F. (1977). Role of gonadotrophin releasing hormones in the reproductive behaviour of female ring doves (*Streptopelia risoria*). *Journal of Endocrinology*, **74**, 37–45.

Clark, C. R. (1987). *Steroid hormone receptors: their intracellular localisation.* Ellis Horwood, Chichester.

Clark, J. T., Kalra, P. S., and Kalra, S. P. (1985). Neuropeptide Y stimulates feeding but inhibits sexual behavior in rats. *Endocrinology*, **117**, 2435–42.

Clarke, I. J. and Cummins, J. T. (1982). The temporal relationship between gonadotropin-releasing hormone (GnRH) and luteinizing hormone in ovariectomized ewes. *Endocrinology*, **111**, 1737–9.

Clarke, I. J. and Cummins, J. T. (1984). Direct pituitary effects of estrogen and progesterone on gonadotropin secretion in the ovariectomized ewe. *Neuroendocrinology*, **39**, 267–74.

Clarke, I. J., Burman, K., Funder, J. W., and Findlay, J. K. (1981). Estrogen receptors in the neuroendocrine tissues of the ewe in relation to breed season and stage of the estrous cycle. *Biology of Reproduction*, **24**, 323–31.

Clarke, I. J., Thomas, G. B., Yao, B., and Cummins, J. T. (1987). GnRH secretion throughout the ovine estrous cycle. *Neuroendocrinology*, **46**, 82–8.

Clarke, I. J., Cummins, J. T., Crowder, M. E., and Nett, T. M. (1988). Pituitary receptors for gonadotropin-releasing hormone in relation to changes in pituitary and plasma gonadotropins in ovariectomized hypothalamo/pituitary-disconnected ewes. II. A marked rise in receptor number during the acute feedback effects of estradiol. *Biology of Reproduction*, **39**, 349–54.

Clegg, M. T., Santolucito, J. A., Smith, J. D., and Ganong, W. F. (1958). The effect of hypothalamic lesions on sexual behaviour and estrous cycles in the ewe. *Endocrinology*, **62**, 790–96.

Coppings, R. J. and Malven, P. V. (1976). Biphasic effect of estradiol on mechanisms regulating LH release in ovariectomized sheep. *Neuroendocrinology*, **21**, 146–56.

Cottingham, S. L. and Pfaff, D. (1986). Interconnectedness of steroid hormone-binding neurons: existence and implications. In *Current topics in neuroendocrinology*, Vol. 7., (ed. D. Ganten and D. Pfaff), pp. 223–49. Springer Verlag, Berlin.

Cumming, I. A., Buckmaster, J. M., Blockey, M. A. B., Goding, J. R., Winfield, C. G., and Baxter, R. W. (1973). Constancy of interval between luteinizing hormone release and ovulation in the ewe. *Biology of Reproduction*, **9**, 24–9.

Curlewis, J. D., Naylor, A. M., Rhind, S. M., and McNeilly, A. S. (1991). Effects of β-endorphin on pulsatile luteinizing hormone and prolactin secretion during the follicular phase in the ewe. *Journal of Neuroendocrinology*, **3**, 123–6.

Currie, W. D., Evans, A. C. O., Ravindra, J. P., and Rawlings, N. C. (1993). Progesterone blockage of the positive feedback effects of 17β-oestradiol in the ewe. *Animal Reproduction Science*, **30**, 281–7.

Dees, W. L., Sorensen, A. M., Jr, Kemp, W. M., and McArthur, N. H. (1981). Immunohistochemical localization of gonadotropin-releasing hormone (GnRH) in the brain and infundibulum of the sheep. *Cell and Tissue Research*, **215**, 181–91.

Domanski, E., Przekop, F., and Skubiszewski, B. (1972*a*). Interaction of progesterone and estrogens on the hypothalamic center controlling estrous behavior in sheep. *Acta Neurobiologica Experientia*, **32**, 763–6.

Domanski, E., Przekop, F., and Skubiszewski, B. (1972*b*). The role of the anterior region of the medial basal hypothalamus in the control of ovulation and sexual behavior in sheep. *Acta Neurobiologica Experientia*, **32**, 753–62.

Domanski, E., Przekop, F., Skubiszewski, B., and Wolinska, E. (1975). The effect and the site of action of indolamines on the hypothalamic centres involved in the control of LH release and ovulation in sheep. *Neuroendocrinology*, **17**, 265–73.

Dorsa, D. M., Smith, E. R., and Davidson, J. M. (1981). Endocrine and behavioral effects of continuous exposure of male rats to a potent luteinizing hormone-releasing hormone (LHRH) agonist: evidence for central nervous system actions of LHRH. *Endocrinology*, **109**, 729–35.

Dubois, M. P. and Barry, J. (1974). Repartition comparée de trois neurofacteurs hypothalamiques; LHRF, SRIF et neurophysisine dans l'hypothalamus et l'éminence médiane: étude en immunofluorescence. *Annales d'Endocrinologie*, **35**, 663–4.

Ebling, F. J. P., Lincoln, G. A., Martin, G. B., and Taylor, P. L. (1987). LHRH and β-endorphin in the hypothalamus of the ram in relation to photoperiod and reproductive activity. *Domestic Animal Endocrinology*, **4**, 149–56.

Fabre-Nys, C. and Martin, G. B. (1991). Hormonal control of proceptive and receptive sexual behavior and the preovulatory LH surge in the ewe: re-assessment of the artificial cycle model. *Hormones and Behavior*, **25**, 295–312.

Fabre-Nys, C. and Venier, G. (1987). Development and use of a method for quantifying female sexual behaviour in ewes. *Applied Animal Behaviour Science*, **17**, 289–304.

Fabre-Nys, C. J. and Venier, G. (1989). Quantitative analysis of oestrous behaviour through the breeding season in two breeds of sheep. *Animal Reproduction Science*, **21**, 37–51.

Fabre-Nys, C., Martin, G. B., and Venier, G. (1993*a*). Analysis of the hormonal control of sexual behaviour and the preovulatory LH surge in the ewe: roles of quantity of estradiol and duration of its presence. *Hormones and Behavior*, **27**, 108–21.

Fabre-Nys, C., Poindron, P., and Signoret, J-P. (1993*b*). Reproductive behaviour. In *Reproduction in domesticated animals*, Vol 89, (ed. G. J. King), pp. 147–94. Elsevier, Amsterdam.

Fabre-Nys, C., Blache, D., Kendrick, K., and Hinton, M. (1994). Microdialysis measurement of monoamines from mediobasal hypothalamus of ovariectomized ewes during oestrus. *Brain Research*, **649**, 282–96.

Fairclough, R. J., Smith, J. F., and Peterson, A. J. (1976). Passive immunization against oestradiol-17β and its effect on luteolysis, oestrus and ovulation in sheep. *Journal of Reproduction and Fertility*, **48**, 169–71.

Feder, H. H. (1981). Estrous cyclicity in mammals. In *Neuroendocrinology of reproduction*, (ed. N. T. Adler), pp. 243–68. Plenum, New York.

Fletcher, I. C. and Lindsay, D. R. (1971). Effect of rams on the duration of oestrous behaviour in ewes. *Journal of Reproduction and Fertility*, **25**, 253–9.

Fuentes, V. O. (1989). Effect of naloxone, nalbuphine, progesterone and pregnant mare's serum gonadotrophin on the sexual behaviour of ewes. *Veterinary Record*, **124**, 274–6.

Geschwind, I. I. and Dewey, R. (1968). Dynamics of luteinizing hormone (LH) secretion in the cycling ewe: a radioimmunoassay study. *Proceedings of the Society for Experimental Biology and Medicine*, **129**, 451–5.

Glass, J. D., Mastran, T., and Nett, T. M. (1986). Effects of estradiol and progesterone on the gonadotropin-releasing hormone (GnRH)-immunoreactive neuronal system of the anestrous ewe. *Brain Research*, **381**, 336–44.

Goding, J. R., Catt, K. J., Brown, J. M., Kaltenbach, C. C., Cumming, I. A., and Mole, B. J. (1969). Radioimmunoassay for ovine luteinizing hormone: secretion of luteinizing hormone during estrus and following estrogen administration. *Endocrinology*, **85**, 133–42.

Goodman, R. L. (1988). Neuroendocrine control of the ovine estrous cycle. In *The physiology of reproduction*, Vol 2., (ed. E. Knobil and J. Neill), pp. 1929–69. Raven Press, New York.

Goodman, R. L., Legan, S. J., Ryan, K. D., Foster, D. L., and Karsch, F. J. (1981). Importance of variations in behavioural and feedback actions of oestradiol to the control of seasonal breeding in the ewe. *Journal of Endocrinology*, **89**, 229–40.

Gorski, J., Welshons, W. V., Sakai, D., Hansen, J., Walent, J., Kassis, J., et al. (1986). Evolution of a model of estrogen action. *Recent Progress in Hormone Research*, **42**, 297–329.

Greene, G. L., Nolan, C., Engler, J. P., and Jensen, E. V. (1980). Monoclonal antibodies to human estrogen receptor. *Proceedings of the National Academy of Sciences of the USA*, **77**, 5115–9.

Gregg, D. W. and Nett, T. M. (1989). Direct effects of estradiol-17β on the number of gonadotropin-releasing hormone receptors in the ovine pituitary. *Biology of Reproduction*, **40**, 288–93.

Gregg, D. W., Allen, M. C., and Nett, T. M. (1990). Estradiol-induced increase in number of gonadotropin-releasing hormone receptors in cultured ovine pituitary cells. *Biology of Reproduction*, **43**, 1032–6.

Hauger, R. L., Karsch, F. J., and Foster, D. L. (1977). A new concept for the control of the estrous cycle of the ewe based on the temporal relationships between luteinizing hormone, estradiol and progesterone in peripheral serum and evidence that progesterone inhibits tonic LH secretion. *Endocrinology*, **101**, 807–17.

Herbison, A. E., Robinson, J. E., and Skinner, D. C. (1993). Distribution of estrogen receptor-immunoreactive cells in the preoptic area of the ewe: co-localization with glutamic acid decarboxylase but not luteinizing hormone-releasing hormone. *Neuroendocrinology*, **57**, 751–9.

Herriman, I. D., Harwood, D. J., Mallinson, D. R., and Heitzman, R. J. (1979). Plasma concentrations of ovarian hormones during the oestrous cycle in the sheep and cow. *Journal of Endocrinology*, **81**, 61–4.

Hess, D. L., Wilkins, R. H., Moossy, J., Chang, J. L., Plant, J. T., McCormack, J. T., *et al.* (1977). Estrogen-induced gonadotropin surges in decerebrated female rhesus monkeys with medial basal hypothalamic peninsulae. *Endocrinology*, **101**, 1264–71.

Hoffman, G. E., Melnyk, V., Hayes, T., Bennet-Clarke, C., and Fowler, E. (1978). Immunocytology of LHRH neurons. In *Brain endocrine interaction Vol* III. Neural hormones and reproduction, (ed. D. E. Scott, G. P. Kozlowski, and A. Weind), pp. 67–82. Karger, Basel.

Horton, R. J. E., Francis, H., and Clarke, I. J. (1989). Seasonal and steroid-dependent effects on the modulation of LH secretion in the ewe by intracerebroventricularly administered beta-endorphin or naloxone. *Journal of Endocrinology*, **122**, 509–17.

Jackson, G. L. (1975). Blockage of estrogen-induced of luteinizing hormone by reserpine and potentiation of synthetic gonadotrophin-releasing hormone by estrogen in the ovariectomized ewe. *Endocrinology*, **97**, 1300–7.

Jackson, G. L. (1977). Effect of adrenergic blocking drugs on secretion of luteinizing hormone in the ovariectomized ewe. *Biology of Reproduction*, **16**, 543–8.

Jackson, G. L., Kuehl, D., McDowell, K., and Zalesky, A. (1978). Effect of hypothalamic deafferentation on secretion of luteinizing hormone in the ewe. *Biology of Reproduction*, **18**, 808–19.

Joubert, D. M. (1962). Sex behaviour of purebred and crossbred Merino and Blackhead Persian ewes. *Journal of Reproduction and Fertility*, **32**, 41–49.

Kaplitt, M. G., Kleopoulos, S. P., Pfaff, D. W., and Mobbs, C. V. (1993). Estrogen increases HIP-70/PLC-α messenger ribonucleic acid in the rat uterus and hypothalamus. *Endocrinology*, **133**, 99–104.

Karsch, F. J., Legan, S. J., Hauger, R. L., and Foster, D. L. (1977). Negative feedback action of progesterone on tonic luteinizing hormone secretion in the ewe: dependence on the ovaries. *Endocrinology*, **101**, 800–6.

Karsch, F. J., Legan, S. J., Ryan, K. D., and Foster, D. L. (1980). Importance of estradiol and progesterone in regulating LH secretion and oestrous behavior during the sheep estrous cycle. *Biology of Reproduction*, **23**, 404–13.

Karsch, F. J., Foster, D. L., Bittman, E. L., and Goodman, R. L. (1983). A role for estradiol in enhancing luteinizing hormone pulse frequency during the follicular phase of the estrous cycle of sheep. *Endocrinology*, **113**, 1333–9.

Karsch, F. J., Bittman, E. L., Foster, D. L., Goodman, R. L., Legan, S. J., and Robinson, J. E. (1984). Neuroendocrine basis of seasonal breeding in sheep. *Recent Progress in Hormone Research*, **40**, 185–232.

Kasa-Vubu, J. Z., Dahl, G. E., Evans, N. P., Thrun, L. A., Moenter, S. M., Padmanabhan, V. *et al.* (1992). Progesterone blocks the estradiol-induced gonadotropin discharge in the ewe by inhibiting the surge of gonadotropin-releasing hormone. *Endocrinology*, **131**, 208–12.

Kaynard, A. H. (1984). Evidence to support a neural site of action for progesterone blockade of the LH surge in the ewe. *Biology of Reproduction*, **30**, Suppl. 1, 34.

Kendrick, K. M. (1991). Microdialysis in large unrestrained animals: neuroendocrine and behavioural studies of acetylcholine, amino acid, monoamine and neuropeptide release in the sheep. In *Microdialysis in the neurosciences* (ed. J. B. Justice and T. E. Robinson), pp. 327–48. Elsevier, Amsterdam.

Kendrick, K. M. and Dixson, A. F. (1985). Luteinizing hormone releasing hormone enhances proceptivity in a primate. *Neuroendocrinology*, **41**, 449–53.

Kendrick, K. M., Fabre-Nys, C., Blache, D., Goode, J., and Broad, K. M. (1993). The role of oxytocin release in the mediobasal hypothalamus of the sheep in relation to female sexual receptivity. *Journal of Neuroendocrinology*, **5**, 13–21.

Khorram, O., Pau, F. K-Y., and Spies, H. G. (1987). Bimodal effects of neuropeptide Y on hypothalamic release of gonadotrophin-releasing hormone in conscious rabbits. *Neuroendocrinology*, **45**, 290–7.

Knobil, E. (1980). The neuroendocrine control of the menstrual cycle. *Recent Progress in Hormone Research*, **36**, 53–88.

Kow, L. M. and Pfaff, D. W. (1988). Behavioural effects of neuropeptides: some conceptual considerations. In *Peptide hormones: effects and mechanisms of action*, Vol. 1. (ed. A. Negro-Vilar and P. M. Conn), pp. 141–64. CRC Press, Boca Raton.

Kozlowski, G. P. and Zimmerman, E. A. (1974). Localization of gonadotrophin-releasing hormone (GnRH) in the sheep and the mouse brain. *Anatomical Record*, **178**, 396.

Krey, L. C., Butler, W. R., and Knobil, E. (1975). Surgical disconnection of the medial basal hypothalamus and pituitary function in the rhesus monkey. I. Gonadotrophin secretion. *Endocrinology*, **96**, 1073–87.

Kuljis, R. and Advis, JP. (1989). Immunocytochemical and physiological evidence of a synapse between dopamine-and luteinizing hormone releasing hormone-containing neurons in the ewe median eminence. *Endocrinology*, **124**, 1579–81. Laws, S. C., Webster, J. C., and Miller, W. L. (1990). Estradiol alters the effectiveness of gonadotropin-releasing hormone (GnRH) in ovine pituitary cultures—GnRH receptors versus responsiveness to GnRH. *Endocrinology*, **127**, 381–6.

Lehman, M. N. and Karsch, F. J. (1993). Do gonadotropin-releasing hormone, tyrosine hydroxylase-, and β-endorphin-immunoreactive neurons contain estrogen receptors? A double-label immunocytochemical study in Suffolk ewe. *Endocrinology*, **133**, 887–95.

Lehman, M. N., Robinson, J. E., Karsch, F. J., and silverman, A. J. (1986). Immunocytochemical localization of luteinizing hormone-releasing hormone (LHRH) pathways in the sheep brain during anestrus and the mid-luteal phase of the estrous cycle. *Journal of Comparative Neurology*, **244**, 19–35.

Lehman, M. N., Ebling, J. P., Moenter, S. M., and Karsch, F. J. (1993). Distribution of estrogen receptor-immunoreactive cells in the sheep brain. *Endocrinology*, **133**, 876–86.

Lehman, M. N., Karsch, F. J., and Silverman, A. J. (1988). Potential sites of interaction between catecholamines and LHRH in the sheep brain. *Brain Research Bulletin*, **20**, 49–58.

Lincoln, G. A. (1979). Differential control of luteinizing hormone and follicle-stimulating hormone by luteinizing hormone releasing hormone in the ram. *Journal of Endocrinology*, **80**, 133–40.

Malven, P. V. and Coppings, R. J. (1977). Brain sites stimulatory to release of luteinizing hormone: comparative effects of localized estrogenic and electrical stimuli in conscious sheep. *Brain Research*, **125**, 175–81.

Martin, G. B. (1984). Factors affecting the secretion of luteinizing hormone in the ewe. *Biological Reviews*, **59**, 1–87.

Martin, G. B. and Scaramuzzi, R. J. (1983). The induction of oestrus and ovulation in seasonally anovular ewes by exposure to rams. *Journal of Steroid Biochemistry*, **19**, 107–10.

Martin, G. B., Scaramuzzi, R. J., and Henstridge, J. D. (1983). Effects of oestradiol, progesterone and androstenedione on the pulsatile secretion of luteinizing hormone in ovariectomized ewes during spring and autumn. *Journal of Endocrinology*, **96**, 181–93.

Martin, G. B., Wallace, J. M., Taylor, P. L., Fraser, H. M., Tsonis, C. G., and McNeilly, A. S. (1986). Roles of inhibin and gonadotrophin-releasing hormone in the control of gonadotrophin secretion in the ewe. *Journal of Endocrinology*, **111**, 287–96.

Martin, G. B., Thomas, G. B., Terqui, M., and Warner, P. (1987). Pulsatile LH secretion during the preovulatory surge in the ewe: experimental observations and theoretical considerations. *Reproduction, Nutrition, Développement*, **27**, 1023–40.

Martin, G. B., Price, C. A., Thiéry, J-C., and Webb, R. (1988). Interactions between inhibin, oestradiol and progesterone in the control of gonadotrophin secretion in the ewe. *Journal of Reproduction and Fertility*, **82**, 319–28.

Mauer, R. E., Revenal, P., Johnson, E. S., Moyer, R. H., Hirata, A., and White, W. F. (1972). Levels of luteinizing hormone in sera of ewes near the time of estrus as determined by radio-immunoassay. *Journal of Animal Science*, **34**, 88–92.

McDonnell, S. M., Diehl, N. K. Garcia, M. C., and Kenney, R. M. (1989). Gonadotropin releasing hormone (GnRH) affects precopulatory behavior in testosterone-treated geldings. *Physiology and Behavior*, **45**, 145–9.

McKenzie, F. F. and Terrill, C. E. (1937). Estrus, ovulation, and related phenomena in the ewe. *Missouri Agricultural Experiment Station (University of Missouri) Research Bulletin*, **264**.

Mobbs, C. V., Harlan, R. E., Burrous, M. R., and Pfaff, D. W. (1988). An estradiol-induced protein synthesized in the ventral medial hypothalamus and transported to the midbrain central gray. *Journal of Neuroscience*, **8**, 113–8.

Mobbs, C., Fink, G., Johnson, M., Welch, W., and Pfaff, D. (1989). Similarity of an estrogen-induced protein and a luteinizing hormone releasing hormone-induced protein. *Molecular and Cellular Endocrinology*, **62**, 297–306.

Mobbs, C. V., Fink, G., and Pfaff, D. W. (1990*a*). HIP-70: a protein induced by estrogen in the brain and LHRH in the pituitary. *Science*, **247**, 1477–9.

Mobbs, C. V., Fink, G., and Pfaff, D. W. (1990*b*). HIP-70: an isoform of phosphoinositol-specific phospholipase C-α. *Science*, **249**, 566.

Moenter, S. M., Caraty, A., and Karsch, F. J. (1990). The estradiol-induced surge of gonadotropin-releasing hormone in the ewe. *Endocrinology*, **127**, 1375–84.

Moenter, S. M., Caraty, A., Locatelli, A., and Karsch, F. J. (1991). Pattern of gonadotropin-releasing hormone (GnRH) secretion leading up to ovulation

in the ewe: existence of a preovulatory GnRH surge. *Endocrinology*, **129**, 1175–82.
Moenter, S. M., Brand, R. M., Midgley, A. R., and Karsch, F. J. (1992*a*). Dynamics of gonadotropin-releasing hormone release during a pulse. *Endocrinology*, **130**, 503–10.
Moenter, S. M., Brand, R. C., and Karsch, F. J. (1992*b*). Dynamics of gonadotropin-releasing hormone (GnRH) secretion during the GnRH surge: insights into the mechanism of GnRH surge induction. *Endocrinology*, **130**, 2978–84.
Moenter, S. M., Karsch, F. J., and Lehman, M. N. (1993). *Fos* expression during the estradiol-induced gonadotropin-releasing hormone (GnRH) surge of the ewe: Induction in GnRH and other neurons. *Endocrinology*, **133**, 896–903.
Moore, N. W. and Robinson, T. J. (1957). The behavioural and vaginal response of the spayed ewe to oestrogen injected at various times relative to the injection of progesterone. *Journal of Endocrinology*, **15**, 360–5.
Morali, G. and Beyer, C. (1979). Neuroendocrine control of mammalian estrous behavior. In *Endocrine control of sexual behavior* (ed. C. Beyer), pp. 33–75. Raven Press, New York.
Moss, R. L. (1977). Role of hypophysiotropic neurohormones in mediating neural and behavioral events. *Federation Proceedings*, **36**, 1978–83.
Moss, R. L. and McCann, S. M. (1973). Influence of mating behavior in rats by luteinizing hormone-releasing factor. *Science*, **181**, 177–9.
Naylor, A. M., Porter, D. W. F., and Lincoln, D. W. (1989*a*). Inhibitory effect of central LHRH on LH secretion in the ovariectomized ewe. *Neuroendocrinology*, **49**, 531–6.
Naylor, A. M., Porter, D. W. F., and Lincoln, D. W. (1989 *b*). Naloxone does not affect the luteinizing hormone-releasing hormone-induced inhibition of luteinizing hormone secretion in sheep. *Journal of Neuroendocrinology*, **1**, 227–31.
Niswender, G. D., Roche, J. F., Foster, D. L., and Midgley, A. R. (1968). Radioimmunoassay of serum levels of luteinizing hormone during the cycle and early pregnancy in ewes. *Proceedings of the Society for Experimental Biology and Medicine*, **129**, 901–4.
Norman, R. L., Gliessman, P., Lindstrom, S. A., Hill, J., and Spies, H. G. (1982). Reinitiation of ovulatory cycles in pituitary stalk-sectioned Rhesus monkeys: evidence for a specific hypothalamic message for the preovulatory release of luteinizing hormone. *Endocrinology*, **111**, 1874–82.
Parsons, S. D. and Hunter, G. L. (1967). Effect of the ram on duration of oestrus in the ewe. *Journal of Reproduction and Fertility*, **14**, 61–70.
Pelletier, J. and Caraty, A. (1981). Characterisation of cytosolic 5α-DHT and 17β-estradiol receptors in the ram hypothalamus. *Journal of Steroid Biochemistry*, **14**, 603–11.
Pelletier, J., Kann, G., Dolais, J., and Rosselin, G. (1968). Dosage radioimmunologique de l'hormone lutèinisante plasmatique chez le mouton: comparaison avec le dosage biologique de LH par la diminution de l'acide ascorbique ovarien, et exemple d'application aux mesures de la LH saguine chez la brebis. *Comptes Rendus Hebdomadaires des Séances, Academie des Sciences, (Paris)*, **266**, 2352–4.

Perry, B. N. and Lopez, A. (1978). The binding of ^{3}H-labelled oestradiol-and progesterone-receptor complexes to hypothalamic chromatin of male and female sheep. *Biochemical Journal*, **176**, 873–83.

Pfaff, D. W. (1973). Luteinizing hormone-releasing factor potentiates lordosis behavior in hypophysectomized ovariectomized female rats. *Science*, **182**, 1148–9.

Pfaff, D. W. (1980). *Estrogens and brain function*. Springer-Verlag, New York.

Pfaff, D. W. and Schwartz-Giblin, S. (1988). Cellular mechanisms of female reproductive behaviors. In *The physiology of reproduction*, Vol. II, (ed. E. Knobil and J. Neill), pp. 1487–569. Raven Press, New York.

Polkowska, J. (1981). Immunocytochemistry of luteinizing hormone releasing hormone (LHRH) and gonadotropic hormones in the sheep after anterior deafferentation of the hypothalamus. *Cell and Tissue Research*, **220**, 637–49.

Polkowska, J., Dubois, M-P., and Domanski, E. (1980). Immunocytochemistry of luteinizing hormone releasing hormone (LHRH) in the sheep hypothalamus during various reproductive stages. *Cell and Tissue Research*, **208**, 327–41.

Polkowska, J., Dubois, M. P., and Jutisz, M. (1987). Maturation of luteinizing hormone-releasing hormone (LHRH) and somatostatin (SRIF) neuronal systems in the hypothalamus of growing lambs. *Reproduction, Nutrition, Développement*, **27**, 627–39.

Przekop, F. (1978). Effect of anterior deafferentation of the hypothalamus on the release of luteinizing hormone (LH) and the reproductive function in sheep. *Acta Physiological Polonica*, **29**, 393–407.

Prezekop, F. and Domanski, E. (1980). Abnormalities in the seasonal course of oestrous cycles in ewes after lesions of the suprachiasmatic area of the hypothalamus. *Journal of Endocrinology*, **85**, 481–6.

Radford, H. M. (1967). The effect of hypothalamic lesions on reproductive activity in sheep. *Journal of Endocrinology*, **39**, 415–22.

Radford, H. M. (1979). The effect of hypothalamic lesions on estradiol-induced changes in LH release in the ewe. *Neuroendocrinology*, **28**, 307–12.

Rawlings, N. C., Kennedy, S. W., and Hendricks, D. M. (1978). Effect of active immunization of the cyclic ewe against oestradiol-17β. *Journal of Endocrinology*, **76**, 11–9.

Roberts, J. L., Dutlow, C. M., Jakubowski, M., Blumm, M., and Millar, R. P. (1989). Estradiol stimulates preoptic area-anterior hypothalamic proGnRH-GAP gene expression in ovariectomized rats. *Molecular Brain Research*, **6**, 127–34.

Robertson, H. A. and Rakha, A. M. (1965). The timing of the neural stimulus which leads to ovulation in the sheep. *Journal of Endocrinology*, **32**, 383–6.

Robinson, J. E. and Kendrick, K. M. (1992). Inhibition of luteinizing hormone in the ewe by progesterone: associated changes in the release of gamma-aminobutyric acid and noradrenaline in the preoptic areas measured by intracranial microdialysis. *Journal of Neuroendocrinology*, **4**, 231–6.

Robinson, J. E., Kendrick, K. M., and Lambart, C. E. (1991). Changes in the release of gamma-aminobutyric acid and catecholamines in the preoptic-septal area prior to and during the preovulatory surge of luteinizing hormone in the ewe. *Journal of Neuroendocrinology*, **3**, 393–400.

Robinson, J. E., Brown, D., Chapman, C., and Yates, J. (1993). Luteinizing hormone-releasing hormone (LHRH) alters GABA release in the preoptic area (POA) of the ewe. *Journal of Reproduction and Fertility, Abstract series*, **11**, 47.

Robinson, T. J. (1954a). The necessity for progesterone with estrogen for the induction of recurrent estrus in the ovariectomized ewe. *Endocrinology*, **55**, 403–8.

Robinson, T. J. (1954b). Relationship of oestrogen and progesterone in oestrous behaviour of the ewe. *Nature*, **173**, 878.

Robinson, T. J. (1965). Accumulation d'oestradiol et hexoestradiol tritiés dans le cerveau et quelques autres tissus chez la brebis. *Annales de Biologie Animale, Biochimie et Biophysique*, **5**, 341–51.

Robinson, T. J., Moore, N. V., and Binet, F. E. (1956). The effect of duration of progesterone pretreatment on the response of the spayed ewe to oestrogen. *Journal of Endocrinology*, **14**, 1–7.

Ronchi, E., Krey, L. C., and Pfaff, D. W. (1992). Steady state analysis of hypothalamic GnRH mRNA levels in male Syrian hamsters: influences of photoperiod and androgen. *Neuroendocrinology*, **55**, 146–55.

Roy, E. J. and McEwen, B. S. (1977). An exchange assay for oestrogen receptors in cell nuclei of the adult rat brain. *Steroids*, **30**, 657–69.

Scaramuzzi, R. J. and Land, R. B. (1978). Oestradiol levels in sheep plasma during the oestrous cycle. *Journal of Reproduction and Fertility*, **53**, 167–71.

Scaramuzzi, R. J., Tillson, S. A., Thorneycroft, I. H., and Caldwell, B. V. (1971). Action of exogenous progesterone and estrogen on behavioural estrus and luteinizing hormone levels in the ovariectomized ewe. *Endocrinology*, **88**, 1184–9

Scaramuzzi, R. J., Lindsay, D. R., and Shelton, J. N. (1972). Effect of repeated oestrogen administration on oestrous behaviour in ovariectomized ewes. *Journal of Endocrinology*, **52**, 269–78.

Scaramuzzi, R. J., Adams, N. R., Baird, D. T., Campbell, B. K., Downing, J. A., Findlay, J. K., et al. (1993). A model for follicle selection and the determination of ovulation rate in the ewe. *Reproduction, Fertility and Development*, **5**, 459–78.

Schally, A. V., Arimura, A., Baba, Y., Nair, R. M. G., Matsuo, H., Redding,T. W., et al.(1971). Isolation and properties of the FSH and LH-releasing hormone. *Biochemical and Biophysical Research Communications*, **43**, 393–9.

Schillo, K. K., Leshin, L. S., Kuehl, D., and Jackson, G. L. (1985). Simultaneous measurement of luteinizing hormone-releasing hormone and luteinizing hormone during estradiol-induced luteinizing hormone surges in the ovariectomized ewe. *Biology of Reproduction*, **33**, 644–52.

Schindler, H. and Amir, D. (1972). Length of oestrus, duration of phenomena related to oestrus, and ovulation time in the local fat-tailed Awassi ewe. *Journal of Agricultural Science (Cambridge)*, **78**, 151–6.

Scott, C. J. and Clarke, I. J. (1993). Inhibition of luteinizing hormone secretion in ovariectomized ewes during the breeding season by gamma-aminobutyric acid (GABA) is mediated by GABA-A receptors, but not GABA-B receptors. *Endocrinology*, **132**, 1789–96.

Selmanoff, M., Shut, C., Petersen, S. L., Barraclough, C. A., and Zoeller, R. T. (1991). Single cell levels of hypothalamic messenger ribonucleic acid encoding luteinizing hormone-releasing hormone in intact, castrated, and hyperprolactinemic male rats. *Endocrinology*, **128**, 459–66.

Shivers, B. D., Harlan, R. E. Morrell, J. I. and Pfaff, D. W. (1983). Immunocytochemical localization of luteinizing hormone-releasing hormone in male and female rat brains—quantitative studies on the effect of gonadal steroids. *Neuroendocrinology*, **36**, 1–12.

Shivers, B. D., Harlan, R. E., Hejtmancik, J. F., Conn, P. M., and Pfaff, D. W. (1986). Localization of cells containing LHRH-like mRNA in rat forebrain using *in situ* hybridization. *Endocrinology*, **118**, 883–5.

Short, R. V., McDonald, M. F., and Rowson, L. E. A. (1963). Steroids in the ovarian venous blood of ewes before and after gonadotrophic stimulation. *Journal of Endocrinology*, **26**, 155–69.

Signoret, J. P. (1969). Action de la progestérone sur le comportement d'oestrus induit par le benzoate d'oestradiol chez la truie ovariectomisée. *Annales de Biologie Animale, Biochimie et Biophysique*, **9**, 361–8.

Signoret, J. P. (1970). Action d'implants de benzoate d'oestradiol dans l'hypothalamus sur le comportement d'oestrus de la brebis ovariectomisée. *Annales de Biologie Animale, Biochimie et Biophysique*, **10**, 549–66.

Skinner, D. C., Herbison, A. E., and Robinson, J. E. (1992). Immunocytochemical identification of oestrogen receptors in the ovine pars tuberalis: localization within gonadotrophs. *Journal of Neuroendocrinology*, **4**, 659–62.

Stabenfeldt, G. H., Holt, J. A., and Ewing, L. L. (1969). Peripheral plasma progesterone levels during the ovine estrous cycle. *Endocrinology*, **85**, 11–4.

Stone, G. M., Murphy, L., and Miller, B. G. (1978). Hormone receptor levels and metabolic activity in the uterus of the ewe: regulation by oestradiol and progesterone. *Australian Journal of Biological Science*, **31**, 395–403.

Tang, B. Y. and Adams, N. R. (1978). Changes in oestradiol-17β binding in the hypothalami and pituitary glands of persistently infertile ewes previously exposed to oestrogenic subterranean clover: evidence of alterations to oestradiol receptors. *Journal of Endocrinology*, **78**, 171–7.

Thiéry, J. C. and Martin, G. B. (1991). Neurophysiological control of the secretion of gonadotrophin-releasing hormone and luteinizing hormone in the sheep—a review. *Reproduction, Fertility and Development*, **3**, 137–73.

Thiéry, J-C. and Pelletier, J. (1981). Multiunit activity in the anterior median eminence and adjacent areas of the hypothalamus of the ewe in relation to LH secretion. *Neuroendocrinology*, **32**, 217–24.

Thiéry, J. C., Pelletier, J., and Signoret, J. P. (1978). Effect of hypothalamic deafferentation on LH and sexual behaviour in ovariectomized ewe under hormonally induced oestrous cycle. *Annales de Biologie Animale, Biochimie et Biophysique*, **18**, 1413–26.

Thorburn, G. D., Bassett, J. M., and Smith, I. D. (1969). Progesterone concentration in the peripheral plasma of sheep during the oestrous cycle. *Journal of Endocrinology*, **45**, 459–69.

Tillet, Y. (1987). Immunocytochemical localization of serotonin-containing neurons in the myelencephalon, brainstem and diencephalon of the sheep. *Neuroscience*, **23**, 501–27.

Tillet, Y. and Thibault, J. (1989). Catecholamine-containing neurons in the sheep brainstem and diencephalon: immunohistochemical study with tyrosine hydroxylase (TH) and dopamine-β-hydroxylase (DBH) antibodies. *Journal of Comparative Neurology*, **290**, 69–104.

Tillet, Y., Caldani, M., and Batailler, M. (1989). Anatomical relationships of monoaminergic and neuropeptide Y-containing fibres with luteinizing hormone-releasing hormone systems in the preoptic area of the sheep brain: immunohistochemical studies. *Journal of Chemical Neuroanatomy*, **2**, 319–26.

Tillet, Y., Batailler, M., and Thibault, J. (1993). Neuronal projections to the medial preoptic area of the sheep, with special reference to monoaminergic afferents: immunohistochemical and retrograde tract tracing studies. *Journal of Comparative Neurology*, **330**, 195–220.

Van der Westhuysen, J. M., Van Niekerk, C., and Hunter, G. L. (1970). Duration of oestrus and time of ovulation in sheep: effect of synchronisation, season and ram. *Agroanimalia*, **2**, 131–8.

Wallace, J. M., Martin, G. B., and McNeilly, A. S. (1988). Changes in the secretion of LH pulses, FSH and prolactin during the preovulatory phase of the oestrous cycle of the ewe and the influence of treatment with bovine follicular fluid during the luteal phase. *Journal of Endocrinology*, **116**, 123–35.

Wehrenberg, W. B., Corder, R., and Gaillard, R. C. (1989). A physiological role for neuropeptide-Y in regulating the estrogen/progesterone induced luteinizing hormone surge in ovariectomized rats. *Neuroendocrinology*, **49**, 680–2.

Weick, R. F. and Stobie, K. M. (1992). Vasoactive intestinal peptide inhibits the steroid-induced LH surge in the ovariectomized rat. *Journal of Endocrinology*, **133**, 433–7.

Welshons, W. V., Lieberman, M. E., and Gorski, J. (1984). Nuclear localization of unoccupied oestrogen receptors. *Nature*, **307**, 747–9.

Wheatley, I. S. and Radford, H. M. (1969). Luteinizing hormone secretion during the oestrous cycle of the ewe as determined by radio-immunoassay. *Journal of Reproduction and Fertility*, **19**, 211–4.

Wheaton, J. E. (1979). Regional brain content of luteinizing hormone-releasing hormone in sheep during the estrous cycle, seasonal anestrus and after ovariectomy. *Endocrinology*, **104**, 839–44.

Wheaton, J. E., Martin, S. K., Swanson, L. V., and Stormshak, F. (1972). Changes in hypothalamic biogenic amines and serum LH in the ewe during the estrous cycle. *Journal of Animal Science*, **35**, 801–4.

Wheaton, J. E., Martin, S. K. and Stormshak, F. (1975). Estrogen and L-dihydroxyphenylalanine changes in hypothalamic biogenic amines and serum LH in the ewe. *Journal of Animal Science*, **40**, 1185–91.

Wheeler, A. G. and Land, R. B. (1977). Seasonal variation in oestrus and ovarian activity of Finnish Landrace, Tasmanian Merino and Scottish Blackface ewes. *Animal Production*, **24**, 363–76.

Wildt, L., Marshall, G., and Knobil, E. (1980). Control of the rhesus monkey menstrual cycle: permissive role of hypothalamic gonadotropin-releasing hormone. *Science*, **207**, 1371–5.

Wise, M. E., Glass, J. D., and Nett, T. M. (1986). Changes in the concentration of hypothalamic and hypophyseal receptors for estradiol in pregnant and postpartum ewes. *Journal of Animal Science*, **62**, 1021–28.

Xia, L., Van Vugt, D., Alston, E. J., Luckhaus, J., and Ferin, M. (1992). A surge of gonadotropin-releasing hormone accompanies the estradiol-induced gonadotropin surge in the rhesus monkey. *Endocrinology*, **131**, 2812–20.

Yang, K., Haynes, N. B., Lamming, G. E., and Brooks, A. N. (1988). Ovarian steroid involvement in endogenous opioid modulation of LH secretion in mature ewes during the breeding and non-breeding seasons. *Journal of Reproduction and Fertility*, **83**, 129–39.

8 The gonadotrophin-releasing hormone receptor: structural determinants and regulatory control

STUART C. SEALFON and ROBERT P. MILLAR

I   Introduction

II Molecular cloning of the mammalian GnRH receptor

III Primary structure of the mammalian GnRH receptor

IV GnRH receptor gene structure and variant transcripts

V   Towards an understanding of tertiary structure

VI GnRH receptor binding pocket

VII GnRH receptor regulation

VIII Receptor desensitization: absence of the carboxy-terminal tail

IX Conclusions

I   INTRODUCTION

The gonadotrophin-releasing hormone receptor plays a pivotal role in the orchestration of the reproductive cycle. The decapeptide gonadotrophin-releasing hormone (GnRH) is generated in neurones of the medial basal hypothalamus by cleavage of a larger precursor and secreted in a pulsatile pattern into the hypophyseal portal vasculature. GnRH binds to specific high-affinity gonadotroph receptors to stimulate the release and biosynthesis of the gonadotrophins leuteinizing hormone (LH) and follicle-stimulating hormone (FSH), which in turn promote gonadal steroid synthesis and gametogenesis.

Regulation of the level and responsiveness of the GnRH receptor is critical to the maintenance of normal reproductive function and to the effectiveness of GnRH analogues in a variety of clinical applications. Peptide analogues of GnRH have proved to be efficacious in treating gonadal hormone-dependent disorders such as precocious puberty, endometriosis, uterine leiomyoma, and prostatic cancer (for review see Filicori and Flamigni 1988; Casper 1991; Conn and Crowley 1991; Barbieri 1992; Gordon and Hodgen 1992).

Two areas of extensive investigation over the two decades since the landmark determination of the sequence of GnRH by the laboratories of Schally and Guillemin have involved probing the structural requirements of GnRH analogues and the mechanisms underlying receptor regulation. Through the synthesis and study of thousands of peptide GnRH analogues, considerable progress has been made in clarifying the requirements of particular GnRH moieties for binding, and agonist and antagonist activity. Radioreceptor ligand studies have provided insight into the factors contributing to receptor regulation. Elucidating the precise interactions between GnRH and the receptor, however, has been difficult using biochemical approaches in the absence of precise knowledge about the receptor's structure. Similarly, the lack of the receptor cDNA left considerable gaps in our understanding of the regulatory control of GnRH receptor expression.

The recent molecular cloning of the GnRH receptor, first in the mouse and subsequently in four more mammalian species (see below) sets the stage for rapid progress in understanding the structure of this receptor, its interaction with ligands and its genetic and non-genetic regulatory control. A full discussion of GnRH receptor-meditated intracellular signalling events can be found in several reviews (Dan *et al.* 1990; Naor 1990; Stojilkovic and Catt 1992; Stojilkovic *et al.* 1992). This review will focus on the present state of knowledge concerning GnRH receptor structure and regulation of receptor expression.

II MOLECULAR CLONING OF THE MAMMALIAN GnRH RECEPTOR

The GnRH receptor was first cloned from the αT3-1 gonadotrope cell line using a homology cloning strategy based on the polymerase chain reaction (PCR) (Tsutsumi *et al.* 1992). The αT3-1 line, although not the only GnRH receptor-expressing gonadotroph cell line reported (see Hurbain *et al.* 1990), has been extensively characterized in many laboratories and, in addition to serving as a source of RNA for cloning the receptor, has proved invaluable in investigating gonadotroph physiology and regulation. αT3-1 cells were developed from transgenic mice by targeting expression of the SV40 T-antigen oncogene using the promoter of the glycoprotein hormone α-subunit gene (Windle *et al.* 1990). These mice developed pituitary tumours from which this cell line, which expresses high levels of the GnRH receptor, was isolated. Efficient heterologous expression of the mammalian GnRH receptor in *Xenopus* oocytes using αT3-1 RNA confirmed that this cell line would be a suitable source for cloning the receptor (Sealfon *et al.* 1990a).

The mouse receptor cDNA was cloned using highly degenerate oligomers designed against conserved transmembrane domains of G protein-coupled receptors (Tsutsumi *et al.* 1992). Molecular cloning of more than 100 G protein-coupled receptors, and analysing their hydrophobicity has revealed that the primary sequences of these critical membrane proteins, contain seven putative transmembrane domains. Proteins with seven transmembrane domains can be grouped according to nucleotide and amino acid similarity into three classes: the metabotropic glutamate receptors (Tanabe *et al.* 1992); the secretin–calcitonin–parathyroid hormone class (Ishihata *et al.* 1991); Lin *et al.* 1991; Abou *et al.* 1992), which also includes the recently isolated corticotrophin-releasing hormone receptor (Chen, *et al.* 1993); and the large G protein-coupled receptor superfamily (Probst *et al.* 1992) of which the GnRH receptor is a member.

Within the transmembrane domains of this superfamily are several highly conserved amino acid motifs, and this has facilitated the cloning of a large number of members by low stringency library screening, for example the dopamine D_2 receptor (Bunzow *et al.* 1988), and by PCR (Libert *et al.* 1989). The GnRH receptor was cloned by synthesizing oligonucleotides containing consensus sequences for the third and sixth transmembrane domains of G protein-coupled receptors, performing PCR with αT3-1 cell RNA and subcloning and sequencing the products obtained.

One shortcoming of a homology-based cloning strategy is that it is not directed, and it may not identify the receptor sought. In fact PCR has led to the cloning of a large number of orphan receptors for which the ligand and identity cannot be determined (see, for example, Schiffmann *et al.* 1990; Eidne *et al.* 1991). In identifying the GnRH receptor clone, the clones isolated were rapidly screened by sequencing and by testing any novel sequences isolated for their ability to block GnRH receptor expression in *Xenopus* oocytes. One of the clones isolated was virtually identical to the published rat endothelin receptor sequence (Arai *et al.* 1990), notable in view of a report that endothelin is an LH secretagogue in rat pituitary cell cultures (Stojilkovic *et al.* 1990). One cDNA fragment isolated, when tested by oligonucleotide hybrid arrest in oocytes, was found to block the expression of the GnRH receptor and was used to isolate a full length clone. The receptor showed appropriate ligand binding and signal transduction when expressed in *Xenopus* oocytes (Tsutsumi *et al.* 1992) and in COS-1 cells (Zhou *et al.*

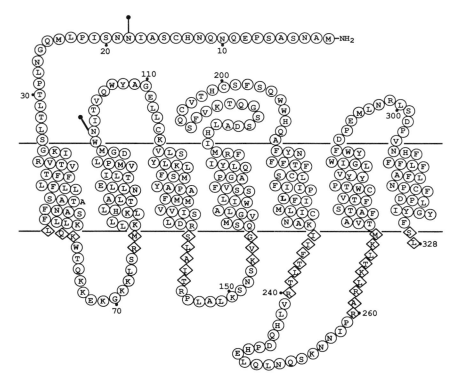

Fig. 8.1 Diagram of the human GnRH receptor. Potential N-linked glycosylation sites are indicated. Residues enclosed in diamonds from putative amphipathic α-helical intracellular domains.

1994). Subsequently, the αT3-1 GnRH receptor sequence was confirmed by two other groups who identified the receptor cDNA via mammalian cell line expression cloning (Reinhart et al. 1992; Perrin et al. 1993). This sequence was used in several laboratories to isolate the GnRH receptor cDNAs of rat (Eidne et al. 1992; Kaiser et al. 1992; Perrin et al. 1993), human (Karkar et al. 1992; Chi et al. 1993), sheep (Brooks et al. 1993; Illing et al. 1993), and cow (Karkar et al. 1993) (Figs 8.1 and 8.2).

The identification of the GnRH receptor as a member of the G protein-coupled receptor superfamily is consistent with previous studies which suggested that the GnRH receptor interacts with a G protein (Limor et al. 1989; Perrin et al. 1989). Recent studies with αT3-1 cells have demonstrated that the G protein involved in coupling to phospholipase C in this cell line is pertussis-toxin insensitive, being G_q and/or G_{11} (Anderson et al. 1992; Hsieh and Martin 1992).

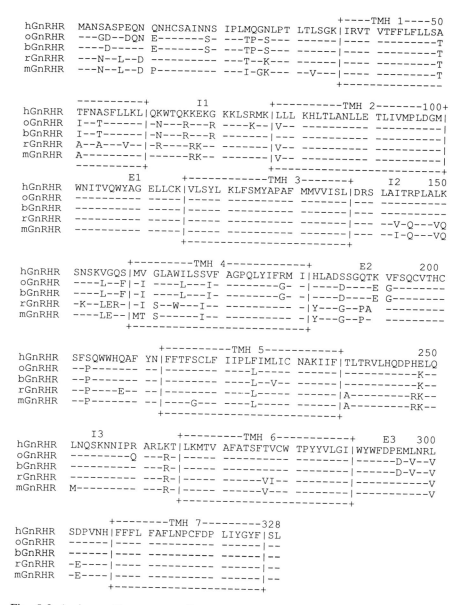

Fig. 8.2 Amino acid sequence alignment of the cloned human (h), ovine (o), bovine (b), rat (r), and murine (m) mammalian GnRH receptors. *Dashes* indicate identity with the human receptor. Approximate extent of the seven transmembrane helix (TMH) domains are enclosed in boxes. I and E refer to putative intracellular and extracellular domains, respectively.

III PRIMARY STRUCTURE OF THE MAMMALIAN GnRH RECEPTOR

A representation of the putative topology of the human GnRH receptor sequence is presented in Fig. 8.1 and an alignment of the five mammalian GnRH receptors cloned to date is shown in Fig. 8.2. The receptor is composed of a single polypeptide chain. Hydrophobicity analysis of its primary sequence confirms the presence of seven putative hydrophobic domains with an extracellular amino-terminus and an intracellular carboxy-terminus. Although direct structural information has only been obtained for bacteriorhodopsin and rhodopsin (Henderson *et al.* 1990; Schertler *et al.* 1993), the transmembrane domains of G protein-coupled receptors are believed to be α-helical and arranged around a hydrophilic core. The predicted amino acid sequences for the GnRH receptors are more than 85 per cent conserved throughout mammalian evolution and are nearly identical within the transmembrane domains. The mouse and rat receptors are 327 amino acids in length whereas the cow, sheep, and human receptors are 328 amino acids long, containing an additional residue in the second extracellular domain.

Most G protein-coupled receptors have consensus glycosylation sites and several receptors have been found to be glycosylated at these sites (Rands *et al.* 1990). Biochemical studies of the GnRH receptor have suggested it is a sialic acid residue-containing glycoprotein (Hazum 1982; Schvartz and Hazum 1985). The cow, sheep, and human receptors contain two potential sites for N-linked glycosylation (Asn-Xaa-Ser/Thr), one in the amino terminus and one in the first extracellular domain. The GnRH receptors of rodents contain an additional potential glycosylation site in the amino terminus. Many intracellular serine and threonine residues could serve as phosphorylation sites involved in receptor responsiveness or intracellular trafficking. Most G protein-coupled receptors contain single conserved cysteines in the first and second extracellular loops that may form a disulphide bond that stabilizes the structure of the functional protein. Mutation of these conserved cysteine residues disrupts the function of rhodopsin, muscarinic receptors, and β-adrenoceptors (Dixon *et al.* 1987; Karnik *et al.* 1988; Fraser 1989; Savarese *et al.* 1992). The GnRH receptor contains a cysteine in the first extracellular domain, two cysteines in the second extracellular loop, and one in the amino-terminal domain. Experimental studies will be required to resolve which of these cysteines form disulphide bonds required for proper receptor folding.

Insight into the potential function of particular residues can be achieved by analysing residues which vary by ligand subtype and which are conserved among most G protein-coupled receptors. A receptor can be viewed as having four functional properties: discriminative binding of a ligand; ligand-induced conformation change; discriminative binding of a particular G protein; and activation of the G protein. Residues that are critical for binding specificity are likely to vary between receptor subclasses and be conserved among receptors with related ligands. For example, a third transmembrane domain aspartate residue is conserved only among receptors for the cationic amines adrenaline, noradrenaline, dopamine, serotonin, and acetylcholine. This receptor locus is postulated to form an ionic interaction with the ligand for these receptors (Strader *et al.* 1987, 1988; Fraser *et al.* 1990; Ho *et al.* 1992). The GnRH receptor has a lysine at this position which may be involved in ligand binding (Zhou and Sealfon unpublished data). Residues conserved among receptors which share common G proteins may interact with specific G proteins. It has been shown that a tyrosine residue at the end of the fifth transmembrane domain which is shared by phospholipase C-coupled muscarinic receptors (but not the GnRH receptor) is critical for phosphatidylinositol coupling of the M_3 muscarinic receptor (Bluml *et al.* 1994). On the other hand, residues which serve to maintain a receptor structure needed for signal transduction through the receptor (ligand-induced conformational change) are likely to be conserved among all G protein-coupled receptors. Several GnRH receptors residues are conserved among other members of the superfamily, including a proline in many transmembrane domains, an asparagine in helix 1, the Phe-Xaa-Xaa-Cys-Trp-Xaa-Pro motif in transmembrane helix 6, and the Asp-Arg at the end of transmembrane helix 3 (Fig. 8.3).

Several features conserved among the G protein-coupled receptors are altered in the mammalian GnRH receptors cloned. These include the surprising lack of an intracellular carboxy-terminal domain, the modification of the Asp-Arg-Tyr sequence of the proximal second intracellular domain to Asp-Arg-Ser and the interchange of a conserved aspartate and asparagine in helix 2 and helix 7. A mutation of Asp-Arg-Ser to Asp-Arg-Ala, generated to test the importance of the serine in this motif, caused no apparent change in receptor binding or activation (Millar *et al.* unpublished data). The other unique features appear to have functional significance and will be discussed below in light of recent experimental results.

IV GnRH RECEPTOR GENE STRUCTURE AND VARIANT TRANSCRIPTS

Only the structure of the mouse gene has been mapped so far (Zhou and Sealfon 1994). In contrast to the genes of many G protein-coupled receptors, which are intronless and are believed to have arisen by retroposition (Brosius 1991), the coding region of the mouse GnRH receptor is distributed over three exons spanning more than 22 kilobases (Fig. 8.4). Exons 1, 2, and 3 encode, respectively, nucleotides +1 to +522, +523 to +739,

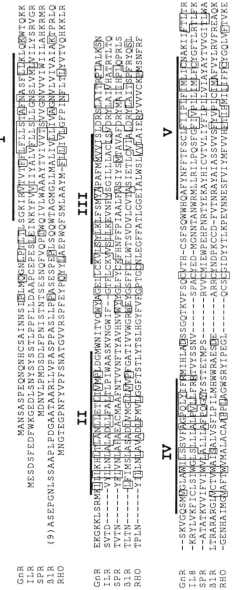

Fig. 8.3 Sequence alignment of the human GnRH receptor (GnR) and the mammalian interleukin-8 (ILR; Murphy and Tiffany 1991), substance P (SPR; Hershey and Krause 1990), β_1-adrenergic receptor (β1R; Frielle et al. 1987), and rhodopsin (RHD; Nathans and Hogness 1984) sequences. Identical residues are contained within boxes.

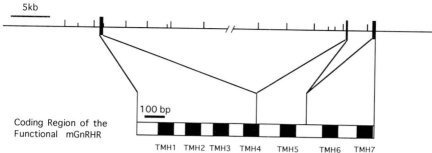

Fig. 8.4 Genomic organization of the mouse GnRH receptor. Exons are represented by filled boxes and introns by thin lines. The filled boxes in the coding region of the receptor cDNA represent the putative transmembrane helices (TMH). Hatch marks indicate XbaI sites.

and +740 to +981 of the open reading frame of the cDNA. Southern blot analysis with mouse, rat, and human DNA is consistent with the presence of a single gene (Zhou and Sealfon 1994, and unpublished data). In addition to the functional mouse receptor cDNA, several alternatively spliced mouse transcripts which do not encode functional receptors have been identified (Zhou and Sealfon 1994). The alternative transcripts found all include exon 1 but lack either exon 2 or 3. The alternative transcripts form a minority of the cDNAs isolated from an αT3–1 cell library and the biological function, if any, of the proteins encoded by these cDNAs is not known.

V TOWARDS AN UNDERSTANDING OF TERTIARY STRUCTURE

The cloning of the GnRH receptor makes possible site-directed mutagenesis studies designed to probe the three-dimensional structure of the protein and to investigate the determinants of ligand binding and receptor activation. Direct structural studies of proteins with seven transmembrane domains have been limited to high resolution electron microscopy of bacteriorhodopsin (Henderson *et al.* 1990) and a lower resolution view of rhodopsin using the same approach (Schertler *et al.* 1993). While bacteriorhodopsin has been used as a template to develop models of the three-dimensional structures of G protein-coupled receptors (Findlay and Eliopoulos 1990; Dahl *et al.* 1991; Hibert *et al.* 1991; Edvardsen *et al.* 1992; Trumpp *et al.* 1992), the relationship of the arrangement of the helices of bacteriorhodopsin to that of the G protein-coupled receptor superfamily is uncertain (Pardo *et al.* 1992). Rhodopsin is more closely related to the superfamily (Fig. 8.3). Two arrangements of the helices of rhodopsin which are mirror images of each other have been proposed to be consistent with the electron microscopic studies (Baldwin 1993). Theoretical three-dimensional models have been developed from the primary sequence of several neurotransmitter receptors using computational approaches based on molecular dynamics and quantum-mechanical considerations (Maloney Huss and Lybrand 1992; Weinstein 1992; Zhang and Weinstein 1993). One recent model for the serotonin 5-HT$_2$ receptor uses molecular dynamic simulations to propose a mechanism of activation for the receptor (Zhang and Weinstein 1993). Using similar approaches, a preliminary three-dimensional model of the GnRH receptor has recently been reported (Zhou *et al.* 1994).

Mutational studies that illuminate the structure of the receptor are critical to the development of a reliable model. The proximity of helix 2 and helix 7 of the GnRH receptor has been proposed based on the results of a reciprocal mutation of two highly conserved G protein-coupled receptor loci (Zhou *et al.* 1994). Residues which are highly conserved among the superfamily, which includes receptors with different ligands and G proteins, are likely to be essential structural determinants of receptor function. Two of these highly conserved residues are a helix 2 aspartate and a helix 7 asparagine (see Fig. 8.3 and Probst *et al.* 1992). A surprising feature of the GnRH receptor is that the residues at these loci are not conserved and, in fact, appear to be interchanged. This inversion suggested that the GnRH receptor may represent a natural double-revertant mutation and that an interaction between these two residues is essential for normal receptor structure and function. This hypothesis was investigated by site-directed mutagenesis. Mutating the Asn^{87} in helix 2 to an aspartate, which is found in virtually all other G protein-coupled receptors, eliminated detectable ligand binding, presumably by interfering with normal folding of the receptor. Introducing a second mutation in transmembrane helix 7, changing Asp^{318} to Asn, which recreates the pattern seen in other G protein-coupled receptors (Probst *et al.* 1992), led to restoration of a receptor with high affinity agonist and antagonist binding. This restoration of binding by a reciprocal mutation supports the proximity of helix 2 and helix 7. Site-directed mutagenesis in rhodopsin also supports the proximity of two residues, Glu^{90} in helix 2 and Lys^{296} in helix 7 (Rao *et al.* 1994). The experimental support of precise points of contact between different helices significantly reduces the degrees of freedom for development of a model of the receptor structure. The iterative process of mutational testing and computational modeling should ultimately lead to a valid three-dimensional model of the GnRH receptor.

Interestingly, the double mutant receptor, Asp^{87} Asn^{318} has markedly reduced coupling to the production of inositol phosphate. Similarly, the single mutation in helix 7, Asp^{318} to Asn, generates a receptor that demonstrates wild-type binding affinities but is also poorly coupled to signal transduction. The differences in coupling of the wild-type and mutant receptors suggest that receptor activation requires other loci on the receptor to interact with one or both of these residues, an arrangement that is not recreated in the reciprocal Asp^{87} Asn^{318} or Asn^{318} receptor mutants. Mutations of the transmembrane helix 2 locus homologous to Asn^{87} in several neurotransmitter receptors leads to altered or diminished receptor coupling (Chung *et al.* 1988; Fraser *et al.* 1990; Surprenant *et al.* 1992). The proximity of this locus with the helix 7 locus suggests that this interaction forms a component of the helix–helix interactions required for the conformational changes underlying receptor activation. In addition to the effects on coupling induced by mutation of the conserved helix 2 aspartate (Asn^{87} in the GnRH receptor), mutation of this locus has been found to induce a variety of alterations including reduced agonist affinity (Strader *et al.* 1987; Chung *et al.* 1988; Fraser 1989; Neve *et al.* 1991; Wang *et al.* 1991), and loss of modulation of binding by pH (Neve *et al.* 1991), sodium (Neve *et al.* 1991; Horstman *et al.* 1992), and by non-hydrolysable GTP analogues (Chung *et al.* 1988; Surprenant *et al.* 1992). While the functional changes reported in different receptors are diverse, these varied findings are all consistent with an alteration in the native or allosterically modulated structure of these receptors associated with the loss of the acidic aspartate side-chain at this position. The proposed proximity of this locus with the transmembrane helix 7 site studied facilitates the development of structural hypotheses for these pharmacological alterations. For example, the decrease in receptor binding observed by sodium or altered pH found in several G protein-coupled receptors is likely to represent modulation of one or more intramolecular interactions within the receptor, possibly the helix 2 and helix 7 sites studied in the GnRH receptor.

VI GnRH RECEPTOR BINDING POCKET

A driving force of investigations into the structure of the GnRH receptor is the hope that they will provide the basis for the rational design of novel analogues. One crucial question is to define the binding pocket for GnRH and the precise sites of interaction of receptor and ligand. Mutagenesis, deletion, and ligand cross-linking studies suggest variability of the location of the binding pocket and ligand interaction sites among different classes of G protein-coupled receptors. Three patterns emerge constituting the neurotransmitter, glycoprotein hormone, and peptide receptors. The best studied is the binding pocket of the opsins and the neurotransmitter receptors. The transmembrane domains appear to form a hydrophilic pocket for ligand binding within the transmembrane domains (for review see Strader *et al.* 1989*b*; Hulme *et al.* 1990; Probst et al. 1992). The glycoprotein hormone receptors, including receptors for TSH, FSH, and LH, have unique structures which contain an enormous amino-terminal domain which encompasses the high affinity ligand-binding site (Loosfelt et al. 1989; Tsai-Morris et al. 1990; Nagayama et al. 1991). For the peptide G protein-coupled receptors, the high affinity binding site appears to require both extracellular and transmembrane residues. High affinity peptide binding to the substance P receptor, for example, requires residues in the amino-terminus (Fong et al. 1992), extracellular loops (Fong et al. 1992; Fong et al. 1993), and transmembrane domains (Huang et al. 1994). Given the size of GnRH, one can assume that a similar pattern of extracellular and transmembrane-domain binding determinants will be observed, a hypothesis that is supported by mutagenesis experiments. For example, deletion of the amino-terminus of the receptor leads to a loss of detectable binding, as does mutation of particular loci of transmembrane domain (our unpublished observations).

Most of the data about the binding site of G protein-coupled receptors comes from site-directed mutagenesis experiments. One of the difficulties in interpreting the results of these studies is that a given mutation can alter affinity in two ways: by changing either the site of interaction with the ligand, or the structure of the binding pocket. Thus, finding that a mutation decreases or eliminates binding does not indicate that the residue mutated is a direct ligand interaction site. For example, the transmembrane helix 2 Asn to Asp mutant could be interpreted as affecting a site of GnRH interaction with the receptor because of the observation of a loss of binding, a conclusion which is probably incorrect. This difficulty can be circumvented, at least in part, by identifying intramolecular contacts through double mutations, as was undertaken for the helix 2 and helix 7 sites of the GnRH receptor. While less straightforward, the same type of analysis can be applied to ligand–receptor interactions through careful study of the structure– activity relations of various analogues with mutant and wild-type receptors. A seminal study using this approach is the determination of the helix 5 sites of agonist docking to the β-adrenoceptor (Strader et al. 1989a).

Modelling studies of GnRH and extensive studies of the pharmacology and bioactivity of GnRH analogues provide important data about the ligand side-chains most likely to interact with the receptor. The GnRH sequence pGlu-His-Trp-Ser-Tyr-Gly-Leu-Arg-Pro-Gly-NH$_2$ has been empirically modified in every position in an attempt to develop more potent agonists and antagonists. The activity of the several thousand analogues reported before 1986 is reviewed by Karten and Rivier (1986). Certain salient properties emerge. Agonist activity requires the amino- and carboxy-terminal residues whereas binding alone requires only carboxy-terminal residues. The residues most critical for agonist binding and activity are the lactam ring of pGlu^1, the aromatic side-chain of Trp^3, and the carboxy-terminal amide of Gly-$NH_2{}^{10}$. Although *des*-His^2-GnRH was the first GnRH receptor antagonist developed (Vale *et al.* 1972), His^2 is not essential for agonist activity as analogues with glycine at this position retain some agonist activity (Karten and Rivier 1986; Gupta *et al.* 1993).

The active conformation of GnRH has been investigated by a combination of conformational energy studies, study of conformationally constrained analogues, and modeling based on homology with peptides of known structure (Momany 1976; Struthers *et al.* 1990; Gupta *et al.* 1993). That the amino and carboxy-terminus of GnRH can exist in close proximity is supported by the development of antisera which require both domains for recognition (King and Millar 1982). The increased affinity of D-Ala^6 analogues has been attributed to constraining a β-II type bend involving Tyr^5-Gly^6-Leu^7-Arg^8 which represents the active conformation (for review see Karten and Rivier 1986). Hagler and co-workers have modeled the conformation of a cyclic GnRH antagonist and found limited flexibility of the backbone (Struthers *et al.* 1990). The proposed conformation was supported by NMR spectroscopy which demonstrated the peptide exists in two slowly interconverting forms involving cis–trans isomerization about the Pro-Ala bond of this analogue. This data was used to design novel cyclic analogues with affinity nearly as high as that of non-cyclic antagonists. Using a different approach to model the active structure of GnRH, Gupta and coworkers searched the structure databank for peptides with sequence homology to GnRH. The model developed from this analysis contains a type-III β-bend through a turn involving Trp^3-Ser^4-Tyr^5-Gly^6 in contrast to the type II β-turn proposed by others involving Tyr^5-Gly^6-Leu^7-Arg^8. These workers propose that the guanidino group of Arg^8, the aromatic side-chain of Trp^3, and the amino and carboxy-termini are juxtaposed and interact with the receptor.

The basic Arg^8 of GnRH is critical for high affinity agonist and antagonist binding. Hazum (1987), based on cation competition experiments, suggested that the Arg^8 on GnRH interacts with at least two carboxylic groups on the receptor. Confirming the importance of Arg^8 are substitutions at Arg^8, as in chicken I GnRH (Gln^8GnRH) which tend to reduce significantly the binding to the mammalian receptor (Millar *et al.* 1989). Yet, while Arg^8 is needed for high affinity binding to the mammalian receptor, others have suggested that the role of Arg^8 is to interact with His^2 and Tyr^5 of GnRH to maintain the ligand in an active conformation (Shinitzky and Fridkin 1976; Shinitzky *et al.* 1976; Hazum *et al.* 1977; Milton *et al.* 1983).

The possibility that Arg^8 of GnRH forms an ionic interaction with an acidic residue of its receptor was investigated by site-directed mutagenesis. To find the potential site of ionic interaction of arginine with the receptor, all acidic residues on the receptor were mutated to their isosteric amine and a mutant that failed to discriminate between Arg^8-GnRH and Gln^8-GnRH was sought. One mutant receptor, Glu^{301}-Gln, had lower affinities for mammalian GnRH than for the wild-type receptor. The affinity of this mutant for chicken I GnRH and for other GnRH analogues with an uncharged residue in the eighth position, however, was unchanged or improved. These results support a role of this residue, located in the third extracellular domain, in conferring the preference of the mammalian receptor for Arg^8 GnRH (Flanagan et al. 1994). The cloning of the GnRH receptor from non-mammalian species, which have different structure–activity profiles, will be important in refining and testing our understanding of the receptor's binding pocket. At this stage, the proposed ionic interaction between Arg^8 on GnRH and the third extracellular loop, and the proximity of helix 2 and helix 7 provide an encouraging foundation upon which to begin to develop a predictive model of the interaction of GnRH with the receptor.

VII GnRH RECEPTOR REGULATION

The utility of GnRH analogues in humans and the complexities of their effects have provided impetus for investigations into the mechanisms underlying regulation of the receptor. The availability of the GnRH receptor cDNA has allowed studies of receptor regulation to be extended to analysis of mRNA expression. The level and responsiveness of the receptor is modulated by a complex interplay of GnRH and heterologous factors, predominantly gonadal hormones. In some cases the results reported in different paradigms and experimental systems appear discordant. The guidelines and caveats presented by Conn et al. (1992) to guide the assessment of which intracellular signals mediate LH secretion in the face of conflicting experimental results can be equally well applied to assessing the data on the mechanisms underlying receptor regulation. In essence, finding a correlation is necessary but not sufficient to prove mediation. Many potential loci of regulatory control of receptor expression exist. The possible mechanisms which contribute to alterations of GnRH receptor number include altered biosynthesis (including gene transcription, RNA stability, and translational efficiency), degradation, stabilization, and recycling of receptors between ligand-accessible and inaccessible pools. To add to the complexity, modulation of receptor responsiveness and post-receptor mechanisms are also important (for review see Huckle and Conn 1988; Naor 1990), and regulatory control may vary in the different mammalian species used for model systems. The present understanding of homologous and heterologous regulation of receptor number will be reviewed. Other recent reviews have addressed receptor coupling to LH secretion (Huckle and Conn 1988; Naor 1990; Davidson et al. 1991).

GnRH is secreted into the portal vasculature in a pulsatile pattern. Pituitary portal blood GnRH concentration has been monitored at 30 second intervals in ovariectomized ewes and found to exhibit an average pulse amplitude of 70-fold above baseline levels with a mean duration of 5.5 minutes (Moenter et al. 1992). The rise in GnRH concentration is abrupt and occurs in a square-wave pattern. In addition to the pulse frequency, the wave pattern itself may have physiological significance. In perifused sheep pituitary cells, square-wave GnRH pulses are more efficient at inducing LH release than more gradual increases in GnRH concentration (McIntosh and McIntosh 1985). The importance of the pulsatile pattern of GnRH secretion is well documented in a number of regulatory phenomena. Pulsatile GnRH, for example, modulates the length of LHβ and α-subunit poly(A) tails (Weiss et al. 1992).

Exposure of dispersed rat pituitary cells to low concentrations of GnRH (0.1–2.5 nM) induces a biphasic modulation of GnRH receptor number. Over the first several hours, receptor number is decreased then receptor number recovers to exceed control levels by 6–9 hours (Loumaye and Catt 1982, 1983; Conn et al. 1984). The GnRH receptor is unusual in being capable of being either down-or up regulated by GnRH, depending on the duration and level of GnRH exposure. Whereas low concentrations of GnRH increase receptor number after 6 hours, higher concentrations decrease receptor number at all time points (Loumaye and Catt 1982; Conn et al. 1984). The upregulation of the GnRH receptor requires both protein and RNA synthesis (Loumaye and Catt 1983; Conn et al. 1984), whereas the downregulation induced by high GnRH concentrations does not (Conn et al. 1984).

Results of experiments on the signals which mediate homologous upregulation are discordant. Early studies in two laboratories confirm the requirement of extracellular calcium for upregulation (Loumaye and Catt 1983; Conn et al. 1984). However in a later study using density-shift labeling, the increased rate of GnRH receptor synthesis was found to be independent of extracellular calcium (Braden and Conn 1990). Density-shift labeling quantifies the rate of receptor biosynthesis by photoaffinity labeling of newly synthesized receptors incorporating heavy amino acids. Using the same technique, phorbol esters were found to increase the synthesis of new receptor. However the finding that the effects of GnRH and phorbol ester were additive suggested that homologous stimulation of receptor synthesis is not mediated by protein Kinase C activation (Braden et al. 1991).

How can GnRH receptor upregulation require transcription and protein synthesis, and be dependent on extracellular calcium (Loumaye and Catt 1983; Conn et al. 1984) when new receptor synthesis does not require extracellular calcium (Braden and Conn 1990)? One possible explanation that arises from recent studies on homologous regulation of receptor mRNA could be that the transcription and protein synthesis required for regulation is not that of the GnRH receptor. αT3-1 cells have been shown to demonstrate upregulation of the receptor in response to a 20 minute exposure to 0.1–10 nM of GnRH to a degree comparable to that reported for rat primary cultures (Tsutsumi, et al. 1993). However this regulation is unaccompanied by any change in GnRH receptor mRNA measured by both nuclease protection assay and northern blot analysis. Furthermore, when pure RNA from regulated αT3-1 cells is injected into *Xenopus* oocytes and the number of receptors synthesized assayed by the electrophysiological response to GnRH, the ability of the RNA to direct the expression of receptors in oocytes is regulated in concert with the regulation that occurs in the αT3-1 cells (Tsutsumi et al. 1993). This suggests the regulation of an RNA distinct from the GnRH receptor mRNA which affects the level of receptor expression. Identification of the precise regulatory mechanism at work in these cells may explain the puzzling discordance of the results reported for homologous regulation in rat primary culture.

In contrast to the results reported with αT3-1 cells, in perifused rat pituitary cells, pulsatile GnRH induces a rapid and significant increase in GnRH receptor mRNA determined by northern blot analysis (Kaiser et al. 1993) Similar results have been obtained in rat primary culture by solution hybridization nuclease protection assay (Tsutsumi and Sealfon unpublished data). Perhaps the up regulation of GnRH receptor mRNA found in rat pituitary represents the calcium-independent regulation of biosynthesis previously identified by density shift labeling. The regulation of an RNA distinct from the GnRH receptor mRNA found in αT3-1 cells could correspond to the extracellular calcium-, transcription-, and translation-dependent regulation in the number of receptors. Another possible explanation for the difference in experimental results is that the regulation in the mouse (or in the αT3-1 mouse cell line) and in rat pituitary occurs via distinct mechanisms. Further studies should resolve these questions.

Homologous upregulation has also been detected *in vivo* where GnRH induced upregulation of the receptor has been shown to be dependent on pulse frequency (Katt *et al.* 1985). The signalling involved in LH release can be dissociated from that involved in upregulation of the receptor. Regulation occurs at GnRH concentrations below that required to stimulate LH release (Loumaye and Catt 1983; Conn *et al.* 1984). Leong has proposed the intriguing hypothesis that the pattern of calcium signalling in an individual gonadotrope represents a binary signal encoder (Leong 1991; Leong *et al.* 1991). High levels of GnRH induce a spike–plateau calcium wave which are proposed to induce secretion, whereas low levels of GnRH induce calcium oscillations which are proposed to signal increased GnRH receptor and LH biosynthesis.

As with upregulation, the downregulation of the receptor induced by exposure to high concentrations of GnRH (10^{-7} to 10^{-6}M) is also reported to occur independently of protein Kinase C activation in rat primary cultures (McArdle *et al.* 1987). Prolonged receptor downregulation with exposure to high concentrations of GnRH should be differentiated from the initial decrease in the level of binding found with exposure to low concentrations of GnRH because the underlying mechanisms may differ (Loumaye and Catt, 1983; Conn *et al.* 1984). Investigation of homologous downregulation in αT3-1 cells at 1 μM GnRH exposure demonstrated that downregulation is not accompanied by any detectable change in receptor mRNA levels (Tsutsumi *et al.* 1990). Similar to the results obtained with upregulation in these cells, injection of RNA into *Xenopus* oocytes from cells in which the receptor had been downregulated induced significantly lower levels of receptor expression than RNA from control cells. Determination of the polysome distribution of GnRHR mRNA in downregulated αT3-1 cells shows a shift of the mRNA to smaller polysomes (Tsutsumi *et al.* 1995). These results suggest that the GnRH receptor mRNA in downregulated cells initiates polysome formation less efficiently, raising the possibility that translational modulation contributes to homologous downregulation.

GnRH-dependent receptor regulation has been reported *in vivo* accompanying both castration and lactation. Gonadectomy of either male or female rats induces an increase in GnRH receptor binding capacity. This upregulation is inhibited by an anti-GnRH antibody (Clayton and Catt 1981; Frager *et al.* 1981; for review of other studies see Clayton *et al.* 1985). Lactation and removal from suckling in rats causes a dramatic regulation in GnRH receptor content. After removal of pups, receptor capacity increases fivefold within 24 hours. This upregulation is also blocked by an anti-GnRH antiserum (Smith and Ru-Lee 1989). The cloning of the GnRH receptor has led to the discovery that receptor regulation in castration and lactation is accompanied by concomitant changes in receptor mRNA. Ovariectomy in rats (Kaiser *et al.* 1993) and castration in sheep (Illing *et al.* 1993) induce an upregulation of the receptor mRNA. An increase in GnRH receptor mRNA following pupremoval from lactating rats has been demonstrated by both *in situ* hybridization (Smith and Reinhart 1993) and nuclease-protection assay (Jakubowski 1993).

As for many peptide receptors, there is evidence that the ligand-bound GnRH receptor can be internalized and either degraded or recycled (Wynn *et al.* 1986). Exposure to phorbol esters results in an increase in binding sites detectable by ligand binding or fluorescent labelling at room temperature. The detection of this apparently preformed receptor population is referred to as unmasking (Braden *et al.* 1989; Leblanc *et al.* 1992). These sequestered receptors are uncoupled from phosphatidylinositol hydrolysis (Huckle *et al.* 1989).

GnRH receptor regulation by gonadal hormones has been well characterized in the rat and in the sheep. In rats, activin A increases receptor synthesis (Braden and Conn 1992). Inhibin decreases steady-state GnRH receptor levels in rat primary culture (Wang *et al.* 1988), but does not affect synthetic rate (Braden *et al.* 1990). In GnRH-deficient *hpg* mice, oestradiol doubles the receptor content, and oestradiol and GnRH have an additive effect (Naik *et al.* 1985). In ovine pituitary cultures, oestradiol, progesterone, and inhibin change GnRH receptor responsiveness, as determined by GnRH-stimulated LH release (Huang and Miller 1980, 1984; Batra and Miller 1985, 1986). These changes in responsiveness correlate with changes in receptor content (Gregg *et al.* 1990; Laws *et al.* 1990a,b; Gregg *et al.* 1991). The apparent receptor number varies 30–40-fold between the lowest receptor levels, in the presence of progesterone treatment, and highest, induced by combined oestradiol and inhibin exposure. Using the oocyte expression bioassay, these regulatory changes were found to be accompanied by changes in GnRH receptor mRNA bioactivity (Sealfon *et al.* 1990b). Regulation of GnRH receptor mRNA by oestradiol, inhibin, and activin has been demonstrated by northern blot analysis (Wu *et al.* 1994).

GnRH receptor content varies during the rat oestrous cycle. Peak receptor levels are seen between metestrus and early proestrus with levels falling rapidly during the preovulatory gonadotropin surge (Savoy-Moore *et al.* 1980). GnRH receptor mRNA also varies significantly through the cycle. Measured by PCR, mRNA levels have been reported to reach a maximum on the afternoon of proestrus (Bauer-Dantoin *et al.* 1993).

The studies published to date demonstrate that GnRH receptor regulation is accompanied by appropriate changes in receptor mRNA expression in many experimental paradigms. The lack of mRNA regulation in some paradigms and the discordant effects of extracellular calcium on GnRH receptor regulation and biosynthesis indicate that there remains much to learn about the mechanisms underlying its physiological regulation.

VIII RECEPTOR DESENSITIZATION: ABSENCE OF THE CARBOXY-TERMINAL TAIL

One striking feature of the mammalian GnRH receptor, unique among the hundred-odd receptors cloned to date, is the complete absence of an intracellular carboxy-terminal domain. In most G protein-coupled receptors, this domain contains a cysteine which has been shown to be palmitoylated in several receptors and forms, in effect, a fourth intracellular loop (Ovchinnikov *et al.* 1988; O'Dowd *et al.* 1989; Ng *et al.* 1993). This domain also contains sites involved in phosphorylation-mediated regulation and desensitization of several receptors (Leeb-Lundberg *et al.* 1987; Sibley *et al.* 1987). In the TRH receptor, for example, agonist-induced receptor internalization requires specific domains in the carboxy-terminus (Nussenzveig *et al.* 1993). Consistent with the potential role for this domain in desensitization of receptors, no desensitization of the αT3-1 cell GnRH receptor is observed during 90 minutes exposure to GnRH. In contrast to this, the TRH receptor in GH3 cells shows a rapid desensitization (Sealfon *et al.* 1992; Davidson *et al.* 1994). As desensitization of the GnRH receptor does not appear to be mediated by receptor phosphorylation, desensitization of the gonadotrope on continuous exposure to GnRH *in vivo* is likely to be mediated by modulation of the level of receptor expression, by post-receptor mechanisms, or both.

IX CONCLUSIONS

The cloning of the mammalian GnRH receptor provides the foundation for significant progress in understanding the molecular mechanisms underlying ligand binding, receptor activation, signal transduction, and regulation. Recent studies have provided considerable insight into the structure of this receptor and its regulatory control. A fuller understanding of these processes should have a significant impact on the development of GnRH analogues and their use in clinical medicine.

ACKNOWLEDGMENTS

Supported by NIH Grant RO1 DK46943.

REFERENCES

Abou-Samra, A. B., Jüppner, H., Force, T., Freeman, M. W., Kong, X. F., *et al.* (1992). Expression cloning of a common receptor for parathyroid hormone and parathyroid hormone-related peptide from rat osteoblast-like cells: a single receptor stimulates intracellular accumulation of both cAMP and inositol trisphosphates and increases intracellular free calcium. *Proceedings of the National Academy of Sciences of the USA*, **89**, 2732–6.

Anderson, L., Hoyland, J., Mason, W. T., and Eidne, K. A. (1992). Characterization of the gonadotrophin-releasing hormone in single alphaT3-1 pituitary gonadotroph cells. *Molecular and Cellular Endocrinology*, **86**, 167–75.

Arai, H., Hori, S., Aramori, I., Ohkubo, H., and Nakanishi, S. (1990). Cloning and expression of a cDNA encoding an endothelin receptor. *Nature*, **348**, 730–2.

Baldwin, J. M. (1993). The probable arrangement of the helices in G protein coupled receptors. *Embo Journal*, **12**, 1693–703.

Barbieri, R. L. (1992). Clinical applications of GnRH and its analogues. *Trends in Endocrinolology and Metabolism*, **3**, 30–4.

Batra, S. K. and Miller, W. L. (1985). Progesterone decreases the responsiveness

of ovine pituitary cultures to luteinizing hormone-releasing hormone. *Endocrinology*, **117**, 1436–40.
Batra, S. K. and Miller, W. L. (1986). Progesterone antagonizes the ability of porcine ovarian inhibin to sensitize ovine pituitary cell culture to luteinizing hormone-releasing hormone: dependence on ovaries *in vivo*. *Endocrinology*, **119**, 1933–8.
Bauer-Dantoin, A. C., Hollenberg, A. N., and Jameson, J. L. (1993). Dynamic regulation of gonadotropin-releasing hormone receptor mRNA levels in the anterior pituitary gland during the rat estrous cycle. *Endocrinology*, **133**, 1911–4.
Bluml, K., Mutschler, E., and Wess, J. (1994). Identification of an intracellular tyrosine residue critical for muscarinic receptor-mediated stimulation of phosphatidylinositol hydrolysis. *Journal of Biological Chemistry*, **269**, 402–5.
Braden, T. D. and Conn, P. M. (1990). Altered rate of synthesis of gonadotropin-releasing hormone receptors: effects of homologous hormone appear independent of extracellular calcium. *Endocrinology*, **126**, 2577–82.
Braden, T. D. and Conn, P. M. (1992). Activin-A stimulates the synthesis of gonadotropin-releasing hormone receptors. *Endocrinology*, **130**, 2101–5.
Braden, T. D., Hawes, B. E., and Conn, P. M. (1989). Synthesis of gonadotropin-releasing hormone receptors by gonadotrope cell cultures: both preexisting receptors and those unmasked by protein Kinase-C activators show a similar synthetic rate. *Endocrinology*, **125**, 1623–9.
Braden, T. D., Farnworth, P. G., Burger, H. G., and Conn, P. M. (1990). Regulation of the synthetic rate of gonadotropin-releasing hormone receptors in rat pituitary cell cultures by inhibin. *Endocrinology*, **127**, 2387–92.
Braden, R. D., Bervig, T., and Conn, P. M. (1991). Protein Kinase-C activation stimulates synthesis of gonadotropin-releasing hormone (GnRH) receptors, but does not mediate GnRH-stimulated receptor synthesis. *Endocrinology*, **129**, 2486–90.
Brooks, J., Taylor, P. L., Saunders, P., Eidne, K. A., Struthers, W. J., McNeilly, A. S. (1993). Cloning and sequencing of the sheep pituitary gonadotropin-releasing hormone receptor and changes in expression of its mRNA during the estrous cycle. *Molecular and cellular Endocrinology*, **94**, R23–7.
Brosius, J. (1991). Retroposons—seeds of evolution. *Science*, **251**, 753.
Bunzow, J. R., Van Tol, H. H. M., Grandy, D. K., Albert, P., Salon, J., Christie, M., et al. (1988). Cloning and expression of a rat D2 dopamine receptor cDNA. *Nature*, **336**, 783–7.
Casper, R. F. (1991). Clinical uses of gonadotropin-releasing hormone analogues. *Canadian Medical Association Journal*, **144**, 153–60.
Chen, R., Lewis, K. A., Perrin, M. H., and Vale, W. W. (1993). Expression cloning of a human corticotropin-releasing-factor receptor. *Proceedings of the National Academy of Sciences of the USA*, **90**, 8967–71.
Chi, L., Zhou, W., Prikhozhan, A., Flanagan, C., Davidson, J. S., Golembo, M., et al. (1993). Cloning and characterization of the human GnRH receptor. *Molecular and Cellular Endocrinology*, **91**, R1–6.
Chung, F-Z., Wang, C-D., Potter, P. C., Venter, J. C., and Fraser, C. M. (1988). Site-directed mutagenesis and continous expression of human β-adrenergic receptors. *Journal of Biological Chemistry*, **263**, 4052–5.

Clayton, R. N. and Catt, K. J. (1981). Regulation of pituitary gonadotropin-releasing hormone receptors by gonadal hormones. *Endocrinology*, **108**, 887–95.

Clayton, R. N., Detta, A., Naik, S. I., Young, S., and Charlton, H. M. (1985). Gonadotrophin-releasing hormone receptor regulation in relationship to gonadotrophin secretion. *Journal of Steroid Biochemistry*, **23**, 691–702.

Conn, P. M. and Crowley, W. F. (1991). Gonadotropin-releasing hormone and its analogues. *New England Journal of Medicine*, **324**, 93–103.

Conn, P. M., Rogers, D. C., and Seay, S. G. (1984). Biphasic regulation of the gonadotropin-releasing hormone receptor by receptor microaggregation and intracellular Ca^{2+} levels. *Molecular Pharmacology*, **25**, 51–5.

Conn, P. M., Hawes, B. E., and Janovick, J. A. (1992). Selection of models for the study of GnRH stimulated gonadotropin release prejudices the assignment of roles for mediators and modulators of hormone action. *Molecular and Cellular Endocrinology*, **84**, C33–C37.

Dahl, S. G., Edvardsen, O., and Sylte, I. (1991). Molecular dynamics of dopamine at the D_2 receptor. *Proceedings of the National Academy of Sciences et al. USA*, **88**, 8111–5.

Dan, C. H., Ben, M. D., and Naor, Z. (1990). The gonadotropin-releasing hormone receptor: signals involved in gonadotropin secretion and biosynthesis. *Hormone Research*, **32**, 76–86.

Davidson, J. S., Wakefield, I. K., and Millar, R. P. (1994). Absence of rapid densensitization of the mouse gonadotropin-releasing hormone receptor. *Biochemical Journal*, **300**, 299–302.

Dixon, R. A., Sigal, L. S., Candelore, M. R., Register, R. B., Rands, E., and Strader, C. D. (1987). Structural features required for ligand binding to the beta-adrenergic receptor. *EMBO Journal*, **6**, 3269–75.

Edvardsen, O., Sylte, I., and Dahl, S. G. (1992). Molecular dynamics of serotonin and ritanserin interacting with the 5-HT_2 receptor. *Molecular Brain Research*, **14**, 166–78.

Eidne, K. A., Zabavnik, J., Peters, T., Yoshida, S., Anderson, L., and Taylor, P. L. (1991). Cloning, sequencing and tissue distribution of a candidate G protein-coupled receptor from rat pituitary gland. *FEBS Letters*, **292**, 243–8.

Eidne, K. A., Sellar, R. E., Couper, G., Anderson, L., and Taylor, P. L. (1992). Molecular cloning and characterisation of the rat pituitary gonadotropin-releasing hormone (GnRH) receptor. *Molecular and Cellular Endocrinology*, **90**, R5–9.

Filicori, M. and Flamigni, C. (1988). GnRH agonists and antagonists: current clinical status. *Drugs*, **35**, 63–82.

Findlay, J. and Eliopoulos, E. (1990). Three-dimensional modelling of G protein-linked receptors. *Trends in Pharmacological Sciences*, **11**, 492–9.

Flanagan, C. A., Becker, I. I., Davidson, J. S., Wakefield, I. K., Zhou, W., Sealfon, S. C., *et al.* (1994). Glu301 of the mouse gonadotropin-releasing hormone receptor determines specificity for Arg8 of mammalian gonadotrophin-releasing hormone. *Journal of Biological Chemistry*, **269**, 22636–41.

Fong, T. M., Yu, H., Huang, R. C., and Strader, C. D. (1992). The extracellular domain of the neurokinin-1 receptor is required for high affinity binding of peptides. *Biochemistry*, **31**, 11806–11.

Fong, T. M., Huang R. R. C., Yu, H., and Strader, C. D. (1993). Mapping the ligand binding site of the NK-1 receptor. *Regulatory Peptides*, **46**, 43–8.

Frager, M. S., Pieper, D. R., Tonetta, S. A., Duncan, J. A., and Marshall, J. C. (1981). Pituitary gonadotropin-releasing hormone receptors: effects of castration, steroid replacement, and the role of gonadotropin-releasing hormone in modulating receptors in the rat. *Journal of Clinical Investigations*, **67**, 615–23.

Fraser, C. M. (1989). Site-directed mutagenesis of β-adrenergic receptors: identification of conserved cysteine residues that independently affect ligand binding and receptor activation. *Journal of Biological Chemistry*, **264**, 9266–70.

Fraser, D. M., Wong, C-D., Robinson, D. A., Gocayne, J. D., and Venter, J. C. (1990). Site-directed mutagenesis of m1 muscarinic acetylcholine receptors: conserved aspartic acids play important roles in receptor function. *Molecular Pharmacology*, **36**, 840–7.

Frielle, T., Collins, S., Daniel, K. W., Caron, M. G., Lefkowitz, R. J., and Kobilka, B. K. (1987). Cloning of the cDNA for the human beta 1-adrenergic receptor. *Proceedings of the National Academy of Sciences of the USA*, **84**, 7920–4.

Gordon, K. and Hodgen, G. D. (1992). Evolving role of gonadotropin-releasing hormone antagonists. *Trends in Endocrinology and Metabolism*, **3**, 259–63.

Gregg, D. W., Allen, M. C., and Nett, T. M. (1990). Estradiol-induced increase in number of gonadotropin-releasing hormone receptors in cultured ovine pituitary cells. *Biology of Reproduction*, **43**, 1032–6.

Gregg, D. W., Schwall, R. H., and Nett, T. M. (1991). Regulation of gonadotropin secretion and number of gonadotropin-releasing hormone receptors by inhibin, activin-A, and estradiol. *Biology of Reproduction*, **44**, 725–32.

Gupta, H. M., Talwar, G. P., and Salunke, D. M. (1993). A novel computer modeling approach to the structures of small bioactive peptides: the structure of gonadotropin releasing hormone. *Proteins*, **16**, 48–56.

Hazum, E. (1982). GnRH-receptor of rat pituitary is a glycoprotein: differential effect of neuraminidase and lectins on agonists and antagonists binding. *Molecular and Cellular Endocrinology*, **26**, 217–22.

Hazum, E. (1987). Binding properties of solubilized gonadotropin-releasing hormone receptor: role of carboxylic groups. *Biochemistry*, **26**, 7011–4.

Hazum, E., Fridkin, M., Meidan, R., and Koch, Y. (1977). *FEBS Letters*, **76**, 187–90.

Henderson, R., Baldwin, J., Ceska, T. H., Zemlin, F., Beckmann, E., and Downing, K. (1990). Model for the structure of bacteriorhodopsin based on high resolution electron cryomicroscopy. *Journal of Molecular Biology*, **213**, 899–929.

Hershey, A. D. and Krause, J. E. (1990). Molecular characterization of a functional cDNA encoding the rat substance P receptor. *Science*, **247**, 958–62.

Hibert, M. F., Trumpp-Kallmeyer, S., Bruinvels, A., and Hoflack, J. (1991). Three-dimensional models of neurotransmitter G-binding protein-coupled receptors. *Molecular Pharmacology*, **40**, 8–15.

Ho, B. Y., Karchiin, A., Branchek, T., Davidson, N., and Lester, H. A. (1992). The role of conserved aspartate and serine residues in ligand binding and in function of the 5-HT$_{1A}$ receptor: a site-directed mutation study. *FEBS Letters*, **312**, 259–62.

Horstman, D. A., Brandon, S., Wilson, A. L., Guyer, C. A., Cragoe, E. J., and Limbird, L. E. (1992). An aspartate conserved among G-protein receptors confers allosteric regulation of α$_2$-adrenergic receptors by sodium. *Journal of Biological Chemistry*, **265**, 21590–5.

Hsieh, K-P. and Martin, T. F. J. (1992). Thyrotropin-releasing hormone and gonadotropin-releasing hormone receptors activate phospholipase C by coupling to the guanosine triphosphate-binding proteins Gq and G11. *Molecular Endocrinology*, **6**, 1673–81.

Huang, E. S. and Miller, W. L. (1980). Effects of estradiol-17β on basal and luteinizing hormone releasing hormone-induced secretion of luteinizing hormone and follicle stimulating hormone by ovine pituitary culture. *Biology of Reproduction*, **23**, 124–34.

Huang, E. S. and Miller, W. L. (1984). Porcine ovarian inhibin preparations sensitize cultured ovine gonadotrophs to luteinizing hormone-releasing hormone. *Endocrinology*, **115**, 513–9.

Huang, R-R. C., Yu, H., Strader, C. D., and Fong, T. M. (1994). Interaction of substance P with the second and seventh transmembrane domains of the neurokinin-1 receptor. *Biochemistry*, **33**, 3007–13.

Huckle, W. R. and Conn, P. M. (1988). Molecular mechanism of gonadotropin releasing hormone action. II. The effector system. *Endocrine Reviews*, **9**, 387–95.

Huckle, W. R., Hawes, B. E., and Conn, P. M. (1989). Protein kinase C-mediated gonadotropin-releasing hormone receptor sequestraton is associated with uncoupling of phosphoinositide hydrolysis. *Journal of Biological Chemistry*, **264**, 8619–26.

Hulme, E. C., Birdsall, N. J., and Buckley, N. J. (1990). Muscarinic receptor subtypes. *Annual Reviews of Pharmacology and Toxicology*, **30**, 633–73.

Hurbain, K. I., Berault, A., Noel, N., Polkowska, J., Bohin, A., Jutisz, M., et al. (1990). Gonadotropes in a novel rat pituitary tumor cell line, RC-4B/C: establishment and partial characterization of the cell line. *In Vitro Cell Developmental Biology*, **26**, 431–40.

Illing, N., Jacobs, G., Becker, I. I., Flanagan, C. A., Davidson, J. S., Eales, A., et al. (1993). Comparative sequence analysis and functional characterization of the cloned sheep gonadotropin-releasing hormone receptor reveal differences in primary structure and ligand specificity among mammalian receptors. *Biochemical and Biophysical Research Communications*, **196**, 745–51.

Ishihara, T., Nakamura, S., Kaziro, Y., Takahashi, T., Takahashi, K., and Nagata, S. (1991). Molecular cloning and expression of a cDNA encoding the secretin receptor. *Embo Journal*, **10**, 1635–41.

Jakubowski, M. (1993). Gonadotropin-releasing hormone gene expression during lactation in intact and ovariectomized rats. *Endocrine Society Abstracts*, **75**, 501.

Kaiser, U. B., Zhao, D., Cardona, G. R., and Chin, W. W. (1992). Isolation and characterization of cDNAs encoding the rat pituitary gonadotropin-releasing hormone receptor. *Biochemical and Biophysical Research Communications*, **189**, 1645–52.

Kaiser, U. B., Jakubowiak, A., Steinberger, A., and Chin, W. W. (1993). Regulation of rat pituitary gonadotropin-releasing hormone receptor mRNA levels *in vivo* and *in vitro*. *Endocrinology*, **133**, 931–4.

Karkar, S. S., Musgrove, L. C., Devor, D. C., Sellers, J. C., and Neill, J. D. (1992). Cloning, sequencing and expression of human gonadotropin releasing hormone (GnRH) receptor. *Biochemical and Biophysical Research Communications*, **189**, 289–95.

Karkar, S. S., Rahe, C. H., and Neill, J. D. (1993). Molecular cloning, sequencing and characterizing the bovine receptor for gonadotropin releasing hormone (GnRH). *Domestic Animal Endocrinology*, **10**, 335–42.

Karnik, S. S., Sakmann, J. P., Chen, H. B., and Khorana, H. G. (1988). Cysteine residues 110 and 187 are essential for the formation of correct structure in bovine rhodopsin. *Proceedings of the National Academy of Sciences of the USA*, **85**, 8459–63.

Karten, M. J. and Rivier, J. E. (1986). Gonadotropin-releasing hormone analog design. Structure–function studies toward the development of agonist and antagonists: rationale and perspective. *Endocrine Reviews*, **7**, 44–66.

Katt, J. A., Duncan, J. A., Herbon, L., Barkan, A., and Marshall, J. C. (1985). The frequency of gonadotropin-releasing hormone stimulation determines the number of pituitary gonadtropin-releasing hormone receptors. *Endocrinology*, **116**, 2113–5.

King, J. A. and Millar, R. P. (1982). Structure of chicken hypothalamic LHRH I. *Journal of Biological Chemistry*, **257**, 10722–8.

Laws, S. C., Beggs, M. J., Webster, J. C., and Miller, W. L. (1990*a*). Inhibin increases and progesterone decreases receptors for gonadotropin-releasing hormone in ovine pituitary culture. *Endocrinology*, **127**, 373–80.

Laws, S. C., Webster, J. C., and Miller, W. L. (1990*b*). Estradiol alters the effectiveness of gonadotrophin releasing hormone (GnRH) in ovine pituitary cultures: GnRH receptors versus responsiveness to GnRH. *Endocrinology*, **127**, 381–6.

Leblanc, P., L'Héritier, A., and Kordon, C. (1992). A Ca^{2+} calmodulin dependent kinase rather than protein kinase C is involved in up-regulation of the LHRH receptor.*Biochemical and Biophysical Research Communications*, **183**, 666–71.

Leeb-Lundberg, L. M. F., Cotecchia, S., DeBlasi, A., Caron, M. G., and Lefkowitz, R. J. (1987). Regulation of adrenergic receptor function by phosphorylation. I. Agonist-promoted desensitization and phosphorylation of α-adrenergic receptors coupled to inositol phospholipid metabolism in DDT1 MR-2 smooth muscle cells. *Journal of Biological Chemistry*, **262**, 3098–105.

Leong, D. A. (1991). A model for intracellular calcium signaling and the coordinate regulation of hormone biosynthesis, receptors and secretion. *Cell Calcium*, **12**, 255–68.

Leong, D. A., Thorner, M. O., Lau, S. K., Lyons, C. J., and Berry, D. J. (1991). A potential code of luteinizing hormone-releasing hormone-induced calcium ion responses in the regulation of luteinizing hormone secretion among individual gonadotropes. *Journal of Biological Chemistry*, **266**, 9016–22.

Libert, F., Parmentier, M., Lefort, A., Dinsart, C., VanSande, J., Maenhaut, C., *et al.* (1989). Selective amplification and cloning of four new members of the G protein-coupled receptor family. *Science*, **244**, 569–72.

Limor, R., Schvartz, I., Hazum, E., Ayalon, D., and Naor, Z. (1989). Effect of guanine nucleotides on stimulus secretion coupling mechanism in permeabilized pituitary cells: relationship to gonadotropin releasing hormone action. *Biochemical and Biophysical Research Communications*, **159**, 209–15.

Lin, H. Y., Harris, T. L., Flannery, M. S., Aruffo, A., Kaji, E. H., Gorn, A., *et al.* (1991). Expression cloning of an adenylate cyclase-coupled calcitonin receptor. *Science*, **254**, 1022–4.

Loosfelt, H., Misrahi, M., Atger, M., Salesse, R., Thi, M., Jolivet, A., *et al.* (1989). Cloning and sequencing of porcine LH-hCG receptor cDNA: variants lacking transmembrane domain. *Science*, **245**, 525–8.

Loumaye, E. and Catt, K. J. (1982). Homologous regulation of gonadotrophin-releasing hormone receptors in cultured pituitary cells. *Science*, **215**, 983–5.

Loumaye, E. and Catt, K. J. (1983). Agonist-induced regulation of pituitary receptors for gonadotrophin-releasing hormone: dissociation of receptor recruitment from hormone release in cultured gonadotrophs. *Journal of Biological Chemistry*, **258**, 12002–9.

Maloney Huss, K. and Lybrand, T. P. (1992). A three-dimensional structure of the β^2-adrenergic receptor protein based on computer modelling studies. *Journal of Molecular Biology*, **225**, 859–71.

McArdle, C. A., Gorospe, W. C., Huckle, W. R., and Conn, P. M. (1987). Homologous down-regulation of gonadotropin-releasing hormone receptors and desensitization of gonadotropes: lack of dependence on protein kinase C. *Molecular Endocrinology*, **1**, 420–9.

McIntosh, R. P. and McIntosh, J. E. A. (1985). Dynamic characteristics of luteinizing hormone release from perifused sheep anterior pituitary cells stimulated by combined pulsatile and continuous gonadotropin-releasing hormone. *Endocrinology*, **117**, 169–79.

Millar, R. P., Flanagan, C. A., Milton, R. C. L., and King, J. A. (1989). Chimeric analogues of vertebrate gonadotropin-releasing hormones comprising substitutions of the variant amino acids in positions 5, 7, and 8. *Journal of Biological Chemistry*, **264**, 21007–13.

Milton, R. C. D. L., King, J. A., Badminton, M. N., Tobler, C. J., Lindsey, G. G., Fridkin, M., *et al.* (1983). Comparative structure–activity studies on mammalian Arg8 LH-RH and Gln8 LH-RH by fluorimetric titration. *Biochemical and Biophysical Research Communications*, **111**, 1082–8.

Moenter, S. M., Brand, R. M., Midgley, A. R., and Karsch, F. J. (1992). Dynamics of gonadotropin-releasing hormone release during a pulse. *Endocrinology*, **130**, 503–10.

Momany, F. A. (1976). Conformational energy analysis of the molecule, luteinizing hormone-releasing hormone. 1. Native decapeptide. *Journal of the American Chemical Society*, **98**, 2990–6.

Murphy, P. M. and Tiffany, H. L. (1991). Cloning of complementary DNA encoding a functional human interleukin-8 receptor. *Science*, **253**, 1280–3.

Nagayama, Y., Russo, D., Chazenbalk, G. D., and Rapoport, B. (1991). Seven amino acids (lys-201-lys-211) and 9 amino acids (gly-222 to leu-230) in the

human thyrotropin receptor are involved in ligand binding. *Journal of Biological Chemistry*, **266**, 14926–30.
Naik, S. I., Young, L. S., Charlton, H. M., and Clayton, R. N. (1985). Evidence for a pituitary site of gonadal steroid stimulation of GnRH receptors in female mice. *Journal of Reproduction and Fertility*, **74**, 615–24.
Naor, Z. (1990). Signal transduction mechanisms of Ca^{2+} mobilizing hormones: the case of gonadotropin-releasing hormone. *Endocrine Reviews*, **11**, 326–53.
Nathans, J. and Hogness, D. S. (1984). Isolation and nucleotide sequence of the gene encoding human rhodopsin. *Proceedings of the National Academy of Sciences of the USA*, **81**, 4851–5.
Neve, K. A., Cox, B. A., Henningsen, R. A., Spanoyannis, A., and Neve, R. L. (1991). Pivotal role for aspartate-80 in the regulation of dopamine D_2 receptor affinity for drugs and inhibition of adenylyl cyclase. *Molecular Pharmacology*, **39**, 733–9.
Ng, G. Y., George, S. R., Zastawny, R. L., Caron, M., Bouvier, M., Dennis, M., et al. (1993). Human serotonin1B receptor expression in Sf9 cells: phosphorylation, palmitoylation, and adenylyl cyclase inhibition. *Biochemistry*, **32**, 11727–33.
Nussenzveig, D. R., Heinflink, M., and Gershengorn, M. C. (1993). Agonist-stimulated internalization of the thyrotropin-releasing hormone receptor is dependent on two domains in the receptor carboxyl terminus. *Journal of Biological Chemistry*, **268**, 2389–92.
O'Dowd, B. F., Hnatowich, M., Caron, M. G., Lefkowitz, R. J., and Bouvier, M. (1989). Palmitoylation of the human β_2-adrenergic receptor. *Journal of Biological Chemistry*, **264**, 7564–9.
Ovchinnikov, Y. A., Abdulaev, N. G., and Bogachuk, A. S. (1988). Two adjacent cysteine residues in the C-terminal cytoplasmic fragment of bovine rhodopsin are palmitylated. *FEBS Letters*, **230**, 1–5.
Pardo, L., Ballesteros, J. A., Osman, R., and Weinstein, H. (1992). On the use of the transmembrane domain of bacteriorhodopsin as a template for modeling the three-dimensional structure of guanine nucleotide-binding regulatory protein-coupled receptors. *Proceedings of the National Academy of Sciences of the USA*, **89**, 4009–12.
Perrin, M. H., Haas, Y., Porter, J., Rivier, J., and Vale, W. (1989). The gonadotropin-releasing hormone pituitary receptor interacts with a guanosine triphosphate-binding protein: differential effects of guanyl nucleotides on agonist and antagonist binding. *Endocrinology*, **124**, 798–804.
Perrin, M. H., Bilezikjian, L. M., Hoeger, C., Donaldson, C. J., Rivier, J., Haas, Y., et al. (1993). Molecular and functional characterization of GnRH receptors cloned from rat pituitary and a mouse pituitary tumor cell line. *Biochemical and Biophysical Research Communications*, **191**, 1139–44.
Probst, W. C., Snyder, L. A., Schuster, D. I., Brosius, J., and Sealfon, S. C. (1992). Sequence alignment of the G-protein coupled receptor superfamily. *DNA and Cell Biology*, **11**, 1–20.
Rands, E., Candelore, M. R., Cheung, A. H., Hill, W. S., Strader, C. D., and Dixon, R. A. F. (1990). Mutational analysis of β-adrenergic receptor glycosylation. *Journal of Biological Chemistry*, **265**, 10759–64.
Rao, V. R., Cohen, G. B., and Oprian, D. D. (1994). Rhodopsin mutation

G90D and a molecular mechanism for congenital night blindness. *Nature*, **367**, 639–42.
Reinhart, J., Mertz, L. M., and Catt, K. J. (1992). Molecular cloning and expression of cDNA encoding the murine gonadotropin-releasing hormone receptor. *Journal of Biological Chemistry*, **267**, 21281–4.
Savarese, T. M., Wang, C. D., and Fraser, C. M. (1992). Site-directed mutagenesis of the rat m1 muscarinic acetylcholine receptor: Role of conserved cysteines in receptor function. *Journal of Biological Chemistry*, **267**, 11439–48.
Savoy-Moore, R. T., Schwartz, N. B., Duncan, J. A., and Marshall, J. C. (1980). Pituitary gonadotropin-releasing hormone receptors during the rat estrous cycle. *Science*, **209**, 942–4.
Schertler, G. F., Villa, C., and Henderson, R. (1993). Projection structure of rhodopsin. *Nature*, **362**, 770–2.
Schiffmann, S. N., Libert, F., Vassart, G., Dumont, J. E., and Vanderhaeghen, J. J. (1990). A cloned G protein-coupled protein with a distribution restricted to striatal medium-sized neurons: possible relationship with D_1 dopamine receptor. *Brain Research*, **519**, 333–7.
Schvartz, I. and Hazum, E. (1985). Tunicamycin and neuraminidase effects of luteinizing hormone (LH)-releasing hormone binding and LH release from rat pituitary cells in culture. *Endocrinology*, **116**, 2341–6.
Sealfon, S. C., Gillo, B., Mundamattom, S., Mellon, P. L., Windle, J. J., Landau, E., et al. (1990). Gonadotropin-releasing hormone receptor expression in *Xenopus* oocytes. *Molecular Endocrinology*, **4**, 119–24.
Sealfon, S. C., Laws, S. C., Wu, J. C., Gillo, B., and Miller, W. L. (1990*b*). Hormonal regulation of gonadotropin-releasing hormone receptors and messenger RNA activity in ovine pituitary culture. *Molecular Endocrinology*, **4**, 1980–7.
Sealfon, S. C., Tsutsumi, M., Zhou, W., Chi, L., Flanagan, C., Davidson, J. S., et al. (1992). Insight on ligand binding, signal transduction and desensitization from the cloning of a functional mouse GnRH receptor. *In GnRH, GnRH analogs, gonadotropins and gonadal peptides: the proceedings of the third Organon Round Table Conference* (ed. P. Bouchard, A. Caraty, J. J. T. Coelingh Bennink, and S. Pavlou), pp. 81–94. Paris.
Shinitzky, M. and Fridkin, M. (1976). Structural features of Luliberin (luteinising hormone releasing factor inferred from fluorescent measurements. *Biochemical Biophysica Acta*, **434**, 137–43.
Shinitzky, M., Hazum, E., and Fridkin, M. (1976). Structure–activity relationships of Luliberin substituted at position 8. *iochimica Biophysica Acta*, **453**, 553–7.
Sibley, D. R., Benovic, J. L., Caron, M. G., and Lefkowitz, R. J. (1987). Regulation of transmembrane signaling by receptor phosphorylation. *Cell*, **48**, 913–22.
Smith, M. S. and Reinhart, J. (1993). Changes in pituitary gonadotropin-releasing hormone receptor messenger ribonucleic acid content during lactation and after pup removal. *Endocrinology*, **133**, 2080–4.
Smith, M. S. and Ru-Lee, L. (1989). Modulation of pituitary gonadotropin-releasing hormone receptor during lactation in the rat. *Endocrinology*, **124**, 1456–61.

Stojilkovic, S. S. and Catt, K. J. (1992). Calcium oscillations in anterior pituitary cells. *Endocrine Reviews*, **13**, 256–80.

Stojilkovic, S. S., Merelli, F., Iida, T., Krsmanovic, L. Z., and Catt, K. J. (1990). Endothelin stimulation of cytosolic calcium and gonadotropin secretion in anterior pituitary cells. *Science*, **248**, 1663–6.

Stojilkovic, S. S., Torsello, A., Iida, T., Rojas, E., and Catt, K. J. (1992). Calcium signaling and secretory responses in agonist-stimulated pituitary gonadotrophs. *Journal of Steroid Biochemistry and Molecular Biology*, **41**, 453–67.

Strader, C. D., Sigal, I. S., Register, R. B., Candelore, M. R., Rands, E., and Dixon, R. A. (1987). Identification of residues required for ligand binding to the beta-adrenergic receptor. *Proceedings of the National Academy of Sciences of the USA*, **84**, 4384–8.

Strader, C. D., Sigal, I. S., Candelore, M. R., Rands, E., Hill, W. S., and Dixon, R. A. F. (1988). Conserved aspartate residues 79 and 113 of the β-adrenergic receptor have different roles in receptor function. *Journal of Biological Chemistry*, **263**, 10267–71.

Strader, C. D., Candelore, M. R., Hill, W. S., Sigal, I. S., and Dixon, R. A. (1989a). Identification of two serine residues involved in agonist activation of the beta-adrenergic receptor. *Journal of Biological Chemistry*, **264**, 13572–8.

Strader, C. D., Sigal, I. S., and Dixon, R. A. F. (1989b). Structural basis of β-adrenergic receptor function. *FASEB Journal*, **3**, 1825–32.

Struthers, R. S., Tanaka, G., Koerber, S. C., Solmajer, T., Baniak, E. L., Gierach, L. M., et al. (1990). Design of biologically active, conformationally constrained GnRH antagonists. *Proteins: Structure, Function and Genetics*, **8**, 295–304.

Surprenant, A., Horstman, D. A., Akbarali, H., and Limbird, L. E. (1992). A point mutation of the alpha 2-adrenoceptor that blocks coupling to potassium but not calcium currents. *Science*, **257**, 977–80.

Tanabe, Y., Masu, M., Ishii, T., Shigemoto, R., and Nakanishi, S. (1992). A family of metabotropic glutamate receptors. *Neuron*, **8**, 169–79.

Trumpp, K. S., Hoflack, J., Bruinvels, A., and Hibert, M. (1992). Modeling of G-protein-coupled receptors: application to dopamine, adrenaline, serotonin, acetylcholine, and mammalian opsin receptors. *Journal of Medicinal Chemistry*, **35**, 3448–62.

Tsai-Morris, C. H., Buczko, E., Wand, W., and Dufau, M. L. (1990). Intronic nature of the rat luteinizing hormone receptor gene defines a soluble receptor subspecies with hormone binding activity. *Journal of Biological Chemistry*, **265**, 19385–8.

Tsutsumi, M., Mellon, P. L., Roberts, J. L., and Sealfon, S. C. (1990). GnRH decreases GnRH receptor mRNA activity in a rodent pituitary cell line. *Society for Neuroscience Abstracts*, **16**, 393.

Tsutsumi, M., Zhou, W., Millar, R. P., Mellon, P. L., Roberts, J. L., Flanagan, C. A., et al. (1992). Cloning and functional expression of a mouse gonadotropin-releasing hormone receptor. *Molecular Endocrinology*, **6**, 1163–9.

Tsutsumi, M., Laws, S. C., and Sealfon, S. C. (1993). Homologous up-regulation of the gonadotropin-releasing hormone receptor in alpha T3-1 cells is associated with unchanged receptor messenger RNA (mRNA) levels and altered mRNA activity. *Molecular Endocrinology*, **7**, 1625–33.

Tsutsumi, M., Laws, S. C., Redic, V., and Sealfon, S. C. (1995). Translational regulation of the gonadotropin-releasing hormone receptor in αT3−1 cells. *Endocrinology*, **136**, 1128–36.

Vale, W., Grant, G., Rivier, J., Monahan, M., Amoss, M., Blackwell, R., *et al.* (1972). Synthetic polypeptide antagonists of the hypothalamic luteinizing hormone releasing factor. *Science*, **176**, 933–4.

Wang, O. F., Farnsworth, P. G., Findlay, J. K., and Burger, H. G. (1988). Effect of purified 31K bovine inhibin on specific binding of gonadotropin-releasing hormone to rat anterior pituitary cells in culture. *Endocrinology*, **123**, 2161–6.

Wang, C. D., Buck, M. A., and Fraser, C. M. (1991). Site-directed mutagenesis of alpha$_{2A}$-adrenergic receptors: identification of amino acids involved in ligand binding and receptor activation by agonists. *Molecular Pharmacology*, **40**, 168–79.

Weinstein, H. (1992). Computational simulations of molecular structure, dynamics and signal transduction in biological systems: mechanistic implications for ecological physical chemistry. In *Ecological Physical Chemistry*, (ed. D. Pitea), pp. 1–16. Elsevier, Amsterdam.

Weiss, J., Crowley, W. F., and Jameson, J. L. (1992). Pulsatile gonadotropin-releasing hormone modifies polyadenylation of gonadotropin subunit messenger ribonucleic acids. *Endocrinology*, **130**, 415–20.

Windle, J. J., Weiner, R. I., and Mellon, P. L. (1990). Cell lines of the pituitary gonadotrope lineage derived by targeted oncogenesis in transgenic mice. *Molecular Endocrinology*, **4**, 597–603.

Wu, J. C., Sealfon, S. C., and Miller, W. L. (1994). Gonadal hormones and gonadotropin releasing hormone alter mRNA levels of GnRH receptors in sheep. *Endocrinology*, **134**, 1846–50.

Wynn, P. C., Suarez-Quian, C. A., Childs, G. V., and Catt, K. J. (1986). Pituitary binding and internalization of radioiodinated gonadotropin-releasing hormone agoinst and antagonist ligands *in vitro* and *in vivo*. *Endocrinology*, **119**, 1852–63.

Zhang, D. and Weinstein, H. (1993). Signal transduction by a 5-HT$_2$ receptor: a mechanistic hypothesis from molecular dynamics simulations of the 3-D model of the receptor complexed to ligands. *Journal of Medicinal Chemistry*, **36**, 934–8.

Zhou, W. and Sealfon, S. C. (1994). Structure of the mouse gonadotropin-releasing hormone gene: variant transcripts generated by alternative processing. *DNA and Cell Biology*, **13**, 605–14.

Zhou, W., Flanagan, C., Ballesteros, J. A., Konvicka, K., Davidson, J. S., Weinstein, H., *et al.* (1994). A reciprocal mutation supports helix 2 and helix 7 proximity in the gonadotropin-releasing hormone receptor. *Molecular Pharmacology*, **45**, 165–70.

Index

abortions
  induced 143
  spontaneous 126, 128
acetylcholine receptor (AChR)
  peptide 11–12, 14, 19
acrosome reaction 95–6, 97–8, 102, 103–4, 107
activin 178–82
  FSHβ mRNA regulation 179–81
  GnRH receptor regulation 181–2, 271–2
  immunoregulation 4
  LH regulation 181–2
  paracrine effects in pituitary 184–6
  receptors (ActR) 183–4
  regulation of synthesis 160, 161
  structure 179
adenosine 63, 66
adrenaline 234
β_1-adrenergic receptor 262
adrenergic receptors, sperm 104
age, maternal 128–9, 136
α-subunit (glycoprotein hormones) 161, 187
  gene, regulation by cAMP 168–9
  mRNA
    post-transcriptional modification 167
    regulation by GnRH 165, 166, 167, 168
    regulation by steroid hormones 173, 174, 175, 176–8
  promoter 188, 189
αT3 cells 169, 188, 256–7, 258, 269–70, 273
γ-aminobutyric acid (GABA) neurones 235
amplification, autoantibody 19–20
ampulla of Fallopian tube
  follicular fluid entry 101
  sperm:egg ratios 105, 107, 108–9
  sperm transport to 88, 91, 105–10
angiogenesis 61, 62–3
  clinical relevance 75–7
angiogenic growth factors 63–75, 77
angiogenin 66
angiotropin 66
anoestrus 207
anticipation 142
AOD-1 gene 23
arcuate nucleus 222, 237
artificial insemination 86, 113
athymic nude (nu/nu) mice 6, 7, 8
attractivity 207, 209
autoantibodies
  amplification 19–20
  autoimmune oophoritis 17–20, 23–4

autoimmune orchitis 17, 18, 23–4
autoimmune epididymovasitis 21
autoimmune gastritis 6–9, 20
autoimmune oophoritis 1–24
  animal models 2, 3
  events that can cause 9–12
  experimental, *see* experimental autoimmune oophoritis
  immunogenetic approach 20–3
  regulatory mechanisms preventing 3–9
autoimmune orchitis 1–24
  animal models 2, 3
  experimental, *see* experimental autoimmune (allergic) orchitis
  immunogenetic approach 20–3
  regulatory mechanisms preventing 3–9
autonomic nervous system 104

bacteriorhodopsin 263
BALB/c mice, EAO susceptibility 6, 21, 22
basal metabolic rate (BMR)
  in lactation 40
  measurement 43
  in pregnancy 40, 41, 43–8, 52–4
bek gene product 67
β-subunits, gonadotrophins, *see* follicle-stimulating hormone β-subunit; luteinizing hormone β-subunit
biopsy, embryo 143
birthweights 56
  different populations 44, 54, 55
  humans vs. animals 35
blood–testis barrier 3
BMR, *see* basal metabolic rate
Bphs gene 21, 22
breeding season 207
1-butyryl glycerol 63, 66

cadherins 107
calcium, GnRH receptor regulation 269
calorimetry, whole-body 52–3
capacitation, sperm 102–4
castrated male rats
  GnRH receptor regulation 169, 172, 271
  gonadotrophin regulation 165, 166, 173, 176, 177
catecholamines, sperm responses 104
CD4+CD8− thymocytes 6

CD4+ T cells
  experimental autoimmune oophoritis/
    orchitis and 5, 12–17
  T_H1 phenotype 8, 15, 23
  T_H2 phenotype 8–9
  transfer of autoimmune oophoritis/
    gastritis 8, 9–10
CD5 molecule 8, 9–10
cell-cycle checkpoints, human embryos
  140–1
centromeric alphoid repeat DNA
  probes 147–9
cholecystokinin 238
choriocarcinoma cells 168, 187
chorionic gonadotrophin 161
chromosomal abnormalities 125, 126–41
  cell-cycle checkpoints and 140–1
  clinically recognized pregnancy 128–9
  developmental stage of origin 126–8
  human embryos 130–2
  human gametes 129–30
  molecular cytogenetics 132–40
  preimplantation diagnosis 147–51, 152–3
chromosome painting 151, 152
comparative genomic hybridization 152–3
complement activation, testis 4–5
contraceptives, long-acting 76
counter-current transfer, ovarian
  hormones 98, 99, 100
cows, sperm transport and fertilization
  85–115
cyclic AMP (cAMP), α-subunit gene
  regulation 168–9
cystic fibrosis (CF) 146
cytogenetics
  interphase 132
  molecular 132–40

decidua
  fibroblast growth factors 67
  vascular endothelial growth factor 71–3,
    74, 75
developing countries, energy costs of
  pregnancy 43–54
diabetic NOD mice 20
diet-induced thermogenesis (DIT) 48–9
diploidy, human sperm 133
disomy, human sperm 133
dopaminergic neurones 233–5
doubly labelled water technique 53
Down syndrome, *see* trisomy 21
Duchenne muscular dystrophy 145, 152
dystrophin gene 151

embryos, preimplantation
  biopsy 143

cell-cycle checkpoints 140–1
diagnosis of genetic defects, *see*
  preimplantation diagnosis
FISH analysis 133–40
karyotype analysis 130–2
mosaicism 134, 136, 140–1
sexing 143–5
see also implantation
encephalomyelitis, experimental
  autoimmune 20, 22
endogenous superantigen 6, 8
endometrial bleeding, long-term hormone
  therapies 76
endometrial carcinoma cells 67, 71
endometriosis 76
endometrium
  angiogenesis 62–3, 64
  fibroblast growth factors 67–8, 69, 75
  vascular endothelial growth factor 71, 72,
    73
β-endorphin 237
endosalpinx, sperm adhesion 106–10
endothelin receptor 257
energy balance, human pregnancy 33–56
energy costs of pregnancy
  accretion requirements 39–40, 49
  adaptive strategies 43–54
    beneficial effects 54
    detrimental effects 54–6
  classical assumptions 39–43
  diet-induced thermogenesis 48–9
  maintenance requirements 40–1, 43–8
  maternal fat storage 41, 49
  new prospective studies 43–54
  per unit time 36
  physical activity 41, 52–4
  total 36–7, 45, 46, 49–50
energy expenditure, total (TEE) 52–4
energy intake
  in lactation 37
  in pregnancy 36, 37, 42–3
energy-sparing traits, natural selection 37–9
epidermal growth factor (EGF) 66
epididymovasitis, autoimmune 21
estradiol; estrogen, *see* oestradiol;
  oestrogen
ewes, *see* sheep
experimental autoimmune (allergic) orchitis
  (EAO) 2, 3
  genetics of susceptibility 21–2
  histopathology 16
  pathogenesis 12–20
  suppressor T cells 5–6
experimental autoimmune encephalomyelitis
  20, 22
experimental autoimmune oophoritis 2,
  3
  pathogenesis 12–20

Fallopian tubes
  ampulla, *see* ampulla of Fallopian tube
  intra-mural portion 91
  isthmus, *see* isthmus of Fallopian tube
  ovarian hormonal influences 98–102
  sperm adhesion to epithelium 106–10
  sperm transport 85–115
  temperature gradients 94
fat stores
  deposition in pregnancy 41, 44, 49
  lactation and 56
  pre-pregnancy 44, 48
    humans vs. animals 37, 38
    total metabolic costs of pregnancy and 50, 51
fecundity rate 126
female genital tract, sperm transport 85–115
fertility, *see* infertility
fertilization
  site, *see* ampulla of Fallopian tube
  timing 89
fetal growth rates 36
FGF, *see* fibroblast growth factor
fibroblast growth factor (FGF) 4, 63–9
  acidic (aFGF) 65, 66, 67, 68, 76
  basic (bFGF) 65, 66, 67, 76
  function in human reproduction 67–9, 75, 76
  in human reproductive tract 67
  receptors 65–7
  structure and function 63–5
fibroblast growth factor-7 (FGF-7) 65
fibroblast growth factor-8 (FGF-8) 65
FISH, *see* fluorescent *in situ* hybridization
flg gene product 67
flt gene product 70–1, 75
fluorescent *in situ* hybridization (FISH) 132–40
  centromeric alphoid repeat DNA probes 147–9
  dual, embryo sexing 144–5
  preimplantation embryos 133–40
  sources of error 147, 149
  sperm chromosomes 132–3
  yeast artificial chromosome probes 150, 151
follicle-stimulating hormone (FSH) 160
  actions 160, 161
  α-subunit, *see* α-subunit
  cellular localization 162
  receptors 161, 266
  regulation of secretion 162–3
    autocrine/paracrine 184–6, 187
    by GnRH 164–9
    by gonadal peptides 178–86
    by steroid hormones 171–8
  secretion in ewes 209–11, 213–14, 215–16
  structure 161–2

follicle-stimulating hormone β-subunit (FSHβ) 161
  gene 161
  mRNA
    regulation by activin 180
    regulation by follistatin 180, 182–3
    regulation by GnRH 165, 166, 167, 168
    regulation by inhibin 179–81
    regulation by steroid hormones 173, 174, 175, 176–8
follicular fluid, released 101
follicular phase 208–9
  early 208, 210, 215–17, 239–40
    GnRH secretion 223
    sex steroid/gonadotrophin secretion 215–16
    sexual behaviour 216–17
  late 208–9, 210, 217–18, 240
    GnRH secretion 223
    sex steroid/gonadotrophin secretion 217–18
    sexual behaviour 218
follistatin 178, 182–3
  FSHβ mRNA regulation 180, 182–3
  paracrine effects 184–6
  regulation of synthesis 160, 161
  structure 182
food availability, seasonal variations 34–5
fos protein 226
fragile X syndrome 142
FSH, *see* follicle-stimulating hormone

gametogenesis
  chromosomal abnormalities 126
  de novo mutations 142
  γ-aminobutyric acid (GABA) neurones 235–6
gastritis, autoimmune 6–9, 20
genetic defects 125–53
  origins 126–42
  preimplantation diagnosis 143–53
  see also chromosomal abnormalities; single-gene defects
genomic hybridization, comparative 152–3
germ cells, male
  autoantibodies 17, 18
  protection from autoimmune responses 3–4
  see also sperm
gestation period 35–6
glycosylation sites, GnRH receptor 260
Gnb-3 gene 22
GnRH, *see* gonadotrophin-releasing hormone
gonadal peptides, *see* activin; follistatin; inhibin
gonadotrophin(s)
  cellular localization 162

genes 161
  regulation of expression 159–89, 214
    by GnRH 164–71
    by gonadal peptides 178–86
    new horizons 187–9
    by steroid hormones 171–8
  secretion in ewes 209–11, 213–14, 215–16, 217–18
  structure 161–2
  see also follicle-stimulating hormone; luteinizing hormone
gonadotrophin-releasing hormone (GnRH) 160, 208, 255, 268–9
  analogues 256, 266–8
  GnRH receptor interaction 265–8
  GnRH receptor regulation 268–71
  LH surge regulation 220–1
  neurones 221–6
    control by GnRH 236–7
    controlling ovulation 223–6
    distribution 222–3
    neurotransmitters and 233, 235–8
    oestrogen effects 174–5, 228–9
  pulsatile secretion 164–5, 210–11, 269
  regulation of gonadotrophins 162, 164–71
    genetic elements 168–9
    in vitro studies 166–8
    in vivo studies 165–6
  regulation of its own secretion 236–7
  regulation by oestrogen 173–4, 175
  regulation by progesterone 175
  secretion in ewes 209–11, 213, 214, 215–16
  testosterone interactions 176–8
gonadotrophin-releasing hormone (GnRH) receptor 168–71, 255–73
  binding pocket 265–8
  conserved residues 260, 261
  desensitization 171, 272–3
  gene 168, 261–3
  inhibin and activin actions 181–2, 271–2
  LH surge and 220
  molecular cloning 256–8
  regulation 169–71, 172, 173–4, 268–72
  signal transduction 171, 258
  site-directed mutagenesis 264–5, 266, 267–8
  structure 170, 258, 259, 260–1
  tertiary structure 263–5
gonadotrophs 162
  autocrine/paracrine influences 185–6, 187
  cell lines 187
  levels of regulation of gonadotrophin synthesis 163, 164
G-protein-coupled receptors 257, 260, 263–4, 268
G proteins, GnRH receptor-coupled 171, 258

Graafian follicles 98, 209, 217
granulocyte colony-stimulating factor 66
granulocyte–macrophage colony-stimulating factor 66
granulosa cells 102, 115
growth rates, fetal 36

H^+K^+-ATPase autoantibodies 6
H-2-linked genes 21, 22
heparin 65, 70
heparin sulphate proteoglycan 65
Hofbauer cells 71
hormone-induced protein (HIP-70) 237
hormone replacement therapy 76
Huntington's disease 142
hyaluronic acid 66
(12R)-hydroxyeicosatrienoic acid 66
hyperactivation, sperm 97, 103–4, 105, 107, 108
hypothalamo-pituitary–gonadal axis 160–1
hypothalamus 219–20, 238
  control of ovulation 226
  GnRH actions 236–7
  GnRH neurones 222–3, 224–5
  oestrogen actions 228–9
  oestrogen receptors 224, 227–8
  progesterone actions 230
  progesterone binding sites 229

IgG, experimental autoimmune orchitis 17, 18
immunoglobulin-like domains 67, 70
implantation
  angiogenesis and 63, 75–6
  vascular endothelial growth factor and 73–5
infertility
  autoimmune oophoritis/orchitis 2–3
  under-nourished women 34
inhibin 178–82
  activin receptor binding 183, 184
  FSH regulation 179–81, 214
  GnRH receptor regulation 181–2, 271–2
  immunoregulation 4
  knock-out mice 188–9
  LH regulation 181–2
  paracrine effects 184–6
  regulation of synthesis 160, 161
  structure 179
insemination, artificial 86, 113
interleukin-1-like factor 4
interleukin-8 receptor (ILR) 262
interphase cytogenetics 132
in vitro fertilization (IVF)
  FISH analysis of embryos 134–40

karyotype analysis of embryos 130–2
isthmus of Fallopian tube
  chemical influences 94
  lumen patency 94–5, 110
  ovarian control of sperm progression 98–102
  peri-ovulatory sperm activation/release 95–8
  post-ovulatory changes 110–11
  pre-ovulatory sperm distribution 87–91
  pre-ovulatory sperm storage 91–5
  regulation of sperm ascending 105–10
  viscous secretions 93

keratinocyte growth factor (FGF-7) 65
Klinefelter's syndrome (XXY karyotype) 127, 129, 133

lactation
  energy intake 37
  fat stores and 56
  GnRH receptor regulation 271
  metabolic rate 40
Lesch–Nyhan syndrome 146
leukocytes, sperm disposal 115
Leydig cells 4
LH, see luteinizing hormone
luteal phase 208, 210, 213–15, 239
  GnRH secretion 223
  gonadotrophin secretion 213–14
  sex steroid secretion 213
  sexual behaviour 214–15
luteinizing hormone (LH) 160
  actions 160–1
  α-subunit, see α-subunit
  cellular localization 162
  pulsatile secretion 210, 211
  receptors 161, 216, 268
  regulation of secretion 162–3
    autocrine/paracrine 184–6, 187
    by GnRH 164–9
    by gonadal peptides 178–86
    by steroid hormones 171–8
  secretion in ewes 209–11, 213–14, 215–16
  structure 161–2
  surge
    ewes 210, 211, 217–18
    neural control 220–1, 225, 226, 228
    neurotransmitters controlling 233–6
    regulation by oestrogen 171–5, 217–18, 219, 220–1
    sexual behaviour and 218–19
luteinizing hormone β-subunit (LHβ) 161
  gene 161
    oestrogen regulatory element 174

mRNA
  post-transcriptional modification 167
  regulation by GnRH 165, 166–7, 168
  regulation by steroid hormones 173, 174, 175, 176–8

macrophages
  endometriosis and 76
  vascular endothelial growth factor 71, 73, 74, 75
mammillary bodies 222
maternal age 128–9, 136
mating
  competitive experiments 114
  multiple 88, 91
  timing, sperm transport and 88, 90
median eminence 222–3, 226
meiosis, errors 126, 130, 133
menstruation 62–3, 64
  angiogenesis and 75
  fibroblast growth factors 67–8
  vascular endothelial growth factor 71, 72, 73
metabolic rate, basal, see basal metabolic rate
MHC-linked genes 21, 22
molecular cytogenetics 132–40
molecular mimicry, autoimmune oophoritis 10–12, 23
monoaminergic neurones 233–5
monosomy X 128, 129, 145
  origins 127
mosaicism
  misdiagnosis of genetic defects and 147, 148, 149
  preimplantation embryos 134, 136, 140–1
muscular dystrophy, Duchenne 145, 151–2
myosalpinx, contractile activity 94–5, 96, 105
myotonic dystrophy 142

naloxone 238
neuropeptide Y 238
neurotensin 238
neurotransmitters 233–8
nicotinamide 63, 66
noradrenergic neurones 233–5
nutritional stresses
  pregnancy in under-nourished women 54–6
  women 33–5
nutrition status
  indicators 43, 44
  metabolic costs of pregnancy and 50, 51

oestradiol (E_2)
  counter-current transfer 99
  GnRH receptor regulation 272
  hypothalamo-pituitary actions 160, 161
  implants, ovariectomized ewes 212
  LH surge regulation 217–18, 219, 220–1, 226
  neural sites of action 224–5, 228–9, 238
  oestrus cycle of ewe 210, 211, 213, 216, 217–18
  regulation of gonadotrophins 214
  sexual behaviour and 224–5, 228–9, 230
  sperm transport and 100
  vascular endothelial growth factor and 71
oestrogen 239
  follistatin regulation 185, 186
  neural sites of action 224–5, 228–9
  receptors 174–5
    brain 224, 227–8
  regulation of gonadotrophins 171–5
    negative-feedback effects 173
    positive-feedback effects 173–5
  sexual behaviour and 214–15, 218, 219
  see also oestradiol (E_2)
oestrous behaviour, see sexual behaviour
oestrus, ewe 209, 218–19
oestrus cycle
  endocrine/behavioural events 210
  ewe 207–11
  follicular phase, see follicular phase
  GnRH receptor regulation 272
  neural control 219–20, 231–9
  neurotransmitters and 233–9
  pre-ovulatory events 211–19
  pre-ovulatory phase, see pre-ovulatory phase
  progestational/luteal phase, see luteal phase
okadaic acid 63, 66
oocytes
  autoantibodies 6
  chromosomal abnormalities 129–30
oophoritis, autoimmune, see autoimmune oophoritis
opioids 237
Orch-1 gene 21
Orch-2 gene 21–2
Orch-3 gene 21, 22
orchitis, autoimmune, see autoimmune orchitis
organum vasculosum of the lamina terminalis (OVLT) 222
ovarian failure, premature 3
ovariectomized ewes 211–12
  gonadotrophin regulation 173–4, 220
  neural pathways 226, 229–30, 236
  ovariectomized, hypothalamo–pituitary disconnected (OVX/HPD) sheep 165, 173, 174, 179
ovariectomized rats 173, 236
  follistatin 185, 186
  GnRH receptor regulation 169, 172, 271
ovaries
  angiogenesis 62
  regulation of sperm transport/activation 98–102
oviducts, see Fallopian tubes
ovulation 206
  GnRH neurones controlling 223–6
  neural mechanisms 219–20
  neurotransmitters and 233–9
  rapid sperm transport near 88, 91
  reflex 207
  silent 219
  sperm activation/release from isthmus 95–8
  sperm capacitation and 103–4
  sperm transport after 110–11
  spontaneous 207
  synchrony with sexual behaviour 205–40
    endocrine mechanisms 211–19
    neural mechanisms 219–38
    questions remaining 239–40
oxytocin
  sexual behaviour and 238
  sperm transport and 101, 114

PCR, see polymerase chain reaction
peptidergic neurones 236–8
pertussis toxin 22
phorbol esters 269, 271
physical activity, in pregnancy 41, 52–4
pigs, sperm transport and fertilization 85–115
pituitary cells, cultured
  GnRH actions 166–8, 269, 270
  gonadal peptide actions 179–81
  steroid hormone actions 175, 176–8
pituitary fragments, cultured 167, 174
pituitary gland 162, 220
  follistatin synthesis 185, 186
  inhibin synthesis 185
  oestrogen receptors 227–8
  paracrine/autocrine control of gonadotrophins 185–6, 187
placenta
  angiogenesis 62, 63, 76–7
  fibroblast growth factors 67, 68–9
  vascular endothelial growth factor 71–3, 74, 75
plasminogen activator 70
polymerase chain reaction (PCR)
  diagnosis of single-gene defects 146, 151
  embryo sexing 143–4

sources of error 146–7
polyspermy 105, 110–11, 130
potassium ions (K$^+$), Fallopian tubes 94
pregnancy
  chromosomal abnormalities 128–9
  energy balance 33–56
  loss rates 126
  termination 143
  zoological perspective 35–7
preimplantation diagnosis 143–53
  dual FISH for sex determination 144–5
  early approaches 143–4
  future developments 147–53
  single-gene defects 146, 151–3
  sources of errors 146–7, 148, 149
preimplantation embryos, *see* embryos, preimplantation
preoptic area 220, 226
  oestrogen actions 228–9
  oestrogen receptors 224, 227–8
  progesterone actions 230
  progesterone binding sites 229
preoptic–hypothalamic complex 222–3, 224–5
pre-ovulatory phase 86, 96
  sperm distribution in isthmus 87–91
  sperm storage 91–5
primer extension preamplification 151–3
proceptivity 207, 209
  endocrine control 218, 219
  neural control 225, 228, 229
progesterone 211, 238, 239
  binding sites in brain 229
  counter-current transfer 99
  GnRH receptor regulation 272
  hypothalamo–pituitary actions 160, 161
  implants, ovariectomized ewes 212
  LH surge regulation 217, 219
  oestrus cycle of ewe 210, 211, 213, 215–16, 217
  regulation of gonadotrophins 175–6, 214
  sexual behaviour and 214–15, 219, 225, 229–31
  sites of action in brain 229–31, 238–9
  sperm transport and 96–7, 100, 110
prostaglandin E$_1$ (PGE$_1$) 63, 66
prostaglandin E$_2$ (PGE$_2$) 4, 63, 66
  counter-current transfer 98, 99
prostaglandins
  counter-current transfer 98, 99
  sperm transport and 100–1
pyruvate 94

receptivity 207, 209
  endocrine control 218, 219
  neural control 225, 228, 229
relaxin 101

rhesus monkeys 221
rhodopsin 262, 263–4

seasonal breeding 206, 207
seasonal variations, food availability 34–5
seminal plasma
  exclusion from Fallopian tube 88–91
  myosalpinx activity and 95
serotonergic neurones 233–5
sex determination
  preimplantation embryos 143–5
  prenatal 143
sex steroid hormones
  counter-current transfer 98
  oestrus cycle of ewe 210, 211, 213, 215–16, 217–18
  regulation of gonadotrophins 171–8
  sites of action in brain 226–31
  see also oestrogen; progesterone; testosterone
sexual behaviour 206–7, 209, 240
  early follicular phase 216–17
  GnRH actions 236–7
  late follicular phase 218
  luteal phase 214–15
  neural control 219–20, 224–5, 228–31
  neurotransmitters and 234–5, 237
  synchrony with ovulation, *see* ovulation, synchrony with sexual behaviour
sexual dimorphism 37
sheep
  gonadotrophin gene regulation 165–6, 173–4, 175, 179
  oestrus cycle 207–11
  sperm transport and fertilization 85–115
  synchronization of sexual behaviour and ovulation 205–40
single-gene defects 125
  origins 141–2
  potential for misdiagnosis 146–7, 148
  preimplantation diagnosis 146, 151–3
somatostatin 238
sperm
  acrosome reaction 95–6, 97–8, 102, 103–4, 107
  adhesion to endosalpinx 106–10
  antigens, T cell responses 15
  capacitation 102–4
  fate of non-fertilizing 114–15
  FISH analysis 132–3
  karyotype analysis 129
  peri-ovulatory activation/release 95–8
  reason for large numbers 111–13
  reservoir/storage in Fallopian tubes 91–5
sperm:egg ratios 105, 107, 108–9
sperm motility 104
  ascending to fertilization site 106–10

hyperactivated, *see* hyperactivation, sperm
  ovarian regulation 98–102
  peri-ovulatory reactivation 95
  pre-ovulatory suppression 93, 94
sperm transport 85–115
  competitive mating studies 114
  ovarian regulation 98–102
  peri-ovulatory phase 95–8
  post-ovulatory 110–11
  pre-ovulatory phase 87–91, 96
  rate 86, 87–91
  very rapid 87
substance P receptor (SPR) 262, 268
sulphated glycoprotein 2: 5
superantigen, endogenous 6, 8
suppressor T cells
  preventing autoimmune disease 6–9
  preventing EAO 4, 5
  testis antigen-specific 5–6
suprachiasmatic nucleus 226

T cells
  activation of pathogenic 10–12
  depletion of regulatory 9–10
  induction of autoimmune oophoritis/gastritis 6–9
  Vβ11+ 6, 8
  see also CD4+ T cells; suppressor T cells
temperature gradients, Fallopian tubes 94
testis
  autoantigens
    protection 3–4
    T cell responses 15
  local immunoregulatory environment 4–5
testosterone
  counter-current transfer 99
  hypothalamo–pituitary actions 160, 161
  regulation of gonadotrophins 176–8
    in vitro studies 176–8
    in vivo studies 176, 177
tetraploidy 128, 134, 135
thermogenesis, diet-induced (DIT) 48–9
thymectomized mice, day 3 (D3TX) 4, 6–8, 9–10, 22–3
thymocytes, CD4+CD8− 6
thyroid-stimulating hormone (TSH) 161
  receptor 266
thyrotrophin-releasing hormone (TRH) receptor 171, 272–3
total energy expenditure (TEE) 52–4
transforming growth factor α (TGFα) 66
transforming growth factor β (TGFβ) 4, 66, 76
  receptor superfamily 183, 184
transgenic animals 188–9
  α-subunit gene-expressing 169

inhibin α-subunit knock-out 188–9
  tumour-producing 187–8
transglutaminase 4
translocations
  carriers 128–9, 149–50
  preimplantation diagnosis 149–51
triple X syndrome (XXX karyotype) 126, 127, 129, 133
triploidy 128, 136
trisomies 128
  maternal age and 128, 129
  origins 126, 127, 130
trisomy 13: 129
trisomy 16: 128
  origins 126, 127, 133
  preimplantation diagnosis 149
trisomy 18: 129
  origins 126, 127, 133
  preimplantation diagnosis 149, 150
trisomy 21 (Down syndrome) 128, 129
  origins 126, 127, 133
  preimplantation diagnosis 149
trisomy 22: 128
tumour necrosis factor (TNF) 15, 23
tumour necrosis factor-α (TNFα) 76
Turner's syndrome, *see* monosomy X
tyrosine kinase receptors 65, 70–1

utero-tubal junction, sperm transport 88–91
uterus, sperm transport 87, 88

vascular endothelial growth factor (VEGF) 69–75
  function in human reproduction 73–5, 76
  in human endometrium 71
  in pregnant reproductive tract 71–3
  receptors 70–1, 75
  structure and function 69–70
vascular permeability factor 70
vasoactive intestinal polypeptide 238
Vβ11+ T cells 6, 8
VEGF, *see* vascular endothelial growth factor
von Willebrand factor 64

weight
  gain during pregnancy 44, 45, 50
  pre-pregnancy 44, 45
  seasonal fluctuations 34–5
women, nutritional stresses 33–5

X-linked disorders
  embryo sexing 144–5
  prenatal sex determination 143

XO karyotype, *see* monosomy X
XXX karyotype 126, 127, 129, 133
XXY karyotype (Klinefelter's syndrome) 127, 129, 133

yeast artificial chromosomes 150, 151

zona pellucida 10
  accessory sperm 110, 113, 115
ZP3 mini-autoantigen 10–11, 13, 24
  autoantibody induction 17–20
  molecular mimicry 11–12, 14
  T cell responses 15–17